Flood Risk Change

Flood Risk Change

A Complexity Perspective

Andreas Paul Zischg
University of Bern, Bern, Switzerland

ELSEVIER

Elsevier
Radarweg 29, PO Box 211, 1000 AE Amsterdam, Netherlands
The Boulevard, Langford Lane, Kidlington, Oxford OX5 1GB, United Kingdom
50 Hampshire Street, 5th Floor, Cambridge, MA 02139, United States

ISBN: 978-0-12-822011-5

For information on all Elsevier publications visit our website at https://www.elsevier.com/books-and-journals

Publisher: Candice G. Janco
Acquisitions Editor: Amy M. Shapiro
Editorial Project Manager: Joshua Mearns
Production Project Manager: Kumar Anbazhagan
Cover Designer: Greg Harris

Typeset by TNQ Technologies

To my family, Heike, Jule, and Linus

Contents

Preface

The book emerged from two different experiences, from practical experience in moderating participative planning processes in flood risk management and from the results of my research in flood risk. I worked for several years as a consultant for public administrations that are responsible for flood protection and water resources management. Planning and implementing flood protection measures require to involve persons from different sectors and disciplines. Flood protection measures need space to enlarge rivers. Therefore, land owners—in many cases the agricultural sector—are stakeholders in the planning process. Moreover, stakeholders from environmental protection, hydropower production, tourism, local recreation, monument preservation, city planning, landscape ecology, archeology, landscape esthetics, water resources management, and individual residents may be involved in the planning process. This results in a diversity of perspectives on the topic or on the project and in a variety of interests and expectations for the planning process. Balancing the interest conflicts or target conflicts in participatory planning processes and finding a common sense over the possible solutions with the group to solve the problems are one of the key challenges for experts in flood risk management. Discussions in such diverse groups shed light on the interdependencies between the different sectors and problems. Simple solutions that may solve one single isolated problem affect many other problems, positively or negatively, or may result in unintended consequences or new problems. Therefore, possible solutions for the flood risk management problem have to satisfy also the expectations of other planning targets. This makes such planning processes complex and very challenging. However, when adopting a complexity perspective on this topic, we can approach the problems with a positive attitude. Instead of being an obstacle for implementing fast-forward and simplistic solutions for a narrowly defined problem, diversity becomes the key for finding sustainable and robust solutions. I experienced many times that the inclusion of multiple stakeholders and their perspectives in the planning process led to holistic solutions. In many cases, the commonly elaborated solutions for flood risk management solved also other problems that are connected with the flood risk management problem. The foundations of complex systems research can provide the basis for a new thinking and thus provide help in tackling complex and wicked problems.

With each flood protection project, we adapt our environment and landscape to our needs. Together with natural changes in the environment and climate, these human adaptations shape the relationships between humans and their environment. All three components of flood risk—hazard, exposure, and vulnerability—are changing. This makes that flood risk is changing dynamically in space and time. A successful and sustainable adaptation to future flood hazards is only possible if we consider the main drivers of changing risks. We must know where and how to intervene for keeping flood risks constant in time, or even reducing it, also under increasing hazards, increasing exposure, and increasing vulnerability. Again, the complexity perspective supports the analysis of the dynamic nature of flood risks. Complex systems science provides theoretical foundations for analyzing change, numerosity, feedbacks, nonlinear behavior, emergence, system behavior and system sensitivity, and deep uncertainty.

The book addresses two types of readerships. Experts and engineers working in flood risk management may find some ideas of how to consider and tackle complexity in their projects. Scientists may find ideas and concepts for analyzing flood risk change. In Chapter 1, the topic of flood risk change is introduced. Chapter 2 summarizes and explains the key drivers of flood risk change. Chapter 3 highlights approaches to show how the complex interactions between the coevolving drivers of flood risk change can be disentangled and how the isolated effects of single drivers of change on overall flood risk evolution can be quantified. Chapter 4 discusses the question if rivers and their floodplains can be interpreted as complex adaptive systems. Chapter 5 describes examples of model experiments for analyzing the spatiotemporal dynamics of flood risk change. The last chapter summarizes the conclusions of the model experiments by looking at the topic of flood risk change from the perspective of complex adaptive systems.

The book cover tells the story of changing flood risk in a mountainous river catchment. A previous flood event triggered new flood protection measures, in this case the heightening of the lateral protective walls. Flood risk was lowered for a time, but the question is how long it will take that the following increase in exposure and vulnerability, as well as the increase in hazard intensity and probability due to climatic changes, will increase flood risk again to preadaptation levels. The cover shows also that there are limits to adaptation by a further heightening of the lateral protective walls in the future. The book is a first step into the analysis of the dynamic nature of flood risks. It follows a systemic approach—including environmental, socioeconomic, and sociotechnical factors—for modeling and managing flood risk change. Readers will get a more complete picture of the topic for understanding the complexity of flood risk change, both from human and natural causes of flooding.

Chapter 1

Introduction

Flood risk change: A complexity perspective. This title covers two main topics of this book. The first and main topic is flood risk change. Floods are still one of the most damaging natural hazards, accounting for the majority of all economic losses from natural hazards worldwide (UNISDR: Making Development Sustainable: The Future of Disaster Risk Management, 2015). Risk resulting from floods is defined as a function of the probability of a flood event or scenario and its related extent of damage. However, as we are entering in the era of "Great Acceleration" (Steffen et al., 2015), socioeconomy and the human environment are changing rapidly. With these global changes, every single factor that contributes to flood risks is changing. This results in a highly dynamic change in flood risks.

Looking at how flood risk is changing or evolving in space and time requires a paradigm change. Most of the problems we are currently dealing with result from solutions of problems of the past. Many decisions were taken in the past without foreseeing possible future development paths. The built environment created will last for a long life cycle, and it shapes the adaptation options of future generation. Flood risk managers are still planning structural flood protection measures under a very static perspective. When conducting a risk analysis, we become a snapshot in time and space. We do not know the trajectory and the change rate of the past flood risk evolution. Even less, we can foresee future developments of flood risk.

For this, the second topic of the book title opens a new perspective on flood risks. Instead of considering flood risk as static in time and space, the book aims at showing how flood risk can change. The book intends to summarize the way toward a new understanding of flood risks. This comprehends to look at how the drivers of risk, the factors of risk, and the resulting risks are changing. The changes of the single risk factors—hazard, exposure, and vulnerability—are not independent from each other. Changes in flood hazards are influencing and triggering changes in hazard and vulnerability. Therefore, we can look at flood risk change with a systemic perspective. In analyzing the dynamic changes of flood risk, we can observe many features of complexity, namely the emergence of patterns, feedback mechanisms, numerosity and diversity, order and disorder, nested structures, modularity, multiple scales, hierarchies, nonlinearity, episodical changes, history and memory, path dependency, and adaptive behavior. These characteristics of complex systems

Flood Risk Change. https://doi.org/10.1016/B978-0-12-822011-5.00004-1

can be observed in the evolution of coupled human and natural systems and must be considered in flood risk management. In this book, I therefore want to look at changing flood risks from the complexity perspective. The book is a primer for analyzing and modeling flood risk change. It paves a way toward the development of modeling frameworks that are able to consider selected aspects of complexity. This is absolutely needed to avoid unintended consequences of decisions in current flood risk management in the long-term perspective. The book should give an outlook on how new-generation modeling approaches can represent the complex behaviors of floodplains and coupled human and natural systems in a highly dynamic environment. These tools and methods should help flood risk management practice to tackle with increasingly wicked problems.

Flood risks are now increasingly being analyzed from a dynamic rather than from a static perspective (Mazzorana et al., 2012; Merz et al., 2010). Several studies have addressed changes in natural risks over recent decades and centuries (Himmelsbach et al., 2015; Hufschmidt et al., 2005; Paprotny et al., 2018), and research on climate change and its impacts has focused on future changes in risks (Alfieri et al., 2016, 2017; Arnell & Gosling, 2016; Dottori et al., 2018; Hirabayashi et al., 2013). However, most studies focused solely on the future increase in flood hazard. Only few studies consider both the impacts of climatic changes to river flows and the future dynamics in the elements at risk (Bouwer et al., 2010; Jongman et al., 2012; Liu et al., 2015; Löschner et al., 2017; Winsemius et al., 2016). Closer examinations of the spatiotemporal dynamics and the actual rate of change are rather rare. Knowledge about hazardous processes and their impacts, as well as about the trajectories of flood risk changes, is essential for the sustainable management of flood risks.

Several intertwined natural and anthropogenic drivers influence the spatiotemporal evolution of flood risk. In this book, the following drivers of flood risk change that are related to environmental changes are considered:

- Floods are either caused by direct rainfall on the floodplain (pluvial floods and surface water floods) or rainfall on river catchments resulting in catchment outflow. The latter causes floods in downstream floodplains (riverine floods and lake floods). Consequently, changes in flood processes, i.e., changes in frequency and magnitude of floods in a floodplain, are determined by changing precipitation.
- In mountainous areas, flood hazards are influenced by sediment transport and deposition processes and debris flows. Debris flows are influenced by environmental changes, such as melting of glaciers and permafrost, or changes in weathering processes and mass movements.
- River morphology changes over time, including natural and gradual changes in the river morphology, or disruptive changes by flood events. An important aspect of river morphology changes is anthropogenic

interventions, which are relevant drivers of flood risk in a floodplain, for example, the construction of flood defenses such as levees and dams or river restoration projects. However, the construction of levees as flood protection measures in one floodplain can have adverse effects in the downstream floodplains and can result in flooding trade-offs between upstream and downstream floodplains.

Beside changes in the natural environment, flood risk is also evolving due to changes in the exposed elements at risk and their vulnerability. From this aspect, the following drivers of change are considered here:

- The increase in the elements at risk change due to socioeconomic development. The growth of settlements and thus the increase of residential buildings are related to population growth.
- Infrastructure is increasing in parallel with population growth. This has wider impacts on the socioeconomic system. For example, in economically active areas, floodplains are increasingly occupied by production facilities, as these require relatively flat areas for their construction. With economic development, the elements at risk and the infrastructure in floodplains are increasing both in terms of quantity and monetary value.
- Increasing values at risk compete with opposing drivers of flood risk reduction measures implemented by individuals and the public. Hence, changes in exposure and vulnerability are influenced by governmental interventions and regulations and by the actions of individuals.

The built environment in floodplains, whether the settlement area or the river channel, is subject to changes and coevolutionary dynamics in both society and nature (Di Baldassarre et al., 2013, 2015; Fuchs et al., 2017). As Vitousek et al. (1997) postulated, the human impact on nature is now considerably larger than at any point in history. This is true for the floodplains, as humans are shaping landscapes with the built environment. These impacts of society on nature influences pathways of flood risk change. The spatiotemporal development of these drivers of change in flood risk leads to difficulties in predicting future flood risk. Consequently, recent studies have extended the framework of risk analysis toward a spatiotemporal framework as drivers for flood risk changes are varying in space and time (Ahmad & Simonovic, 2013; Zischg et al., 2018). This book goes one step further and shows examples of model experiments that analyze flood risk change, the drivers of flood risk change, or features of complexity. We look at flood risk from a dynamic perspective. The reader will become hands-on examples of how we can analyze past changes in flood risk, of how we can monitor changing risks, and of how we can model flood risk changes by considering complex interactions between the drivers of change.

Chapter 2 summarizes and explains the key drivers of flood risk change. The chapter begins with definitions of flood risks and explains how flood risk can be calculated from the risk factors hazard, exposure, and vulnerability. It describes how systems can be delimited and represented in risk analyses. The focus is laid on river systems and floodplains. In these areas, flood hazards meet the anthropogenic elements at risk and their vulnerability. River systems and floodplains evolve in time. The chapter describes the main factors that shape the long-term evolution of floodplains and of flood risk. It provides an overview of all factors that are relevant for the increase or decrease of flood risks, ranging from environmental change, climatic change, and socioeconomic change. After describing the isolated effects of the single drivers of change, Chapter 2 ends with a description of coevolution of risk factors. Moreover, an insight into spatiotemporal dynamics of flood risk change is given by explaining where changes in the system components occur and where these become relevant for flood risk change. Herein, upstream–downstream and a river basin–river reach relationship as well as legacy effects in time and space are explained.

Chapter 3 highlights theoretical and methodological approaches to show how the complex interactions between the coevolving drivers of flood risk change can be disentangled and how the isolated effects of single drivers of change on overall flood risk evolution can be quantified. A special focus is laid on the potential of model experiments for analyzing flood risk change. Thus, principles of frameworks for analyzing and modeling flood risk change are explained. The chapter introduces several examples of model experiments that aim at disentangling the effects of selected drivers of change to overall flood risk evolution. A special example shows how flood risks changed within the past 300 years after the early geoengineering project of the Kander River deviation in the Canton of Bern, Switzerland. This case study shows how a cascade of unintended consequences for the society can be caused by a human intervention in the natural river system. Moreover, this case study also points out how difficult a long-term perspective is to keep on flood risk management. The technological advancement of the industrial revolution totally changed the framework of society's values that underline decision-making in flood risk management. The presented case studies can be used to compare the effects of environmental changes and the effects of human interventions on flood risk evolution. It becomes clear that future flood risk evolution can radically be shaped by human adaptation and intervention. One example clearly shows that the implementation of flood prevention measures remarkably drives flood risk reduction. The model experiment bases mainly on counterfactual simulations, i.e., comparing the present-day situation with alternative pathways of natural and societal development. Some of the case studies exemplify how negative feedback effects will contribute to flood risk reduction although climatic changes might increase the intensity and frequency of extreme precipitation events. Herein, sediment supply or sediment deficit determines flood risk change in mountain areas. In the last part, Chapter 3 shows analyses on the effects of land use planning and time lags in the land use regulation on emerging patterns of risk hot spots in the long term. Moreover, it presents examples of the effects of the declining societal memory of past flood events on flood exposure and vulnerability.

Chapter 4 argues if the long-term evolution of rivers and their floodplains can be interpreted as complex adaptive systems. To this end, the main features and characteristics of complex systems are summarized. This chapter sheds light on the nature of the coupled human and natural systems of floodplains and identifies which characteristics of complex systems can be observed on floodplains. The reader will get an answer to the question if flood risk in floodplains is a complicated issue or if floodplains can be considered as a complex system or as a complex adaptive system; or, if rivers and floodplains are nowadays to be considered as sociotechnical systems rather than natural systems. These answers help the reader to understand the emergent phenomena in the long-term development of floodplains. Examples of past flood events will be discussed, and they will be shown how these events influenced policy, technology, and risk awareness. Vice versa, examples will show how past technological innovations shaped rivers and floodplains. The chapter ends with an outlook on possible future trajectories and pathways of flood risk evolution and with a discussion of the sensitivity of floodplains to climatic changes.

Chapter 5 describes examples of model experiments and analyses of spatiotemporal dynamics of flood risk change. It explains five interactive web tools that have been developed by the Mobiliar Lab for Natural Risks of the Oeschger Center for Climate Change Research of the University of Bern in Switzerland. The first tool supports citizen science that aims at collecting geolocalized images and photographs of flood events. This growing database can be visualized and queried in a web-based map application. It contributes to keeping the societal flood memory alive. The second tool describes an interactive web-mapping application that estimates the nation-wide exposure of buildings, residents, work places, and critical infrastructure to floods. It contributes to the understanding of the difference between a hazard-centered and an exposure-centered visualization of flood risks. The third tool goes one step further and provides an interactive model experiment to test the sensitivity of flood damages to urban densification, the implementation of property-level flood protection measures, and structural flood protection measures in the river channel. This tool aims to show how one desired effect of urban planning, namely to prioritize the internal densification of cities by developing unused or abandoned spaces against urban sprawl, potentially can result in an unintended increase of flood risks. It furthermore shows how existing flood risks can be reduced by implementing flood prevention measures by different stakeholders. Fourth, a model experiment is described and implemented in an interactive web-based application that allows to assess the trajectory of flood risk change in a floodplain in the past two centuries. The user can test counterfactual scenarios of urban development and of alternative flood risk management strategies. Moreover, the user can test the effects of different drivers of change to the overall flood risk evolution. This tool contributes to the understanding of the dynamic nature of flood risks in space and time. The fifth tool shows storylines of extreme rainfall events and their impacts. The user of this interactive web application can navigate through the spatiotemporal evolvement of a flood event with a time slider and zoom in and out at different scales. This tool visualizes direct and indirect effects of flood events at an hourly timescale. It helps to understand the complex patterns of rainfall events over a complex mountain topography, how the impacts of the event on society are evolving in space and time, namely from upstream to downstream, and how the spatial footprint of the flood impacts follows the rainfall pattern with a time lag. Chapter 5 is also describing a model experiment that shows how the philosophical school that is behind decision-making in flood risk management can lead to the emergence of patterns in flood risk hot spots. It shows how the consideration or nonconsiderations of social justice aspects in priority setting can potentially guide the distribution of public funds for investments in flood protection toward remote and economically disadvantaged or toward economically advantaged regions. Finally, the last example describes the operationalization of an information system for monitoring flood risk change. Chapter 5 summarizes the principles of model coupling for analyzing and modeling flood risk change. It discusses the advantages or limitations of coupled component models. These models have a potential for considering features of complex systems such as feedbacks, adaptive behavior, structural change, regime change, and nonlinear behavior. The chapter closes with a summary of the lessons learnt during the development of several coupled component models and the implementation of the model experiments that were described in the previous chapters. It closes with a perspective on the use of coupled component models in hydrology and in the simulation of coupled human and natural systems.

The last chapter summarizes the conclusions of the model experiments and analyses described in this book by looking at the topic from the perspective of complex adaptive systems. It discusses recommendations of how to consider and implement some thoughts of complex systems science into flood risk management practice and in the development of next-generation simulation models. This aims at enabling modeler's and decision-makers to analyze the big trends or future pathways of flood risk evolution by considering human behavior in simulations or at least by coupling models for simulating human behavior with models for simulating natural processes. This chapter shows how flood risk management can be enhanced with the perspective of complex systems science. It will show how urgently we need to consider the complex interactions between the drivers of change and complexity in flood risk analyses and decision-making in flood risk management. Finally, the chapter outlines a complexity perspective for governing flood risk change in the 21st century.

With all this, the book scraps the surface of a new perspective and a new set of methods to tackle the complex problems of the Anthropocene and to confront deep uncertainties.

References

Ahmad, S. S., & Simonovic, S. P. (2013). Spatial and temporal analysis of urban flood risk assessment. *Urban Water Journal, 10*(1), 26–49. https://doi.org/10.1080/1573062X.2012.690437

Alfieri, L., Bisselink, B., Dottori, F., Naumann, G., de Roo, A., Salamon, P., Wyser, K., & Feyen, L. (2017). Global projections of river flood risk in a warmer world. *Earth's Future, 5*(2), 171–182. https://doi.org/10.1002/2016EF000485

Alfieri, L., Feyen, L., & Di Baldassarre, G. (2016). Increasing flood risk under climate change: A pan-European assessment of the benefits of four adaptation strategies. *Climatic Change, 136*(3–4), 507–521. https://doi.org/10.1007/s10584-016-1641-1

Arnell, N. W., & Gosling, S. N. (2016). The impacts of climate change on river flood risk at the global scale. *Climatic Change, 134*(3), 387–401. https://doi.org/10.1007/s10584-014-1084-5

Bouwer, L. M., Bubeck, P., & Aerts, J. C. J. H. (2010). Changes in future flood risk due to climate and development in a Dutch polder area. *Global Environmental Change, 20*(3), 463–471. https://doi.org/10.1016/j.gloenvcha.2010.04.002

Di Baldassarre, G., Kooy, M., Kemerink, J. S., & Brandimarte, L. (2013). Towards understanding the dynamic behaviour of floodplains as human-water systems. *Hydrology and Earth System Sciences, 17*(8), 3235–3244. https://doi.org/10.5194/hess-17-3235-2013

Di Baldassarre, G., Viglione, A., Carr, G., Kuil, L., Yan, K., Brandimarte, L., & Blöschl, G. (2015). Debates—Perspectives on socio-hydrology: Capturing feedbacks between physical and social processes. *Water Resources Research, 51*(6), 4770–4781. https://doi.org/10.1002/2014WR016416

Dottori, F., Szewczyk, W., Ciscar, J. C., Zhao, F., Alfieri, L., Hirabayashi, Y., Bianchi, A., Mongelli, I., Frieler, K., Betts, R. A., & Feyen, L. (2018). Increased human and economic losses from river flooding with anthropogenic warming. *Nature Climate Change, 8*(9), 781–786. https://doi.org/10.1038/s41558-018-0257-z

Fuchs, S., Röthlisberger, V., Thaler, T., Zischg, A., & Keiler, M. (2017). Natural hazard management from a coevolutionary perspective: Exposure and policy response in the European Alps. *Annals of the American Association of Geographers, 107*(2), 382–392. https://doi.org/10.1080/24694452.2016.1235494

Himmelsbach, I., Glaser, R., Schoenbein, J., Riemann, D., & Martin, B. (2015). Flood risk along the upper Rhine since AD 1480. *Hydrology and Earth System Sciences Discussions, 12*, 177–211. https://doi.org/10.5194/hessd-12-177-2015

Hirabayashi, Y., Mahendran, R., Koirala, S., Konoshima, L., Yamazaki, D., Watanabe, S., Kim, H., & Kanae, S. (2013). Global flood risk under climate change. *Nature Climate Change, 3*(9), 816–821. https://doi.org/10.1038/nclimate1911

Hufschmidt, G., Crozier, M., & Glade, T. (2005). Evolution of natural risk: Research framework and perspectives. *Natural Hazards and Earth System Science, 5*(3), 375–387. https://doi.org/10.5194/nhess-5-375-2005

Jongman, B., Ward, P. J., & Aerts, J. C. J. H. (2012). Global exposure to river and coastal flooding: Long term trends and changes. *Global Environmental Change, 22*(4), 823–835. https://doi.org/10.1016/j.gloenvcha.2012.07.004

Liu, J., Hertel, T. W., Diffenbaugh, N. S., Delgado, M. S., & Ashfaq, M. (2015). Future property damage from flooding: Sensitivities to economy and climate change. *Climatic Change, 132*(4), 741–749. https://doi.org/10.1007/s10584-015-1478-z

Löschner, L., Herrnegger, M., Apperl, B., Senoner, T., Seher, W., & Nachtnebel, H. P. (2017). Flood risk, climate change and settlement development: A micro-scale assessment of Austrian municipalities. *Regional Environmental Change, 17*(2), 311–322. https://doi.org/10.1007/s10113-016-1009-0

Mazzorana, B., Levaggi, L., Keiler, M., & Fuchs, S. (2012). Towards dynamics in flood risk assessment. *Natural Hazards and Earth System Science, 12*(11), 3571–3587. https://doi.org/10.5194/nhess-12-3571-2012

Merz, B., Hall, J., Disse, M., & Schumann, A. (2010). Fluvial flood risk management in a changing world. *Natural Hazards and Earth System Science, 10*(3), 509–527. https://doi.org/10.5194/nhess-10-509-2010

Paprotny, D., Sebastian, A., Morales-Nápoles, O., & Jonkman, S. N. (2018). Trends in flood losses in Europe over the past 150 years. *Nature Communications, 9*(1). https://doi.org/10.1038/s41467-018-04253-1

Steffen, W., Broadgate, W., Deutsch, L., Gaffney, O., & Ludwig, C. (2015). The trajectory of the anthropocene: The great acceleration. *Anthropocene Review, 2*(1), 81–98. https://doi.org/10.1177/2053019614564785

UNISDR. (2015). Making development sustainable: The future of disaster risk management. In *Global assessment report on disaster risk reduction, 4.2015*. United Nations.

Vitousek, P. M., Mooney, H. A., Lubchenco, J., & Melillo, J. M. (1997). Human domination of Earth's ecosystems. *Science, 277*(5325), 494–499. https://doi.org/10.1126/science.277.5325.494

Winsemius, H. C., Aerts, J. C. J. H., van Beek, L. P. H., Bierkens, M. F. P., Bouwman, A., Jongman, B., Kwadijk, J. C. J., Ligtvoet, W., Lucas, P. L., van Vuuren, D. P., & Ward, P. J. (2016). Global drivers of future river flood risk. *Nature Climate Change, 6*(4), 381–385. https://doi.org/10.1038/nclimate2893

Zischg, A. P., Hofer, P., Mosimann, M., Röthlisberger, V., Ramirez, J. A., Keiler, M., & Weingartner, R. (2018). Flood risk (d)evolution: Disentangling key drivers of flood risk change with a retro-model experiment. *Science of the Total Environment, 639*, 195–207. https://doi.org/10.1016/j.scitotenv.2018.05.056

Chapter 2

Key drivers of flood risk change

The term flood risk is here defined according the general concept of disaster risk of the Intergovernmental Panel on Climate Change IPCC (IPCC, 2012). Disaster risk is defined as the "likelihood over a specified time period of severe alterations in the normal functioning of a community or a society due to hazardous physical events interacting with vulnerable social conditions, leading to widespread adverse human, material, economic, or environmental effects that require immediate emergency response to satisfy critical human needs and that may require external support for recovery." Thus, risk requires a potential hazard (*H*) and the "presence of people; livelihoods; environmental services and resources; infrastructure; or economic, social, or cultural assets in places that could be adversely affected." This is defined as exposure (*E*). Moreover, exposed people, communities, and infrastructure must be vulnerable to the hazard. Vulnerability (*V*) is the "propensity or predisposition of exposed people or infrastructure to be adversely affected" (IPCC, 2012). Risk is composed of these three main components. Thus, risk is a function of hazard, exposure, and vulnerability (UNISDR, 2015) and is often expressed in Eq. (2.1)

$$R = H \times E \times V \tag{2.1}$$

These three main components of risk are not independent. Exposure is connected with the magnitude and probability of the hazardous event. The higher the magnitude of the hazard, the higher is the number of exposed population or infrastructure. However, this interaction is nonlinear and depends on many factors, which will be discussed below. The vulnerability of exposed assets also depends on the frequency and magnitude of the hazard as well as on the characteristics of the exposed objects, and thus there is an interaction between these components of risk. The main reason of this interdependency is mostly due to the effects of social learning. Buildings and infrastructure built near rivers with frequent floods have often a reduced vulnerability because they have been adapted to the hazard after previous events (Elmer et al., 2010; Kienzler et al., 2015). The consideration of three main components in defining risk opens the possibility to consider human interventions for risk reduction. These interventions can range from the installation of concrete hydraulic engineering structures to measures for avoiding urbanization in hazardous areas by land use planning or building

Flood Risk Change. https://doi.org/10.1016/B978-0-12-822011-5.00005-3

flood-proof infrastructure with reduced vulnerability. The first type of interventions targets at reducing the hazard H, either in terms of reducing frequency or magnitude of flood events. The second type of intervention targets at reducing exposure E, and the third type of interventions aims at reducing vulnerability V.

The calculation of risk with or without risk management measures leads to the assessment and quantification of the efficacy of these measures in terms of risk reduction. The comparison of the benefits with the costs of the risk reduction measures allows to analyze the efficiency of risk reduction measures. However, this requires the monetization of risks, i.e., quantifying the risk and the reduced risk after the implementation of risk reduction measures in monetary units. In natural sciences and engineering, risk is quantified as a measure of uncertainty based on the concept of probability. Thus, risk is the probability of a loss within a certain time period. As the loss depends on the magnitude and frequency of the hazard triggering the loss, the probability of occurrence of a hazard event or scenario with a certain magnitude is taken as the probability of the loss. Thus, event risk or scenario risk can be written as shown in Eq. (2.2) (Oberndorfer et al., 2020).

$$R_{i,j} = f\left(p_j, p_{i,j}, A_{i,j}, v_{i,j}\right) \tag{2.2}$$

Where $R_{i,j}$ is the risk dependent on object i and scenario j, p_j is the probability of defined scenario j, $p_{i,j}$ is the probability of exposure of object i to scenario j, A_i is the value of object i affected by scenario j, and $v_{i,j}$ is the vulnerability of object i in dependence on scenario j. The probability of scenario j is the probability of occurrence, mostly simplified by the number of how often the event of this magnitude potentially occurs within a specified time period. This is also termed as the return period of an event. If a hazard event of a certain magnitude is expected to occur once in 100 years, it is called a "1-in-100 years event" or a "100 year event." If the value of object i is expressed in monetary terms and the vulnerability is a dimensionless degree of the object suffering loss, the risk function is a multiplication and results in the probability of a damage event (Eq. 2.3).

$$R_{i,j} = p_j \times p_{i,j} \times A_{i,j} \times v_{i,j} \tag{2.3}$$

The term "loss" here defines the quantifiable aspects of harm and damage that result from a hazard event. In this way, the scalar value of risk includes both probability and the severity of the impact, but in a way that keeps hazard, exposure, and vulnerability distinct and gives rise to a single dimension (Rougier et al., 2013). The term "damage" is used for explicitly focusing on the monetized impacts of hazard events and thus is a subset of the overall loss.

If multiple hazard scenarios are considered in the quantification of risk, risk or the mean expected loss—in many cases in a 1-year-interval—is the sum of the product of each possible loss and its probability. The losses of several hazard scenarios with increasing magnitudes and decreasing occurrence probability form the loss exceedance probability curve (Fig. 2.1). In natural hazards management, this function is often termed as the "hazard footprint" function. The footprint function aggregates the spatial overlay between the footprint(s) of the hazard scenario(s) and the values at risk with their object-specific vulnerabilities. The form and shape of the footprint function show how the loss increases with event magnitude. Fig. 2.1 shows an example of a flood hazard footprint function in terms of exposed population. If risk is derived from the exceedance probability curve or from the footprint function, risk is calculated as the mean loss from the area under the curve. From the risk change perspective, it becomes obvious that risk quantification must include the time period for which the footprint function and the assessed risk value are valid.

Principles of flood risk analysis

As this book deals mainly with flood risks in mountainous or hilly terrain, the procedure of analyzing flood risk is outlined schematically in the following section. The concept is described from a "top-down" perspective, by following the course of the water through the landscape.

The main starting point is the definition of the hazard domain and the spatial delimitation of the system that is considered in the risk analysis. The

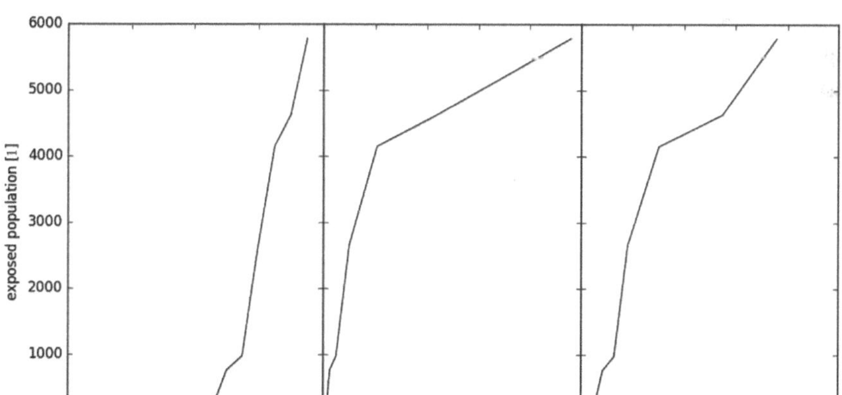

FIGURE 2.1 Hazard footprint. An example of a hazard footprint function showing the relationship between increasing flood magnitude and increasing flood exposure. *Modified from Zischg, A. P., & Bermúdez, M. (2020). Mapping the sensitivity of population exposure to changes in flood magnitude: Prospective application from local to global scale.* Frontiers in Earth Science, 8. *https://doi.org/10.3389/feart.2020.534735.*

first step is the definition of the hazard process. In the case of floods, there are several possibilities to consider. Probably the simplest flood hazard in terms of direct causal effect from heavy rainfall to consider in flood risk analysis is flooding from infiltration excess overflow, i.e., surface runoff leading to surface water floods. Floods by surface runoff result from rainfall events that exceed the infiltration capacity of the soil. In urban environments with a high share of sealed surface and limited sewer capacity, the threshold of rainfall intensity and sum for triggering surface water floods are lower. This process is also termed as pluvial floods. In this case, the spatial delimitation of the system equals the area for which the risk must be calculated (area of interest) or for which risk reduction measures are planned plus the (small) catchments from which surface water is flowing into the area of interest (Fig. 2.2, left). In case of riverine floods, the spatial delimitation of the system considered for risk analysis is the floodplain where the settlements or infrastructure are located (area of interest) plus the upstream watershed or river basin, which contributes to the discharge in the river that is triggering flood events in the area of interest (Fig. 2.2, right). In larger river basins, there can be several areas of interest if more than one floodplain exists. In this case, the areas of interest for risk analysis are nested within the river basin.

In summary, the system delimitation comprises the type of hazard considered in the study, the area of interest where the risk has to be analyzed, and the upstream areas where the hazard processes originate from and form the boundary condition of the inner system. The inner system can be defined as an open system. Hazards can originate and occur within the system itself and can enter into the system from outside at the system boundary. The outer system delimitation, i.e., the upstream catchment area, is thus needed for modeling the hazard in case of floods.

Besides surface water floods and riverine floods, flood hazards are also influenced by other flood processes such as sediment transport, lateral erosion, large wood transport and deposition, as well as debris flow. Especially in

FIGURE 2.2 System delimitation. Spatial delimitation of the system considered in flood risk analysis. Left: the area of interest for risk analysis plus the contributing areas for surface water floods. Right: the area of interest plus the upstream catchment. Background map from map.geo.admin.ch.

mountainous areas, floods and their impacts on infrastructure are likely dominated by sediment transport and deposition. Thus, the definition of the system for which the risk analysis is valid contains a clear statement for which type of hazard the risk analysis is valid. This is part of the system delimitation. This explicit definition of the hazard processes considered and the explicit exclusion of the hazard types that are not considered are necessary for the interpretability of the results. Especially if the risk analysis is termed as "flood risk analysis," misunderstandings or misinterpretations can result from unclear definitions of the flood process type. Another requirement of the system delimitation is a clear statement of which type of exposure is considered in the risk analysis. Is the risk assessment valid for people, buildings, or infrastructure only? Are mobile assets as, for example, cars moving around considered in the analysis? Depending on the type of asset considered, the system can be spatially limited, but it is open to input from outside. One example of an open system is road traffic entering and exiting the floodplains.

In most cases of flood risk analysis, the area of interest (inner system) is a floodplain or a set of floodplains within a river basin. In floodplains, the main factors of flood risk, the flood process, and the values at risk intersect spatially. From a physical perspective, floodplains are defined as areas of land adjacent to and formed by flowing water in times of floods. From a socioeconomic perspective, floodplains provide land for settlement, infrastructure, and other human activities. In the following, the term "floodplain" is used synonymously for the area of interest for flood risk analysis if not otherwise specified.

Once the type of hazard and exposure is defined and the system is spatially delimited, the main hazard input parameters for estimating the probability and magnitude of the hazard events and scenarios are determined. The calculation of flood risk requires a set of events or scenarios with increasing magnitude. This is the so-called "event set." In an ideal case, the case study area has a long record of well-documented flood events. The spatial footprints of these past flood events can be reconstructed (Bomers et al., 2019) and overlaid with the actual or historic dataset of exposure, e.g., houses or traffic infrastructure. If the occurrence probability of the reconstructed flood events can be determined, the historic event set leads to the loss footprint function of the system under investigation. Eventually, risk can be quantified from this footprint function.

However, the event set is very small in many cases, due to the inherent low probability of occurrence of extreme events. In this situation, the event set must be developed by means of simulation models or by a cascade of simulation models. The first step for determining the event set for surface water floods is to analyze the characteristics of rainfall events (Bernet et al., 2019). The meteorological data from nearby located meteorological stations is analyzed with extreme value statistics. An intensity-duration function is derived from the data and a set of rainfall scenarios with increasing magnitude is derived from the extreme value statistics. This set of rainfall scenarios then is used as an input for modeling surface water floods. The hydrodynamic simulation models use the rainfall upon the digital elevation model within the spatial delimitation of the system and convert the rainfall into surface runoff. The water fluxes over the digital elevation model are computed with the shallow water equations or with simplified formulas (Bernet et al., 2018). The output of these hydrodynamic simulation models is a map of the water depths and flow velocities for selected time steps during the time interval of the simulation for each rainfall scenario. In case of modeling surface water floods, the simulation time (temporal system delimitation) equals the duration of the rainfall event.

In contrast to modeling surface water floods, the modeling of riverine floods requires an additional modeling step. A hydrological model is needed to convert the rainfall scenarios over the catchment upstream of the area of interest for the risk analysis (inner system) into water discharge at the interface between the catchment outlet and the outer boundary of the inner system (Felder et al., 2018). The water coming from the catchment and flowing into the area of interest for risk analysis is given as an input into the hydrodynamic model (Felder et al., 2017). In this case, the hydrodynamic model computes the water fluxes within the inner system from the upper boundary condition to the lower boundary condition. Advanced hydrodynamic models are able to simulate both flood processes in parallel, surface water floods, and riverine floods, e.g., LISFLOOD-FP (Bates & De Roo, 2000). The output is the same as for the simulation of surface water floods, namely a set of maps showing water depths and flow velocities for selected time steps during the time interval of the simulation. However, the simulation time period must be extended to consider the time lag between rainfall in the upper catchment and outflow from the catchment and the time lag between the inflow of water into the floodplain and the outflow from the floodplain. The higher the time lag is, the larger is the size of the catchment, and the length of the river reaches through the system. Thus, the simulation time (temporal system delimitation) remarkably exceeds the duration of the rainfall event.

In case of riverine floods, the event set, or the set of scenarios, must not necessarily be derived from rainfall characteristics over the catchment. If hydrological measurements are available nearby the area at interest, the time series of the discharge measurements can be analyzed with extreme values statistics. The occurrence probability of flood peaks of a certain magnitude can be derived directly from these statistics. However, the hydrodynamic simulations of the inundation processes require a hydrograph for the whole length of the simulation period. Thus, if flood probability is derived from discharge time series, the form of the hydrograph must also be derived from the time series (Zischg et al., 2018). Each flood scenario in the event set must be described by the flood peak and the flood volume, i.e., by the shape of the flood hydrograph. This hydrograph is then used as an input for the hydrodynamic model to simulate inundation dynamics and to create flood maps. The hydrological model to convert the rainfall event set in a corresponding set of flood hydrographs is not needed in this case. The spatial delimitation of the system is therefore restricted to the spatial extent of the inner system, i.e., the area of interest for the risk analysis. Moreover, the temporal delimitation of the system can also be restricted to the time period of the flood event in the hydrograph plus the time needed for water traveling through the floodplain within the area of interest.

In summary, the model cascade from rainfall to inundation enables to correlate each rainfall event or flood event of the event set with a related flood map. Thus, the probability of occurrence of the input event or scenario corresponds to the occurrence probability of the flood event, expressed in a hazard footprint. The hazard footprint is described by the extent of the inundated area, the inundation depths, and the flow velocities.

In mountainous areas, flood processes are often influenced by sediment transport and large wood dynamics. Thus, flood maps differ if these processes are considered in hazard mapping or not. The effects of sediment transport dynamics on flood maps can be complex and have consequences for the calculation of damages. A flood event with sediment transport can lead to aggradation of the riverbed, thus leading to a decrease of the transport capacity of the river channel during the flood event. This increases the volume of water exiting the river channel and thus the extent of the flooded areas in the floodplain. Sediment deposited in the floodplain can also lead to higher damages to infrastructure and houses, at least by increasing the cleaning costs after the flood event (Fuchs et al., 2007). The vulnerability of houses to floods is also higher if sediment is involved or deposited around and within the houses. Another example of increasing hazard is the transport of large wood coming from the upstream catchment. Wood pieces come from flooded forest areas and lateral erosion and are transported downstream by the river. At bridge piers, the wood logs can lead to occlusion of the river channel below the bridge and thus to a decreasing of the discharge capacity of this cross section (Ruiz-Villanueva et al., 2014). This in turns leads to an increase of the outflow of water from the river channel into the floodplain and an increase of the flooded area. If these processes have to be expected, the flood hazard maps in many cases have a higher extent as the flood maps of the same occurrence probability that are based on plain water simulations only. The consideration of fluvial sediment transport and large wood dynamics in the elaboration of the hazard maps requires the adaptation of the spatial delimitation of the system to be analyzed because the sources of sediment and large wood transport often originate from river reaches upstream of the areas of interest (Mazzorana et al., 2009).

The next step in this top-down model cascade is the flood impact modeling part. The complexity of the impact modeling increases with the kind of assets and values at risk considered in the analysis. The simplest analysis setup is the spatial overlay of the flood maps with houses and the derivation of monetary damages from the estimated repairing costs. The houses are fixed in space and do not move around as population does. The consideration of population in the risk analysis increases the level of complexity. First, population is mobile and leaves or enters the area of interest. During floods, people move out of the flooded area and relief units move in. The same is valid for cars. The complexity of impact modeling further increases with the consideration of adaptation measures, e.g., the effects of early warning systems on preparedness, or when cascading impacts have to be expected. Examples of cascading impacts are the physical damage to single parts of the electric grid infrastructure leading to a blackout and to subsequent cascading impacts to the economic activities. Another example is the cascading impact if roads are effected widely. Working places and production facilities are not only affected directly by the flood but can also be indirectly affected by road closures that hamper work forces to commute from home to their working place.

Impact modeling is mostly conducted by means of geographical information systems (GIS) because they have the ability to combine the outcomes of the flood models with geodata and other data on socioeconomic activities. The impact analysis is done on a spatially explicit way and the damages or impacts calculated for each object (value at risk) are aggregated at the level of the area of interest. This results in a scalar value of risk for the floodplain. However, the spatiotemporal dynamics of the flood impacts can also be analyzed. Fig. 2.3 shows schematically a model cascade from rainfall to flood impacts.

In summary, there are many factors that influence flood risk and flood risk change. Most of these factors change over time at different timescales and change in space. Thus, the assumption of stationarity and constant conditions

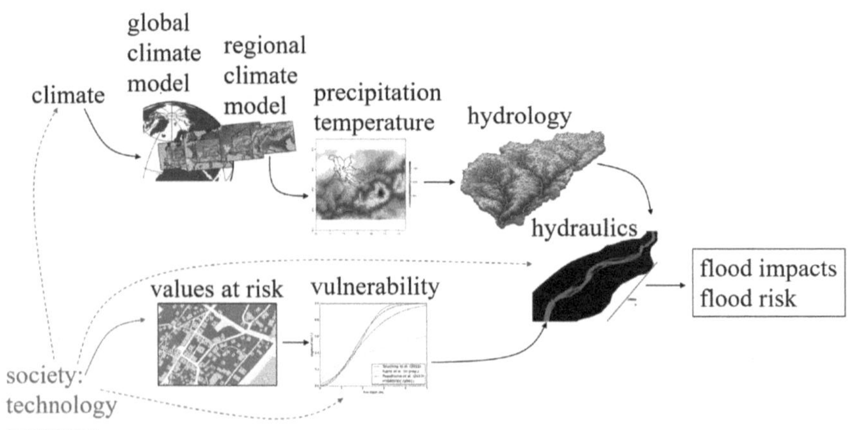

FIGURE 2.3 Model cascade. Model cascade from rainfall to flood impacts.

in the analysis of flood risk is to be discussed. In the following, the main drivers of flood risk change are summarized.

Environmental changes

As stated above, floods are either caused directly by rainfall falling onto the system under investigation (pluvial floods and surface water floods, for example) or by falling onto river catchments resulting in a catchment outflow, with or without sediment and large woods. The latter causes floods in downstream floodplains (riverine floods and lake floods). Thus, the boundary condition of floods in floodplains can either be rainfall, river flow, or both. Consequently, changes in flood processes, i.e., changes in the frequency and magnitude of floods, are determined by these external influencing factors. This opens a number of freedom degrees in changing the hazard footprint. In the following, plausible factors of change in the environment and in society are listed that are possibly influencing flood risk. The list is ordered systematically by the distinction between changes in the upstream catchments that are relevant for the changing the boundary conditions of the risk analysis and changes in the narrow area of interest.

Environmental changes in the upstream catchment

Dams, reservoirs, and regulated lakes

The most obvious factor that is altering the hydrology of a river basin is the construction of dams and reservoirs. Dams are constructed for temporarily holding back the water flow and store water in the reservoir. Reservoirs can have multiple purposes, ranging from hydropower and electricity production to drought management and flood control. The effect of an upstream reservoir on downstream risk is given by the retention effect of the reservoir. A reservoir with the purpose for flood risk reduction is constructed to store the volume of water during the peak flow exceeding the discharge capacity of the downstream river reaches. After the rainfall event, the reservoir is steadily releasing the stored flood water over a longer period. This leads to an attenuation of the flood peak downstream. Depending on the storage volume, the water level in the reservoir must be lowered before the rainfall event to increase the storage capacity for the inflow flood volume. This requires the establishment and functioning of weather forecast and early warning systems. In many river basins, multiple reservoirs exist along the river course. In these river basins, the hydrological regime is remarkably altered. Flood attenuation is one of the main purposes for many of the existing dams and for some major dams of the world currently under construction (ICOLD International Commission on Large Dams, 2021). In total, 7450 large reservoirs worldwide are in operation for the purpose of flood control.

Reservoirs for other purposes as well as regulated lakes also store water during rainfall events und thus have similar effects on flood frequency and magnitude at downstream river reaches. However, the effects of these reservoirs are not as optimized for flood control as reservoirs built for flood control. The attenuation effect of a reservoir or a regulated lake is measured by comparing the flood frequency curve of the inflowing and the outflowing flood hydrograph. This requires discharge measurements upstream and downstream of the reservoir, which are in many cases commercially sensitive and not available. A simplified method for estimating the retention effect is to calculate the flood attenuation index. This is the ratio of dam outflow to inflow at the flood peak (Zhao et al., 2020). The analysis of the effects of reservoirs on hydrology is analyzed by inserting reservoir operation rules in hydrological models (Fleischmann et al., 2019; Meng et al., 2019). Boulange et al. (2021) estimate that dams reduce the number of flood-exposed population downstream of dams by 20% in a climate change scenario if assuming population as constant.

A similar flood attenuation effect arises from natural dams developed by large landslides, debris flows, or rock avalanches that obstruct the main river course of a valley bottom. Until the river erodes the deposits and enlarges the discharge capacity through the deposits, a natural dam delimits the flow through the deposits. In case of floods, a share of the flood water is held back by the natural dam. This leads to backwater flooding of the area upstream of the natural dam and to a reduction of the flood peak downstream. Thus, their effect on flood risk is twofold: flood risk in downstream areas may decrease while in upstream areas risk may increase.

An indirect effect of dams and reservoirs on flood risk is the alteration of sediment transport. Dams are holding back sediment fluxes from upstream catchments. Sediment transport downstream of dams is limited, except during operation phases intended for emptying the reservoir from sediment. Thus, long periods (decades) without sediment input from upstream change river morphology and vegetation. Sediment transport capacity increases and thus riverbed erosion tends to increase. This results in increasing conveyance capacity downstream and thus in reduced probability of flooding in the floodplain. Another effect of flow regulation by reservoir operation is a change in vegetation in and near the riverbed. The decreasing frequency of high-water situations leads to the growth of forests and dense vegetation. This has a negative effect from the flood risk perspective as dense vegetation increases the surface friction, lowers the flow velocity, and ultimately increases water levels during flood situations. This eventually increases the probability of flooding in the floodplain. Moreover, the presence of trees in flood areas increases the availability of large wood recruitment.

In summary, dams and reservoirs have in many cases a flood attenuation effect and thus tend to reduce flood risks downstream of the facility. In rare cases, dams of reservoirs may collapse or may be operated erroneously and trigger widespread flooding downstream with catastrophic consequences (Larrauri & Lall, 2020). This residual risk will increase in the near feature because the majority of existing dams are reaching their lifetime (Perera et al., 2021).

Land use changes at catchment scale

The outflow from a river basin or a mountainous catchment during or after a rainfall event is varying with a high number of factors related to the land cover and land use. The study of the influence of environmental conditions on catchment outflow is a wide field of research in the discipline of hydrology and covers multiple spatial and temporal scales. Vegetation cover (type of vegetation, areal extent, and canopy density) and especially forest cover markedly influence interception during rainfall. A higher forest cover intercepts a relatively higher share of the rainfall and reduces surface runoff. Thus, afforestation and reforestation lead to a reduced flood peak. Prior to a rainfall event, vegetation cover controls basin evapotranspiration and thus controls antecedent conditions of the catchment, e.g., soil moisture and soil saturation. Moreover, forests and natural vegetation are related with natural soils that are not compacted by anthropogenic activities such as some agricultural practices. Therefore, soils in forested areas have a higher infiltration capacity than soils in agricultural land and catchments with a higher share of forested areas have lower flood peaks than catchments with a high share of areas with compacted soils or bare land (Schilling et al., 2014). Forested and reforested areas reduce surface runoff and eventually flood peak but also sediment and nutrient load of rivers (Ouyang et al., 2013). As surface runoff increases over agricultural land and especially over temporarily bare soils, soil erosion increases too, and thus fine sediment is directed toward the water courses and ultimately to the river network.

The intensification of agricultural land does not only lead to soil compaction (Alaoui et al., 2018), but comprises also investments into drainage systems. This infrastructure drains the water from agricultural land in floodplains and transports the water toward the main river course. Depending on the timing of the flood peak during a rainfall event, this additional water from the floodplains can alter the flood peak in the river if synchronized with the peak flow of the main river. The drainage system decreases the retention effect of the floodplains, and a higher share of the rainfall is directly converted into runoff within the time period of the rainfall event. The transformation of wetlands to agricultural areas has a similar effect on flood peak as drainage systems (O'Connell et al., 2007). The retention capacity is lowered and the drainage from these transformed areas is increased (Gulbin et al., 2019). The study of Javaheri and Babbar-Sebens (2014) shows that wetlands can reduce the peak flow in downstream floodplains up to 42%, flood areas up to 55%, and maximum flow velocity up to 15%. Even more important than these land use changes for altering flood peaks is the transformation of agricultural land to settlement and infrastructure with sealed soils (Monk et al., 2019). Sealed soils totally inhibit the infiltration of rainfall water and thus reduce the storage capacity of the soil layer. Moreover, sealed surfaces have low surface frictions. Surface runoff is therefore fast on these surfaces and catchments with a high share of sealed areas react fast.

As land use has changed significantly in the last centuries, nonstationarity in the processes of surface runoff generation and hydrological catchment characteristics must be expected widely. However, the impact of land use changes on stream flow and floods is complex and studies often obtain contradictory results (Rogger, Chirico, et al., 2017).

Changes in the glacial and periglacial environment

Mountain permafrost is characterized by ice-rich sediments and quaternary deposits. The ice layers below the surface inhibit infiltration into deeper parts of the substrate and thus rainfall on permafrost leads to rapid reactions of high-alpine catchments. Locally, also potentially very deep groundwater storage volumes as, for example, debris slope below rock walls and deep sedimentary deposits can be impermeable because of the existence of thin ice layers. Warming of permafrost thus eliminates the ice layers in the long term, i.e., with a long delay and potentially increases the storage capacities of high alpine catchments (Rogger, Agnoletti, et al., 2017). However, permafrost thawing will decrease slope stabilities and thus led to increased landslide activities. These processes provide an increased amount of material for debris flow and sediment transport in high alpine torrents.

Natural climate variability

Flood frequency varies naturally at decadal and centennial time scales. The variability of flood occurrence is related with the long-term variability of the global atmospheric circulation patterns. Paleoflood reconstructions show that periods with higher flood occurrence in the region of the European Alps coincide with warmer summer temperatures (Glur et al., 2013) and lower with cooler summer temperatures. Increased flood frequency is correlated with latitudinal shifts of Atlantic and Mediterranean storm tracks. A known influence of climate variability on flood occurrence are the El Niño and La Niña events, the extremes of ENSO climate variability (Emerton et al., 2017). These events influence river flow and flooding at the global scale. In Central Europe, a high number of floods occurred during and at the end of the little ice in the 19th century (Brönnimann et al., 2019). An increased frequency of flood-prone weather types for the European Alps has been found in the mid-19th century and a decreased frequency in the postwar period, consistent with a climate reconstruction that shows changes in the cyclonic flow over Western Europe.

Also in other parts of Europe, a relation between flood frequency and climate variability has been found (e.g, Ward et al., 2008; Wirth et al., 2013). Periods with a cooler climate are periods with advancing glaciers, a more extensive snow cover during winter, and a higher potential for snow melt influenced floods. Volcanic eruptions can also have a temporal influence on flood occurrence (Rössler & Brönnimann, 2018). Schmocker-Fackel and Naef (2010) analyzed flood-rich and flood-poor periods in Switzerland and found that periods with reduced flood occurrence coincide with periods of minimum solar activity. An overview of regions where flood occurrence is actually increasing or decreasing in Europe is given by Blöschl et al. (2017).

Environmental changes in the floodplain

Environmental changes in the floodplain itself, i.e., in most cases the area of interest for risk analysis, alter both surface water floods and riverine floods. Changes in the floodplain can be gradual or abrupt. They can have natural causes but also anthropogenic causes, ranging from changes in the river morphodynamics and flood regime, to changes in the adjacent vegetation or disruptive changes by flood events, for example, by levee failures. Anthropogenic interventions are more or less the most relevant driver of flood risk in a floodplain; that is, the construction of flood defenses such as levees and dams or river restoration projects. Furthermore, the construction of levees as flood protection measures in one floodplain can have adverse effects in downstream floodplains and thus, result in trade-offs between upstream and downstream floodplains. Moreover, floodplains can be affected by land subsidence due to drainage or groundwater extraction. This results in increasing flood hazards and consequently, increasing flood risk (Carisi et al., 2017). In the following, an overview of environmental changes in the floodplain is given.

Changes in river morphology at reach scale

Themorphologyoftheriverisimportantforfloodriskasitdeterminesthedischargecapacityorconveyancecapacity.Thehigherthetransportcapacity,theloweristheprobabilityofinundationintheadjacentriverfloodplain(Fig.2.4).Thus,floodoccurrence probabilityinthefloodplainisdeterminedmainlyberiverwidthandriverdepth,i.e.,the geometryoftherivercrosssections.

Water flows and sediment supply and transport are inherently unsteady (Church & Ferguson, 2015). Thus, natural rivers change their channel morphology periodically. These changes are mostly driven by lateral erosion and riverbed erosion. Especially flood events can change the river morphology because of the high shear stress on riverbed and riverbanks. Besides riverbed erosion and aggradation during and after flood events, also adjacent vegetation is changed (Guan et al., 2016). Bank erosion leads to the elimination of dense bank vegetation and the hydrodynamic force on vegetated river islands can

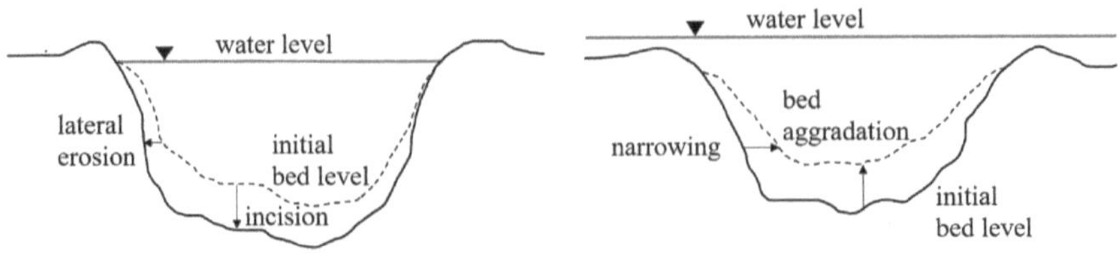

FIGURE 2.4 River morphology. The effect of river channel morphology changes for flood risk change. Fluvial erosion amplifies the river conveyance capacity and reduces inundation probability in the floodplain (left). Bed aggradation reduces the river conveyance capacity and amplifies inundation probability (right). *Modified from Slater, L.J., Singer, M.B., & Kirchner, J.W. (2015). Hydrologic versus geomorphic drivers of trends in flood hazard. Geophysical Research Letters, 42(2), 370–376. https://doi.org/10.1002/2014GL062482.*

lead to the elimination of vegetation within the river channel. This lowers hydraulic friction and leads to an increased flow velocity. Riverbank vegetation and riparian forests have an effect on bank stabilization and the flow regime (Abernethy & Rutherfurd, 2001). Riparian vegetation increases stream bank resistance and critical shear stress (Pollen & Simon, 2005) and reduces flow velocity at the stream bank interface by increasing channel roughness provoking flow energy dissipation through plant deformation. One approach to implement the mechanical effects of roots on bank erosion in modeling is by adapting critical shear stress. Further, the degree of protection depends on the location, root depth, and species composition (Slater et al., 2019) showed that river morphology adjusts to climate variability. In addition, river morphology determines also the water surface elevation in case of floods. The substrate of the river channel determines friction and with this it determines flow velocity and water surface levels. A coarser surface has a higher friction. This lowers flow velocity and heightens water surface elevation. Vice versa, a smooth surface increases flow velocity. Another factor that determines flow conditions in the river channel is the slope.

An important driver of morphology change during floods is the failure and breach of lateral levees. The local change of water flow and sediment at levee breaches triggers erosion and sedimentation processes within the river channel and in the adjacent floodplain. This change obviously modifies the inundation pattern in the floodplain. Flood models used for the elaboration of flood inundation maps rarely consider levee breaches. In this case, the water outflow from the levee breach potentially increases the flooded area. Thus, risk analysis based on flood maps of floodplains with river levees that do not consider levee breaches underestimates risks.

More relevant than natural causes of changing river morphology are anthropogenic causes (Gregory, 2006). In Europe, rivers have been modified since more than 300 years. The reasons for actively modifying river morphology range from making rivers navigable, making use of hydropower generation, to flood protection. More recently, the main motivation for changing river morphology is river revitalization. These anthropogenic interventions are more or less the most relevant driver of flood risk change.

The intention of flood risk reduction by river engineering is to enhance river conveyance capacity. The more water can flow within the river channel, the lesser is the occurrence probability of inundation in the floodplain. This aim can be reached by the construction of lateral levees. These constructions enhance conveyance capacity by heightening the levels of the lateral shores. Thus, the cross-sectional area is increased. However, the heightening of the lateral riverbanks is limited, especially in urban areas. Higher lateral dams require the heightening of the access roads to bridges too with consequences for city planning. Thus, another option is to artificially deepen the riverbed by excavation. This option is limited because it influences the stability of the riverbanks and foundations of existing flood defenses. A third option is to enlarge the river channel to extend conveyance capacity. This is the most ideal situation, but it is usually limited because the lateral borders of rivers in urban areas are occupied by houses or road infrastructure. All three options require a balancing as they can have unintended consequences. Widening the river too much can lead to a decrease of sediment transport capacity and to the aggradation of the riverbed. This is eventually reducing conveyance capacity (Hall et al., 2014). A narrow river channel tends to incise if sediments supply from upstream is limited. The increasing of conveyance capacity has also another effect for flood risk reduction: a wider river has not only a lower probability of inundation in the floodplain but inundation depths are also lower on average because a higher share of the water remains in the channel in comparison to narrower rivers. Structural flood defenses such as lateral levees have an effect downstream. If the inundation of the floodplain is prevented upstream, more water enters the downstream floodplain. This potentially enhances peak discharge downstream. Hence, the effects of river engineering works must be analyzed at larger scale, e.g., at river basin scale (Seher & Löschner, 2018).

Traditional engineering solutions against flooding and lateral mobility of river channels (e.g., straightening, embankment, channel dredging, bank protection) have strongly altered basic functions of hydrosystems, such as the sediment, water, and nutrients exchanges between the channel and the floodplain, or the longitudinal continuity of sediment transport along stream networks (Arnaud et al., 2019; Bravard et al., 1999; Gilvear, 1999). It became also clear that the physical status of rivers has been negatively impacted by other strong human pressures, such as gravel mining and hydroelectricity (Best, 2019). Many human interventions in the river system today are river restoration projects aiming at improving the ecological quality of rivers by dam or dike removal, riverbed widening, artificial sediment replenishment, among others (Wohl et al., 2015). Today, floodplain restoration and giving more space to rivers are actively promoted in flood risk management. These goals of restoring or enhancing catchment and river processes that have been affected by human interventions are also aiming at reducing flood hazards. These activities are summarized by the term "natural flood management" (Dadson et al., 2017).

The draining of floodplains in the past centuries has a long-term effect on the soil subsidence. The drying of former wetlands leads to the compaction of the whole soil layer and thus leads to shrinking. The subsidence of the soil layer in the floodplain in comparison a more stable riverbed increases flood inundation depths and thus can alter flood risk in the long term. Groundwater extraction has similar effects on soil subsidence and consequently on flood risk change (Carisi et al., 2017).

The analysis of the effects of morphological changes on flood risks requires to study the environmental changes in the entire catchment in an integrated way. Only with this perspective, local changes in river morphology and ecosystems can be evaluated in context with upstream catchment processes (Molnar et al., 2002).

Climate changes

Flood processes heavily depend on rainfall. Since rainfall is an important parameter of global and regional climate and climate is expected to change, a change in rainfall intensity and frequency, especially extreme rainfall events, is also expected to alter flood processes (IPCC, 2012). The most important aspect of a warming climate is the increase of water in the atmosphere. According to the Clausius–Clapeyron relation, saturation humidity will increase with higher temperature. This implies that extreme moisture transport will increase in a warmer climate. This can lead to more intensive rainfall or a higher frequency of extreme rainfall events. In mountainous areas, a warming climate influences the altitude level of the 0°C temperature line. The upward shift of this parameter determines the form of precipitation. Above this line, precipitation falls as snow and (more or less) below this line precipitation is falling as rain. The upward shift of the snow line extends the areas where precipitation is falling as rain and thus increases the contributing areas of flood processes. It is expected that this change is also influencing the probability of rainfall on snow cover (Rössler et al., 2014). A warmer atmosphere in mountainous areas potentially leads to the intensification of convective processes and to a higher rainfall intensity during thunderstorms. Flood peaks tend also to shift seasonally (Blöschl et al., 2017). In upstream catchments, climate change will influence soil erosion processes (Li & Fang, 2016). This potentially causes the formation of gully erosion and an increase of sediment transport in downstream river reaches. The increase of temperatures and changing precipitation patterns affect glacial and periglacial areas in the mountains. Glaciers are shrinking and permafrost ice is melting. This leads to an increase of exposed sediments that might become susceptible for erosion and triggering debris flows (Keiler et al., 2010).

Climate change is also expected to change atmospheric circulation patterns. This might lead to changes in regional weather types. This changes spatiotemporal patterns of the rainfall and can play an important role for flood runoff generation (Martius et al., 2020). Depending on the topology of the river network, the timing of stream flow accumulation along the river network driven by spatially distributed rainfall influences peak discharge in the individual river reaches (Zischg, Felder, Mosimann, et al., 2018; Pattison et al., 2014; Zoccatelli et al., 2011). Hence, the synchronization of peak flows in a river network may change with changing rainfall patterns. Flood assessment at the catchment scale is dependent on space—time variability of rainfall within a catchment, hydrological (soil and vegetation) processes leading to local runoff formation, and the propagation through the river network.

Socioeconomic changes

Besides changes in the natural environment, flood risk is also changing due to variations in the exposed elements at risk and their vulnerability. The elements at risk increase with socioeconomic development (Elmer et al., 2012). This increase in exposure directly enters the risk formula (Eq. 2.3) and thus has a direct effect on flood risk change. Settlement growth and thus the increase of residential buildings are related to population growth. Studies have shown that the number of houses that are potentially exposed to flood processes increased by a factor of 7 in the past century (Fuchs et al., 2015). These are long-term changes of population exposure that are shown by the growth of houses. However, population growth increases also the mobility of people and therefore short-term exposure to hazards. By traveling, people temporarily pass hazardous areas that are in many cases avoided for constructing houses. Hence, fluctuation of the exposure of people and mobile assets is an important factor in analyzing natural risks (Keiler et al., 2005).

Infrastructure is increasing in parallel with population growth. This has wider impacts on the socioeconomic system. For example, in economically active areas, floodplains are increasingly occupied by production facilities, as these require relatively flat areas for their construction. With economic development, the elements at risk and the infrastructure in floodplains are increasing both in terms of quantity and monetary value.

Only if population and infrastructure exposed to floods are vulnerable, a flood event results in flood damages. The state of susceptibility to harm from exposure to stresses associated with environmental and social change and from the absence of capacity to adapt mirrors regional socioeconomic characteristics and varies in space and time (Adger, 2006). In flood risk research, physical vulnerability is distinguished from social vulnerability. Physical vulnerability is defined as the degree of damage. The damage is expressed as a share of the total value of an object. The values range from 0 to 1, where 0 means no vulnerability and 1 total damage. Social vulnerability describes the ability or inability of a society to cope with the adverse impacts of a hazard (Cutter, 2003). Both physical and social vulnerability change over time. The experience of flood events often results in a lower social vulnerability against the subsequent flood events. This is the effect of social learning. Homeowners see what can happen and reduce the physical vulnerability of their houses by installing object protection measures (Elmer et al., 2010; Kreibich et al., 2011; Thieken et al., 2007). In relation to the increase in gross domestic product and in population, fatalities and losses do not increase to the same extent and are thus relatively decreasing (Jongman et al., 2015). However, vulnerability can also increase over time. An example of an increasing physical vulnerability is the installation of home automation techniques in houses that result in high damages in case of floods. These damages due to new characteristics of the houses are additional to the other structural damages. Increasing mobility and the dependency of the economy on traffic, e.g., due to the optimization of supply chains, lead to an increasing social vulnerability. Highly interconnected industrial production processes that are spatially distributed are more vulnerable than single production facilities. The more infrastructure is connected, the higher are the probabilities of cascading effects of flood on the functioning of infrastructure (Wang et al., 2018). This interconnectedness of infrastructure is increasing vulnerability. Hence, vulnerability changes with the technological advancement. New technologies allow increasing efficiency and reduce redundancy what in case of widespread flood leads to outage of parts of the production system. These outages cascade through the whole socioeconomic system. Moreover, physical vulnerability changes with legislation. In countries with a strong land use planning sector, the construction of houses in hazard zones is prohibited or is only possible with constraints (Röthlisberger et al., 2017). These constraints aim at reducing vulnerability, e.g., the obligation to construct new buildings in a flood save way. Last, but not the least, vulnerability reduction is target of flood risk management strategies, and thus it changes with a widespread implementation of these strategies.

Coevolution of key drivers of change

In the previous chapters, we have seen that many of the components of the risk formula (Eq. 2.3) change over time. Some changes occur at a large timescale (decades to centuries), whereas others occur at shorter timescales (daily to yearly). The built environment in floodplains, whether the settlement area or the river channel, is subject to changes and coevolutionary dynamics in both society and nature (Di Baldassarre et al., 2013). As Vitousek et al. (1997) postulated, the human impact on nature is now considerably larger than at any point in history. This is also true for the floodplains, as humans are shaping landscapes with the built environment, including the rivers in the Anthroposphere by modifying their courses. These impacts of society on nature influence future risk pathways. The spatiotemporal development of these drivers of change in flood risk leads to difficulties in predicting future flood risk. Consequently, recent studies have extended the framework of risk analysis toward a spatiotemporal framework as drivers for flood risk changes are varying in space and time (Ahmad & Simonovic, 2013; Zischg, Hofer, Mosimann, et al., 2018).

To improve the understanding of flood risk change, the drivers of flood risk change are described from a coevolutionary perspective. This widens the explanatory power of multiple drivers in hazard dynamics, exposure dynamics, and vulnerability dynamics. The coevolutionary framework provides a guideline for analyzing and explaining the linkage between these three main factors of flood risk (Fuchs et al., 2017). Coevolution includes two or more interdependently evolving systems (Gual & Norgaard, 2010). Two systems coevolve when their change in space or time depends on each other. The aim of this framework is to analyze and understand the intertwined changes within the different interacting systems, where coevolutionary dynamics are path-dependent (Kallis, 2007). These dynamics include social adaptation to environmental change and the shaping of the environment by social adaptation. A central theme inherent to the coevolutionary perspective is the analysis of human adaptation to catastrophic flood events. The analysis of coevolution of the drivers of flood risk changes opens the opportunity to analyze the feedback mechanisms, i.e., to analyze how each driver of change influences others (Roe, 2009). The interaction must not necessarily endure over the whole period but can be reactive and abrupt. Coevolution can consist of slow and fast dynamics and can be variable in space and time.

One of the most postulated coevolutionary dynamics is the increase of exposure to extreme floods with the increase of flood protection measures that reduce frequent flooding. For example, Ferdous et al. (2020) observe a higher growth of population density with higher levels of flood protection. This is sometime termed as the "levee paradox" (White, 1945). Thus, human adaptation coevolves with changes in flood frequency. Herein, the interaction is a two-way feedback. A flood event of a certain magnitude may initiate the installation of flood protection measures against flood events of this magnitude. Subsequently, only flood events markedly exceeding this magnitude will cause flood damages and trigger new adaptation (Sivapalan & Blöschl, 2015). If the impacts of flood events are large enough to influence not only the local level but are relevant also for national or international societies, governments at higher hierarchy may adapt their legislative framework, subsidy policies, and risk management policies (Lane, 2014). This has feedback effects to adaptation at community and individual levels.

It can also be that coevolutionary dynamics in the drivers of flood risk change have opposing trends. For example, Boccard (2021) observed two distinct trends in flood impacts: Fatalities from flood events are falling worldwide while financial losses are rising. Increasing values at risk compete with opposing drivers of flood risk reduction measures implemented by individuals and the public. Hence, changes in exposure and vulnerability are influenced by governmental interventions and regulations and by the actions of individuals. Increasing values at risk also coevolve with the observed increase of insured damages. Barredo (2009) thus proposed to normalize the time series of flood damages with the increasing values at risk.

The physical drivers of flood risk change are also intertwined, involving a multitude of interlinked processes. Catchments respond to changes in land use and climate in a nonlinear way. Climate, vegetation, and soil coevolve and result in landscape changes and runoff characteristics of catchments (Gaál et al., 2012). Changes of vegetation cover and type cause changes of hydrological processes affecting flooding by changes of interception, evapotranspiration, and infiltration (Runyan et al., 2012). Ultimately, changes in vegetation influence soil moisture and thus antecedent conditions of catchments. Similar coevolutionary dynamics arise with human modifications of land cover in catchments. Deforestation or afforestation also change the runoff properties of the catchment. A coevolution is also expected with the shrinking of glaciers and the melting of subsurface permafrost ice and the arising of new source areas for debris flows in high mountain catchments.

Davenport et al. (2021) showed that change in precipitation coevolves with changes in damages of pluvial floods. More extreme rainfall does not only increase surface runoff and flooding but also may increase erosion and landslides. Perdigão and Blöschl (2014) showed that changes of spatial patterns of rainfall markedly cause changes in flood risks.

The influence of climate change, as well as that of human interventions on nature, is being studied comprehensively and in a differentiated manner with regard to future flooding scenarios. The contribution of humans to changes in the sediment budget of the world's largest rivers was investigated on a global scale by Syvitski et al. (2005). Human-induced soil erosion increases the sediment load in rivers. At the same time, less sediment is transported to the coasts due to reservoir effects caused by humans (e.g., large dam projects). However, changes in soil erosion with an influence on the bedload balance are not only caused by humans, but also to a large extent by climate change. There are direct and indirect influences of climate change on soil erosion with associated changes in the sediment content of rivers, which can cause both an increase in sediment load and a decrease, depending on the climatic conditions (Li & Fang, 2016). Yang et al. (2018) studied the changes in sediment load in the Yellow River in China and concluded that it has decreased by 60% since 1997 and that climatic changes, in addition to human influence, have contributed a large part to this. An opposite effect was researched by Dibike et al. (2018) in Canada. The future sediment load was modeled with the inclusion of climatic changes in the Lower Athabasca River. The results indicate that an overall increase of up to 50% in sediment load in the river as well as in the deposition area is expected for the next 60 years. Increased sediment load with regard to an increase in extreme events was investigated by Tu et al. (2020). They concluded that floods will increase especially during events with a 200-year return period, resulting in more sediment being transported and deposited.

According to Naylor et al. (2017), river geometry dynamically adapts to changing flow regimes, which can lead to extreme events altering channel capacity and floodplain geometry, thereby changing the risk of future flooding. This form of feedback means that extreme events that occurred in the past represent a modified flood risk for future events due to the geomorphological response (Croke et al., 2015), and Coulthard and Van De Wiel (2007) discuss these morphological changes of river morphology by river bank failures as a manifestation of self-organized criticality. On the one hand, it is known that high-magnitude flood events can erode, transport, and deposit large amounts of sediment, which can reshape the river system and have a corresponding effect on channel capacity (the cross-sectional area of the channel) and thus on flow velocity. On the other hand, it has been shown that increasing flooding is also caused by ongoing geomorphological changes (sedimentation in the channel), which gradually reduce channel capacity. The influence of river morphology and interactions within the riverbed on flood behavior has long been underestimated (Raven et al., 2010). Lane et al. (2007) criticize the lack of consideration of geomorphological processes in flood risk management. It shows how important it is to consider the natural processes of a river. Lateral erosion, for example, can cause high sediment inputs and thus influence river transport and increase flood risk. According to their calculations, a sediment input during 16 months can cause a short-term bed elevation, which leads to about half of the increase in flood area in an event with a return period of about 1−5 years. Slater et al. (2015) complement this research by showing the importance of the coevolution of river morphology change and river conveyance capacity. On average, a 1% decrease in channel velocity or flow cross-sectional area resulted in a 2% increase in flood hazard frequency. These results suggest that shifts in sediment flow, grain size, and/or vegetation may play a significant role in controlling flood hazard frequency. Slater (2016) examined the findings from the United States for England and Wales. She concluded that a 10% decrease (or increase) in channel capacity would result in an increase (or decrease) in flood frequency of about 1.5 days per year on average. The widespread increase in the frequency of flood events was amplified by both hydrological and geomorphological effects.

Important sediment contributors to rivers are tributaries. Picking up on this theme, Bressan et al. (2020) showed that both river geometry and runoff input from tributaries are key factors determining the magnitude and duration of shear stress. Larger tributaries can keep the water depth higher and thus increase the shear stress. Smaller tributaries can add clearer water and thus reduce sediment concentrations. Failure to include tributaries or the use of incorrect hydrographs leads to a large under- or overestimation of sediment volumes.

In summary, there are many possibilities how one driver of flood risk change interacts with other drivers of change. The effects of these intertwined drivers on flood risk have to be disentangled before being analyzed in an isolated way. In the next chapters, methods for disentangling the effects of single drivers on flood risk change are shown.

References

Abernethy, B., & Rutherfurd, I. D. (2001). The distribution and strength of riparian tree roots in relation to riverbank reinforcement. *Hydrological Processes, 15*, 63–79. https://doi.org/10.1002/hyp.152

Adger, W. N. (2006). Vulnerability. *Global Environmental Change, 16*(3), 268–281. https://doi.org/10.1016/j.gloenvcha.2006.02.006

Ahmad, S. S., & Simonovic, S. P. (2013). Spatial and temporal analysis of urban flood risk assessment. *Urban Water Journal, 10*(1), 26–49. https://doi.org/10.1080/1573062X.2012.690437

Alaoui, A., Rogger, M., Peth, S., & Blöschl, G. (2018). Does soil compaction increase floods? A review. *Journal of Hydrology, 557*, 631–642. https://doi.org/10.1016/j.jhydrol.2017.12.052

Arnaud, F., Schmitt, L., Johnstone, K., Rollet, A. J., & Piégay, H. (2019). Engineering impacts on the Upper Rhine channel and floodplain over two centuries. *Geomorphology, 330*, 13–27. https://doi.org/10.1016/j.geomorph.2019.01.004

Barredo, J. I. (2009). Normalised flood losses in Europe: 1970-2006. *Natural Hazards and Earth System Science, 9*(1), 97–104. https://doi.org/10.5194/nhess-9-97-2009

Bates, P. D., & De Roo, A. P. J. (2000). A simple raster-based model for flood inundation simulation. *Journal of Hydrology, 236*(1–2), 54–77. https://doi.org/10.1016/S0022-1694(00)00278-X

Bernet, D. B., Trefalt, S., Martius, O., Weingartner, R., Mosimann, M., Röthlisberger, V., & Zischg, A. P. (2019). Characterizing precipitation events leading to surface water flood damage over large regions of complex terrain. *Environmental Research Letters, 14*(6). https://doi.org/10.1088/1748-9326/ab127c

Bernet, D. B., Zischg, A. P., Prasuhn, V., & Weingartner, R. (2018). Modeling the extent of surface water floods in rural areas: Lessons learned from the application of various uncalibrated models. *Environmental Modelling and Software, 109*, 134–151. https://doi.org/10.1016/j.envsoft.2018.08.005

Best, J. (2019). Anthropogenic stresses on the world's big rivers. *Nature Geoscience, 12*(1), 7–21. https://doi.org/10.1038/s41561-018-0262-x

Blöschl, G., Hall, J., Parajka, J., Perdigão, R. A. P., Merz, B., Arheimer, B., Aronica, G. T., Bilibashi, A., Bonacci, O., Borga, M., Čanjevac, I., Castellarin, A., Chirico, G. B., Claps, P., Fiala, K., Frolova, N., Gorbachova, L., Gül, A., Hannaford, J., … Živković, N. (2017). Changing climate shifts timing of European floods. *Science, 357*(6351), 588–590. https://doi.org/10.1126/science.aan2506.

Boccard, N. (2021). Analysis of trends in disaster risk. *International Journal of Disaster Risk Reduction, 53*. https://doi.org/10.1016/j.ijdrr.2020.101989

Bomers, A., Schielen, R. M. J., & Hulscher, S. J. M. H. (2019). Application of a lower-fidelity surrogate hydraulic model for historic flood reconstruction. *Environmental Modelling and Software, 117*, 223–236. https://doi.org/10.1016/j.envsoft.2019.03.019

Boulange, J., Hanasaki, N., Yamazaki, D., & Pokhrel, Y. (2021). Role of dams in reducing global flood exposure under climate change. *Nature Communications, 12*(1). https://doi.org/10.1038/s41467-020-20704-0

Bravard, J. P., Landon, N., Peiry, J. L., & Piégay, H. (1999). Principles of engineering geomorphology for managing channel erosion and bedload transport, examples from French rivers. *Geomorphology, 31*(1–4), 291–311. https://doi.org/10.1016/S0169-555X(99)00091-4

Bressan, F., Mantilla, R., Schilling, K. E., Palmer, J. A., & Weber, L. (2020). Hydrologic-hydraulic modeling of sediment transport along the main stem of a watershed: Role of tributaries and channel geometry. *Hydrological Sciences Journal, 65*(2), 183–199. https://doi.org/10.1080/02626667.2019.1687897

Brönnimann, S., Frigerio, L., Schwander, M., Rohrer, M., Stucki, P., & Franke, J. (2019). Causes of increased flood frequency in central Europe in the 19th century. *Climate of the Past, 15*(4), 1395–1409. https://doi.org/10.5194/cp-15-1395-2019

Carisi, F., Domeneghetti, A., Gaeta, M. G., & Castellarin, A. (2017). Is anthropogenic land subsidence a possible driver of riverine flood-hazard dynamics? A case study in Ravenna, Italy. *Hydrological Sciences Journal, 62*(15), 2440–2455. https://doi.org/10.1080/02626667.2017.1390315

Church, M., & Ferguson, R. I. (2015). Morphodynamics: Rivers beyond steady state. *Water Resources Research, 51*(4), 1883–1897. https://doi.org/10.1002/2014WR016862

Coulthard, T. J., & Van De Wiel, M. J. (2007). Quantifying fluvial non linearity and finding self organized criticality? Insights from simulations of river basin evolution. *Geomorphology, 91*(3–4), 216–235. https://doi.org/10.1016/j.geomorph.2007.04.011

Croke, J., Denham, R., Thompson, C., & Grove, J. (2015). Evidence of self-organized criticality in riverbank mass failures: A matter of perspective? *Earth Surface Processes and Landforms, 40*(7), 953–964. https://doi.org/10.1002/esp.3688

Cutter, S. L. (2003). The vulnerability of science and the science of vulnerability. *Annals of the Association of American Geographers, 93*(1), 1–12. https://doi.org/10.1111/1467-8306.93101

Dadson, S. J., Hall, J. W., Murgatroyd, A., Acreman, M., Bates, P., Beven, K., Heathwaite, L., Holden, J., Holman, I. P., Lane, S. N., O'Connell, E., Penning-Rowsell, E., Reynard, N., Sear, D., Thorne, C., & Wilby, R. (2017). A restatement of the natural science evidence concerning catchment-based "natural" flood management in the UK. *Proceedings of the Royal Society A: Mathematical, Physical and Engineering Sciences, 473*(2199). https://doi.org/10.1098/rspa.2016.0706

Davenport, F., Burke, M., & Diffenbaugh, N. S. (2021). Contribution of historical precipitation change to US flood damages. *Proceedings of the National Academy of Sciences, 118*(4). https://doi.org/10.1073/pnas.2017524118. e2017524118.

Di Baldassarre, G., Kooy, M., Kemerink, J. S., & Brandimarte, L. (2013). Towards understanding the dynamic behaviour of floodplains as human-water systems. *Hydrology and Earth System Sciences, 17*(8), 3235–3244. https://doi.org/10.5194/hess-17-3235-2013

Dibike, Y., Shakibaeinia, A., Eum, H. I., Prowse, T., & Droppo, I. (2018). Effects of projected climate on the hydrodynamic and sediment transport regime of the lower Athabasca River in Alberta, Canada. *River Research and Applications, 34*(5), 417–429. https://doi.org/10.1002/rra.3273

Elmer, F., Hoymann, J., Düthmann, D., Vorogushyn, S., & Kreibich, H. (2012). Drivers of flood risk change in residential areas. *Natural Hazards and Earth System Science, 12*(5), 1641–1657. https://doi.org/10.5194/nhess-12-1641-2012

Elmer, F., Thieken, A. H., Pech, I., & Kreibich, H. (2010). Influence of flood frequency on residential building losses. *Natural Hazards and Earth System Science, 10*(10), 2145–2159. https://doi.org/10.5194/nhess-10-2145-2010

Emerton, R., Cloke, H. L., Stephens, E. M., Zsoter, E., Woolnough, S. J., & Pappenberger, F. (2017). Complex picture for likelihood of ENSO-driven flood hazard. *Nature Communications, 8*. https://doi.org/10.1038/ncomms14796

Felder, G., Gómez-Navarro, J. J., Zischg, A. P., Raible, C. C., Röthlisberger, V., Bozhinova, D., Martius, O., & Weingartner, R. (2018). From global circulation to local flood loss: Coupling models across the scales. *Science of the Total Environment, 635*, 1225–1239. https://doi.org/10.1016/j.scitotenv.2018.04.170

Felder, G., Zischg, A., & Weingartner, R. (2017). The effect of coupling hydrologic and hydrodynamic models on probable maximum flood estimation. *Journal of Hydrology, 550*, 157–165. https://doi.org/10.1016/j.jhydrol.2017.04.052

Ferdous, M. R., Di Baldassarre, G., Brandimarte, L., & Wesselink, A. (2020). The interplay between structural flood protection, population density, and flood mortality along the Jamuna River, Bangladesh. *Regional Environmental Change, 20*(1). https://doi.org/10.1007/s10113-020-01600-1

Fleischmann, A., Collischonn, W., Paiva, R., & Tucci, C. E. (2019). Modeling the role of reservoirs versus floodplains on large-scale river hydrodynamics. *Natural Hazards, 99*(2), 1075–1104. https://doi.org/10.1007/s11069-019-03797-9

Fuchs, S., Heiss, K., & Hübl, J. (2007). Towards an empirical vulnerability function for use in debris flow risk assessment. *Natural Hazards and Earth System Science, 7*(5), 495–506. https://doi.org/10.5194/nhess-7-495-2007

Fuchs, S., Keiler, M., & Zischg, A. (2015). A spatiotemporal multi-hazard exposure assessment based on property data. *Natural Hazards and Earth System Sciences, 15*(9), 2127–2142. https://doi.org/10.5194/nhess-15-2127-2015

Fuchs, S., Röthlisberger, V., Thaler, T., Zischg, A., & Keiler, M. (2017). Natural hazard management from a coevolutionary perspective: Exposure and policy response in the European Alps. *Annals of the American Association of Geographers, 107*(2), 382–392. https://doi.org/10.1080/24694452.2016.1235494

Gaál, L., Szolgay, J., Kohnová, S., Parajka, J., Merz, R., Viglione, A., & Blöschl, G. (2012). Flood timescales: Understanding the interplay of climate and catchment processes through comparative hydrology. *Water Resources Research, 48*(4). https://doi.org/10.1029/2011WR011509

Gilvear, D. J. (1999). Fluvial geomorphology and river engineering: Future roles utilizing a fluvial hydrosystems framework. In *Geomorphology* (Vol. 31, pp. 229–245). https://doi.org/10.1016/S0169-555X(99)00086-0, 1–4.

Glur, L., Wirth, S. B., Büntgen, U., Gilli, A., Haug, G. H., Schär, C., Beer, J., & Anselmetti, F. S. (2013). Frequent floods in the European Alps coincide with cooler periods of the past 2500 years. *Scientific Reports, 3*. https://doi.org/10.1038/srep02770

Gregory, K. J. (2006). The human role in changing river channels. *Geomorphology, 79*(3–4), 172–191. https://doi.org/10.1016/j.geomorph.2006.06.018

Gual, M. A., & Norgaard, R. B. (2010). Bridging ecological and social systems coevolution: A review and proposal. *Ecological Economics, 69*(4), 707–717. https://doi.org/10.1016/j.ecolecon.2008.07.020

Guan, M., Carrivick, J. L., Wright, N. G., Sleigh, P. A., & Staines, K. E. H. (2016). Quantifying the combined effects of multiple extreme floods on river channel geometry and on flood hazards. *Journal of Hydrology, 538*, 256–268. https://doi.org/10.1016/j.jhydrol.2016.04.004

Gulbin, S., Kirilenko, A. P., Kharel, G., & Zhang, X. (2019). Wetland loss impact on long term flood risks in a closed watershed. *Environmental Science and Policy, 94*, 112–122. https://doi.org/10.1016/j.envsci.2018.12.032

Hall, J., Arheimer, B., Borga, M., Brázdil, R., Claps, P., Kiss, A., Kjeldsen, T. R., Kriaučiūnienė, J., Kundzewicz, Z. W., Lang, M., Llasat, M. C., Macdonald, N., McIntyre, N., Mediero, L., Merz, B., Merz, R., Molnar, P., Montanari, A., Neuhold, C., … Blöschl, G. (2014). Understanding flood regime changes in Europe: A state-of-the-art assessment. *Hydrology and Earth System Sciences, 18*(7), 2735–2772. https://doi.org/10.5194/hess-18-2735-2014.

ICOLD International Commission on Large Dams. (2021). *World register of dams*. https://www.icold-cigb.org/GB/world_register/general_synthesis.asp.

IPCC. (2012). *Managing the risks of extreme events and disasters to advance climate change adaptation: Special report of the Intergovernmental Panel on Climate Change.*

Javaheri, A., & Babbar-Sebens, M. (2014). On comparison of peak flow reductions, flood inundation maps, and velocity maps in evaluating effects of restored wetlands on channel flooding. *Ecological Engineering, 73*, 132–145. https://doi.org/10.1016/j.ecoleng.2014.09.021

Jongman, B., Winsemius, H. C., Aerts, J. C. J. H., Coughlan de Perez, E., van Aalst, M. K., Kron, W., & Ward, P. J. (2015). Declining vulnerability to river floods and the global benefits of adaptation. *Proceedings of the National Academy of Sciences, 112*(18), E2271–E2280. https://doi.org/10.1073/pnas.1414439112

Kallis, G. (2007). When is it coevolution? *Ecological Economics, 62*(1), 1–6. https://doi.org/10.1016/j.ecolecon.2006.12.016

Keiler, M., Knight, J., & Harrison, S. (2010). Climate change and geomorphological hazards in the eastern European Alps. In *Philosophical transactions of the royal society A: Mathematical, physical and engineering sciences* (Vol. 368, pp. 2461–2479). Royal Society. https://doi.org/10.1098/rsta.2010.0047, 1919.

Keiler, M., Zischg, A., Fuchs, S., Hama, M., & Stötter, J. (2005). Avalanche related damage potential- Changes of persons and mobile values since the mid-twentieth century, case study Galtür. *Natural Hazards and Earth System Science, 5*(1), 49–58. https://doi.org/10.5194/nhess-5-49-2005

Kienzler, S., Pech, I., Kreibich, H., Müller, M., & Thieken, A. H. (2015). After the extreme flood in 2002: Changes in preparedness, response and recovery of flood-affected residents in Germany between 2005 and 2011. *Natural Hazards and Earth System Sciences, 15*(3), 505–526. https://doi.org/10.5194/nhess-15-505-2015

Kreibich, H., Seifert, I., Thieken, A. H., Lindquist, E., Wagner, K., & Merz, B. (2011). Recent changes in flood preparedness of private households and businesses in Germany. *Regional Environmental Change, 11*(1), 59–71. https://doi.org/10.1007/s10113-010-0119-3

Lane, S. N. (2014). Acting, predicting and intervening in a socio-hydrological world. *Hydrology and Earth System Sciences, 18*(3), 927–952. https://doi.org/10.5194/hess-18-927-2014

Lane, S. N., Tayefi, V., Reid, S. C., Yu, D., & Hardy, R. J. (2007). Interactions between sediment delivery, channel change, climate change and flood risk in a temperate upland environment. *Earth Surface Processes and Landforms, 32*(3), 429–446. https://doi.org/10.1002/esp.1404

Larrauri, P. C., & Lall, U. P. (2020). Assessing the exposure of critical infrastructure and other assets to the climate induced failure of aging dams in the.

Li, Z., & Fang, H. (2016). Impacts of climate change on water erosion: A review. *Earth-Science Reviews, 163*, 94–117. https://doi.org/10.1016/j.earscirev.2016.10.004

Martius, O., Viviroli, D., Rössler, O., Zischg, A., Röthlisberger, V., Aemisegger, F., Schulte, L., Stuber, M., Pena, J. C., Ruiz-Villanueva, V., & Molnar, P. (2020). Understanding flood triggering mechanisms and flood risk changes.

Mazzorana, B., Zischg, A., Largiader, A., & Hübl, J. (2009). Hazard index maps for woody material recruitment and transport in alpine catchments. *Natural Hazards and Earth System Science, 9*(1), 197–209. https://doi.org/10.5194/nhess-9-197-2009

Meng, X., Chang, J., Wang, X., Wang, Y., & Wang, Z. (2019). Flood control operation coupled with risk assessment for cascade reservoirs. *Journal of Hydrology, 572*, 543–555. https://doi.org/10.1016/j.jhydrol.2019.03.055

Molnar, P., Burlando, P., & Ruf, W. (2002). Integrated catchment assessment of riverine landscape dynamics. *Aquatic Sciences, 64*(2), 129–140. https://doi.org/10.1007/s00027-002-8061-1

Monk, W. A., Compson, Z. G., Choung, C. B., Korbel, K. L., Rideout, N. K., & Baird, D. J. (2019). Urbanisation of floodplain ecosystems: Weight-of-evidence and network meta-analysis elucidate multiple stressor pathways. *Science of the Total Environment, 684*, 741–752. https://doi.org/10.1016/j.scitotenv.2019.02.253

Naylor, L. A., Spencer, T., Lane, S. N., Darby, S. E., Magilligan, F. J., Macklin, M. G., & Möller, I. (2017). Stormy geomorphology: Geomorphic contributions in an age of climate extremes. *Earth Surface Processes and Landforms, 42*(1), 166–190. https://doi.org/10.1002/esp.4062

Oberndorfer, S., Sander, P., & Fuchs, S. (2020). Multi-hazard risk assessment for roads: Probabilistic versus deterministic approaches. *Natural Hazards and Earth System Sciences, 20*(11), 3135–3160. https://doi.org/10.5194/nhess-20-3135-2020

O'Connell, E., Ewen, J., O'Donnell, G., & Quinn, P. (2007). Is there a link between agricultural land-use management and flooding? *Hydrology and Earth System Sciences, 11*(1), 96–107. https://doi.org/10.5194/hess-11-96-2007

Ouyang, Y., Leininger, T. D., & Moran, M. (2013). Impacts of reforestation upon sediment load and water outflow in the Lower Yazoo River Watershed, Mississippi. *Ecological Engineering, 61*, 394–406. https://doi.org/10.1016/j.ecoleng.2013.09.057

Pattison, I., Lane, S. N., Hardy, R. J., & Reaney, S. M. (2014). The role of tributary relative timing and sequencing in controlling large floods. *Water Resources Research, 50*(7), 5444–5458. https://doi.org/10.1002/2013WR014067

Perdigão, R. A. P., & Blöschl, G. (2014). Spatiotemporal flood sensitivity to annual precipitation: Evidence for landscape-climate coevolution. *Water Resources Research, 50*(7), 5492–5509. https://doi.org/10.1002/2014WR015365

Perera, D., Smakhtin, V., Williams, S., North, T., & Curry, A. (2021). *Ageing water storage infrastructure: An emerging global risk. UNU-INWEH report series* (Vol. 11).

Pollen, N., & Simon, A. (2005). Estimating the mechanical effects of riparian vegetation on stream bank stability using a fiber bundle model. *Water Resources Research, 41*(7), 1–11. https://doi.org/10.1029/2004WR003801

Raven, E. K., Lane, S. N., & Bracken, L. J. (2010). Understanding sediment transfer and morphological change for managing upland gravel-bed rivers. *Progress in Physical Geography, 34*(1), 23–45. https://doi.org/10.1177/0309133309355631

Roe, G. (2009). Feedbacks, timescales, and seeing red. *Annual Review of Earth and Planetary Sciences, 37*, 93–115. https://doi.org/10.1146/annurev.earth.061008.134734

Rogger, M., Agnoletti, M., Alaoui, A., Bathurst, J. C., Bodner, G., Borga, M., Chaplot, V., Gallart, F., Glatzel, G., Hall, J., Holden, J., Holko, L., Horn, R., Kiss, A., Kohnová, S., Leitinger, G., Lennartz, B., Parajka, J., Perdigão, R., … Blöschl, G. (2017). Land use change impacts on floods at the catchment scale: Challenges and opportunities for future research. *Water Resources Research, 53*(7), 5209–5219. https://doi.org/10.1002/2017WR020723.

Rogger, M., Chirico, G. B., Hausmann, H., Krainer, K., Brückl, E., Stadler, P., & Blöschl, G. (2017). Impact of mountain permafrost on flow path and runoff response in a high alpine catchment. *Water Resources Research, 53*(2), 1288–1308. https://doi.org/10.1002/2016WR019341

Rössler, O., & Brönnimann, S. (2018). The effect of the Tambora eruption on Swiss flood generation in 1816/1817. *Science of the Total Environment, 627*, 1218–1227. https://doi.org/10.1016/j.scitotenv.2018.01.254

Rössler, O., Froidevaux, P., Börst, U., Rickli, R., Martius, O., & Weingartner, R. (2014). Retrospective analysis of a nonforecasted rain-on-snow flood in the Alps-A matter of model limitations or unpredictable nature? *Hydrology and Earth System Sciences, 18*(6), 2265–2285. https://doi.org/10.5194/hess-18-2265-2014

Röthlisberger, V., Zischg, A. P., & Keiler, M. (2017). Identifying spatial clusters of flood exposure to support decision making in risk management. *Science of the Total Environment, 598*, 593–603. https://doi.org/10.1016/j.scitotenv.2017.03.216

Rougier, J. (2013). Quantifying hazard losses. In J. Rougier, R. S. J. Sparks, & L. Hill (Eds.), *Risk and uncertainty assessment for natural hazards* (pp. 19–39). Cambridge University Press.

Ruiz-Villanueva, V., Bodoque, J. M., Díez-Herrero, A., & Bladé, E. (2014). Large wood transport as significant influence on flood risk in a mountain village. *Natural Hazards, 74*(2), 967–987. https://doi.org/10.1007/s11069-014-1222-4

Runyan, C. W., D'Odorico, P., & Lawrence, D. (2012). Physical and biological feedbacks of deforestation. *Reviews of Geophysics, 50*(4). https://doi.org/10.1029/2012RG000394

Schilling, K. E., Gassman, P. W., Kling, C. L., Campbell, T., Jha, M. K., Wolter, C. F., & Arnold, J. G. (2014). The potential for agricultural land use change to reduce flood risk in a large watershed. *Hydrological Processes, 28*(8), 3314–3325. https://doi.org/10.1002/hyp.9865

Schmocker-Fackel, P., & Naef, F. (2010). Changes in flood frequencies in Switzerland since 1500. *Hydrology and Earth System Sciences, 14*(8), 1581–1594. https://doi.org/10.5194/hess-14-1581-2010

Seher, W., & Löschner, L. (2018). Balancing upstream–downstream interests in flood risk management: Experiences from a catchment-based approach in Austria. *Journal of Flood Risk Management, 11*(1), 56–65. https://doi.org/10.1111/jfr3.12266. Blackwell Publishing Inc.

Sivapalan, M., & Blöschl, G. (2015). Time scale interactions and the coevolution of humans and water. *Water Resources Research, 51*(9), 6988–7022. https://doi.org/10.1002/2015WR017896

Slater, L. J. (2016). To what extent have changes in channel capacity contributed to flood hazard trends in England and Wales? *Earth Surface Processes and Landforms, 41*(8), 1115–1128. https://doi.org/10.1002/esp.3927

Slater, L. J., Khouakhi, A., & Wilby, R. L. (2019). River channel conveyance capacity adjusts to modes of climate variability. *Scientific Reports, 9*(1). https://doi.org/10.1038/s41598-019-48782-1

Slater, L. J., Singer, M. B., & Kirchner, J. W. (2015). Hydrologic versus geomorphic drivers of trends in flood hazard. *Geophysical Research Letters, 42*(2), 370–376. https://doi.org/10.1002/2014GL062482

Syvitski, J. P. M., Vörösmarty, C. J., Kettner, A. J., & Green, P. (2005). Impact of humans on the flux of terrestrial sediment to the global coastal ocean. *Science, 308*(5720), 376–380. https://doi.org/10.1126/science.1109454

Thieken, A. H., Kreibich, H., Müller, M., & Merz, B. (2007). Coping with floods: Preparedness, response and recovery of flood-affected residents in Germany in 2002. *Hydrological Sciences Journal, 52*(5), 1016–1037. https://doi.org/10.1623/hysj.52.5.1016

Tu, T., Ercan, A., Carr, K. J., Kavvas, M. L., Trinh, T., Ishida, K., Nosacka, J., & Brown, K. (2020). Coupling hydroclimate-hydraulic-sedimentation models to estimate flood inundation and

sediment transport during extreme flood events under a changing climate. *Science of the Total Environment, 740*. https://doi.org/10.1016/j.scitotenv.2020.140117
UNISDR. (2015). Making development sustainable: The future of disaster risk management. In *Global assessment report on disaster risk reduction* (Vol. 4)United Nations, 2015.
Vitousek, P. M., Mooney, H. A., Lubchenco, J., & Melillo, J. M. (1997). Human domination of Earth's ecosystems. *Science, 277*(5325), 494–499. https://doi.org/10.1126/science.277.5325.494
Wang, W., Yang, S., Hu, F., Stanley, H. E., He, S., & Shi, M. (2018). An approach for cascading effects within critical infrastructure systems. *Physica A: Statistical Mechanics and Its Applications, 510*, 164–177. https://doi.org/10.1016/j.physa.2018.06.129
Ward, P. J., Renssen, H., Aerts, J. C. J. H., Van Balen, R. T., & Vandenberghe, J. (2008). Strong increases in flood frequency and discharge of the River Meuse over the late Holocene: Impacts of long-term anthropogenic land use change and climate variability. *Hydrology and Earth System Sciences, 12*(1), 159–175. https://doi.org/10.5194/hess-12-159-2008
White, G. (1945). *Human adjustment to floods*. USA: Department of Geography, University of Chicago.
Wirth, S. B., Glur, L., Gilli, A., & Anselmetti, F. S. (2013). Holocene flood frequency across the Central Alps - solar forcing and evidence for variations in North Atlantic atmospheric circulation. *Quaternary Science Reviews, 80*, 112–128. https://doi.org/10.1016/j.quascirev.2013.09.002
Wohl, E., Lane, S. N., & Wilcox, A. C. (2015). The science and practice of river restoration. *Water Resources Research, 51*(8), 5974–5997. https://doi.org/10.1002/2014WR016874
Yang, X., Sun, W., Li, P., Mu, X., Gao, P., & Zhao, G. (2018). Reduced sediment transport in the Chinese Loess Plateau due to climate change and human activities. *Science of the Total Environment, 642*, 591–600. https://doi.org/10.1016/j.scitotenv.2018.06.061
Zhao, G., Bates, P., & Neal, J. (2020). The impact of dams on design floods in the conterminous US. *Water Resources Research, 56*(3). https://doi.org/10.1029/2019WR025380
Zischg, A. P., Felder, G., Mosimann, M., Röthlisberger, V., & Weingartner, R. (2018). Extending coupled hydrological-hydraulic model chains with a surrogate model for the estimation of flood losses. *Environmental Modelling and Software, 108*, 174–185. https://doi.org/10.1016/j.envsoft.2018.08.009
Zischg, A. P., Felder, G., Weingartner, R., Quinn, N., Coxon, G., Neal, J., Freer, J., & Bates, P. (2018). Effects of variability in probable maximum precipitation patterns on flood losses. *Hydrology and Earth System Sciences, 22*(5), 2759–2773. https://doi.org/10.5194/hess-22-2759-2018
Zischg, A. P., Hofer, P., Mosimann, M., Röthlisberger, V., Ramirez, J. A., Keiler, M., & Weingartner, R. (2018). Flood risk (d)evolution: Disentangling key drivers of flood risk change with a retro-model experiment. *Science of the Total Environment, 639*, 195–207. https://doi.org/10.1016/j.scitotenv.2018.05.056
Zoccatelli, D., Borga, M., Viglione, A., Chirico, G. B., & Blöschl, G. (2011). Spatial moments of catchment rainfall: Rainfall spatial organisation, basin morphology, and flood response. *Hydrology and Earth System Sciences, 15*(12), 3767–3783. https://doi.org/10.5194/hess-15-3767-2011

Chapter 3

Disentangling drivers of change

Flood risk is composed of multiple factors, summarized by the three main factors: hazard, exposure, and vulnerability. These three main factors of risk vary in space and time. The changes of these main risk factors are influenced by a high number of drivers of change. As the previous chapter has shown, drivers of change comprise changes in the environment as well as changes in society. The drivers of change are furthermore interconnected, they coevolve with each other. Therefore, causes and effects of flood risk change are not always clearly distinguishable by statistical inference methods. An example is an expected correlation between the number of houses constructed annually in a floodplain and the increase in flood risk. Flood risk is not increasing linearly with the number of exposed houses but with the number of exposed houses that are vulnerable to flooding and with the exposure to different hazard intensities. If a land use regulation foresees the obligatory installation of object protection measures such as sealing the openings of the house to avoid water entering into the house up to a specific flood water level, the relationship between the number of newly constructed houses and the flood risk in the floodplain is nonlinear. Moreover, statistical inference methods require specific data. In this example, the increase of flood risk in a growing settlement area is not directly deductible from data, e.g., from house registers or from insurance claim data.

Moreover, the main aim of analyzing flood risk change is to learn about how the different factors drive the increase or decrease of flood risk. Knowing the key drivers of risk and their changes in space and time enables decision-makers to target the most promising levers for flood risk reduction. Thus, flood risk reduction measures can be targeted at the most influencing driver of change. The key drivers of change heavily depend on local context. While in some floodplains, settlement growth might be the most important driver of change, climate change, or other environmental changes might outpace societal drivers of flood risk change. The request to learn about the most influencing driver of change requires to look at the different drivers of change individually. This requires a method that enables the analysis of the effects of single risk drivers on overall flood risk in an isolated mode. The interdependencies between the drivers of change must be artificially eliminated

Flood Risk Change. https://doi.org/10.1016/B978-0-12-822011-5.00006-5

when the isolated effects of key drivers of change on flood risk evolution are being studied. Thus, a method for analyzing the drivers of change and their effects on the overall flood risk evolution must enable a disentangling of the effects of single drivers of change.

The method for disentangling the effects of isolated drivers of change on flood risk evolution must furthermore enable the analysis of the effects of coevolving drivers of change on overall flood risk change. The method therefore must be modular, i.e., it must allow the variation of more than one driver of change in space and time and must allow to consider their interdependence. Therefore, the modules for simulating the drivers of change must be coupled to allow for consideration of coevolutionary dynamics. This requirement is a precondition for analyzing and modeling feedback mechanisms. Finally, the analysis framework should enable to consider the cumulative effects of all drivers of change if we want to find out if and how the drivers of change, positive or negative, balance each other out or not.

The same requirements for the analysis framework are also valid for a modeling framework that aims at projecting the effects of future changes of risk drivers on future flood risk evolution. Hence, an analysis and modeling framework must fulfill the following criteria:

- Disentangling of drivers of flood risk change: the effects of single drivers of change on overall flood risk evolution can be analyzed and modeled in an isolated way.
- Detection of key drivers of change: the isolated effects of the single drivers of change on overall flood risk change can be compared and weighted.
- Analyzing feedback and coevolutionary dynamics: the effects of mutual influence between drivers of change on overall flood risk change can be analyzed and modeled.
- Quantifying the cumulative effects of changes: the balancing of the positive and negative effects of risk drivers on flood risk change can be analyzed and modeled.

In the following, a framework for analysis and modeling of flood risk change is presented.

Analysis and modeling framework

From the requirements listed above, it becomes clear that the analysis and modeling framework must be set up as a model experiment. The principal idea of the model experiment is to create an artificial or virtual environment that enables the testing of hypotheses as well as to analyze the effects of parameter variations on flood impacts. As flood risk is composed of many processes and factors, the modeling approach must consider a certain level of complexity. This can be completed using different ways (Kelly et al., 2013; Werner & McNamara, 2007) as described in the following.

Probably the most common approach in modeling coupled human and natural systems is the development of **integrated assessment models**. In these models, all processes and submodules are incorporated into one model. Integrated assessment models have a common spatial and temporal resolution for all submodules and processes. This is one of the main limitations of integrated models, as disaggregation and aggregation approaches are needed to convert different processes to the same spatial and temporal scale. This in turn makes validation and calibration more complicated. Furthermore, the uncertainties are assumed to propagate through the submodels (Voinov & Shugart, 2013). Earth system models (Donges et al., 2020) are the most prominent representatives of this kind of models. These are global climate models with the added capability to explicitly represent biogeochemical processes that interact with the physical climate, thereby altering its response to climate forcing, such as that associated with human-caused emissions of greenhouse gases (Flato, 2011). Giorgi and Avissar (1997) stated that the spatial resolution of earth system models is too coarse to capture local heterogeneity effects. Thus, most processes need to be parametrized or considered as subgrid processes. In addition, surface heterogeneity can lead to dynamic errors in coarse models. This is especially important for the representation of water fluxes (Clark et al., 2015; Nazemi & Wheater, 2015). Several attempts have extended earth system models with the ability to consider the actions of humans on natural systems (Donges et al., 2017, 2020; Müller-Hansen et al., 2017; Palmer & Smith, 2014). An overview and future directions of integrated human—earth system modeling was produced by Calvin and Bond-Lamberty (2018).

Another approach for modeling coupled human and natural systems is the **system dynamics** approach. This is a computer simulation problem-solving approach with a foundation in the concept of system feedback used to gain insight into real-world system behavior (Neuwirth, 2017). System dynamics models are based on the first computational experiments of Forrester (1969) and on Luhmann's system theory (Luhmann, 1987). These approaches have been used for conceptualizing human—flood interactions (Di Baldassarre et al., 2013), vulnerability analyses (Rougé et al., 2015) modeling the feedbacks between flooding and economic growth (Grames et al., 2015), and analyzing upstream—downstream trade-offs in the internalization and externalization of flood risks (Ciullo et al., 2019; Roos et al., 2017). However, system dynamic models are, in most cases, lumped models. The ground processes and earth surface processes, that is, the spatiotemporal dynamics of hazard and exposure, are not modeled in a spatially explicit way. Thus, the outcome is quantitative but aggregated into a lumped variable. Only a few recent studies addressed the spatial discretization of system dynamic models (Neuwirth, 2017; Wingo et al., 2017). In flood risk research, system dynamic models either focus on structural changes in flood risks in general (Neuwirth et al., 2015), the management of flood risk (Simonovic, 2009), or disaster management (Simonovic, 2011). These approaches provide a potential for system conceptualization and thus a holistic analysis of floodplains. However, methods are still needed for incorporating physically based process models and linking them with other modules toward more complex models.

A third alternative is the coupling of different models. **Coupled component models** are composed of specialized disciplinary models representing parts or processes of a studied system. Coupled models integrate submodels to form a model chain that represents a whole system (Blair & Buytaert, 2016). Coupled component models are hypothesized to be flexible regarding the level of integration and relatively transparent because the submodels have mostly been validated in their specific discipline (Andreas Zischg, 2018). However, the submodels often use different spatial and temporal scales, so the bridging of different scales in model coupling is challenging. Another advantage is that coupled component models can potentially combine both lumped and spatially explicit models. The coupling of models aims to represent the full chain of natural processes from atmospheric and meteorological processes to hydrologic and hydraulic processes and finally to the drivers of flood risk change. The coupling of disciplinary models also offers a high potential for approaches that synthesize relevant scientific progress in the different disciplines of flood risk. Under a reductionist philosophy, all the subdisciplines of hydrology and flood risk research have made relevant progress in developing reliable models for specific and narrow purposes. From a universalist view, this fragmented knowledge has to be compiled and incorporated to meet the requirements of holistic thinking in solving our current problems, which are often interconnected, complex, and interdisciplinary. Thus, a more philosophical hypothesis is that coupled component modeling frameworks have the potential to unify specialist knowledge and to bring together experts or knowledge from different disciplines for holistically analyzing human and natural systems. After (Tobin & Montz, 1994), natural risks result from the interactions between humans and their environment. In this sense, coupled component models allow to simulate "coupled human—natural systems" (CHANS). The CHANS approach focuses on simulations of changes in systems by considering feedback mechanisms between human activities and the natural environment (Liu, Dietz, Carpenter, Alberti, et al., 2007; Liu, Dietz, Carpenter, Folke, et al., 2007; Werner & McNamara, 2007).

The remainder of this chapter will focus on the use of coupled component models and model coupling approaches for disentangling the effects of drivers of change on flood risk evolution and for modeling complex interactions between different drivers of flood risk change. A modular structure for modeling flood risk change is therefore proposed.

The modular structure must at least comprise a climate/meteorological model to generate extreme rainfall scenarios, a hydrological model to simulate river runoff, a hydrodynamic model to simulate inundation processes, and an impact model to simulate the effects of floods on society. Depending on the research questions, the model structure can be extended with a morphological model or models that simulate the effects of flood risk management actions and risk reduction measures on flood risk change. Fig. 3.1 conceptually shows the modules of model experiment for analyzing flood risk change.

The proposed modular structure allows multiple variations of input parameters and even the exchange of one model with another. These two characteristics enable sensitivity analyses and uncertainty analyses. In the following, the potential of this analysis and modeling approach is highlighted.

Simulations driven by global and regional climate models

An often used approach in the research on the effects of climate change on flood risk is to run a climate model for a certain period without climate forcing and with different scenarios of climate forcing, i.e., pathways of economic development and related greenhouse gas emissions. This results in different timeseries of weather parameters. In the case of flood risk analysis, these are mainly the parameters rainfall and temperature. A hydrological model simulates river flows continuously over a selected period with these parameters, and the resulting flood events are analyzed using extreme value statistics. The hydrodynamic model simulates the inundations in the floodplains and the flood impact model analyzes the impacts of the flood events extracted from the timeseries of river flows. This results in timeseries of flood impacts. The probabilities of flood impact, i.e., the flood risk, can be calculated from the extreme value statistics of the rainfall timeseries. Examples of this type of modeling approach are given in Dottori et al. (2018), Hirabayashi et al. (2013), IPCC (2012), Pappenberger et al. (2012), Ward et al. (2013), Winsemius et al. (2013). These models provide global information concerning the regions in which river flow and thus flood risk is likely to increase or decrease.

The comparison of the two different timeseries of flood impacts or sets of flood events extracted from the river flow timeseries can lead to the analysis of the effects of climatic changes on flood impacts. The change in flood risk can be quantified by analyzing the changes of occurrence probability and frequency of extreme rainfall events and the related changes of flood impacts. The other modules of the model chain are being held constant in this approach. This means that the model experiment assumes constant river morphology, constant settlements and infrastructure (no economic growth assumption), or constant vulnerabilities of exposed assets over the simulation period. Doing so, the effects of the climate forcing's to flood impacts and flood risk can be analyzed independently from the variations of all other parameters that drive flood risk change. This leads to the attribution of the climate change effect to flood risk by quantifying the fraction of attributable risk (Kreibich et al., 2019).

The mentioned studies are mostly focused on global-scale flood risk analysis. However, the presented approach must be adapted when analyzing flood risks at a regional scale. Modeling floods at a global scale is based on

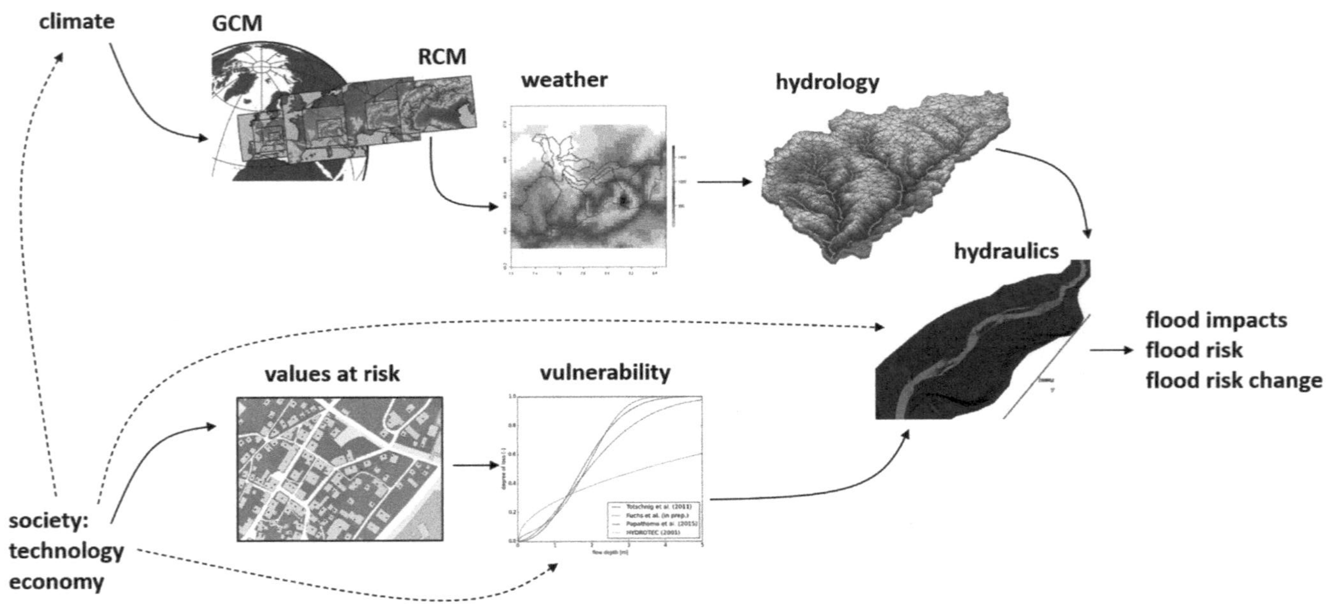

FIGURE 3.1 Model experiment for analyzing flood risk change. Modular structure of the model experiment for analyzing flood risk change and disentangling the drivers of change.

simplified models at a coarse spatial and temporal scale and on several assumptions that are used to parametrize the effects of local-scale processes (Ward et al., 2015). In contrast, modeling floods on a regional and local scale requires specific data and high spatial resolution (Felder et al., 2018). The outputs of the global climate model simulations must be downscaled to a spatial and temporal scale that can be used as input for a hydrological model running at a regional scale. In the process of downscaling global climate model simulations to regional and local scale, cross-scale issues may arise when studying the impact of climatic change on floods and flood risk (Fleischmann et al., 2019; Mateo et al., 2017). Global climate models provide the basis for flood models, but the outputs of these models need to be downscaled to continental or regional scales with regional climate models. Only the downscaled model results can be fed into regional- and local-scale hydrological and inundation models to provide flood projections (Felder et al., 2018; Gusyev et al., 2016). The downscaling can be based on statistical or dynamical approaches. However, the use of model cascades that are driven by global and regional climate models is controversial. The precipitation output of such a downscaling process often exhibits significant biases (Smith et al., 2014), and methods to correct these biases are therefore required (Christel Prudhomme et al., 2002). Moreover, uncertainties propagate through the coupled hydrologic and hydraulic models (Grimaldi et al., 2019; Steinschneider et al., 2015; Trigg et al., 2016). Significant uncertainties are also associated with return period estimates under current and future climatic conditions (Winsemius et al., 2013), depending on the length of observation and the periods simulated, and thus on the sample size used for extreme value statistics (Smith et al., 2014; Trigg et al., 2016). Hence, the capacity of this approach to identify and analyze extreme flood events is still under discussion (Felder et al., 2018). To overcome these uncertainties in flood risk change attribution by using this model cascade that is driven by global climate models, several other approaches have been developed recently.

Sensitivity analysis and scenario-neutral approaches

One of these alternative approaches is the sensitivity analysis approach or the so-called scenario-neutral approach (Keller, Rössler, et al., 2019; Prudhomme et al., 2010). Herein, the modeling approach is some kind inverted. Instead of running climate simulations with specific assumptions of climate forcing that provide weather patterns and parameters to a hydrological model for simulating river runoff, monitored timeseries of environmental parameters are perturbed. For example, the measured temperatures and rainfall intensities are increased or decreased step by step. The perturbed timeseries are fed into the model chain. This results in timeseries of perturbed river flows (Broderick et al., 2019; Culley et al., 2021; Guo et al., 2017, 2018) or in timeseries of perturbed flood impacts (Keller, Rössler, et al., 2019). The comparison of the perturbed timeseries of flood impacts with the original timeseries finally leads to the identification of the most sensitive input parameter and to the sensitivity of the studied system to changing environmental or climate boundary conditions. This approach is interesting from the complexity perspective. It allows us to analyze if the increase in flood impacts is linear or nonlinear with the increasing precipitation intensities. Depending on the local terrain characteristics and the general flood exposure situation, a certain increase in rainfall intensities and sum can lead to nonlinear increases in flood impacts, e.g., the relative increase of impacts is much higher or lower than the relative increase in rainfall. Moreover, this approach can also be used to study emergent phenomena or the rise of tipping points, i.e., rainfall thresholds that lead to a sudden collapse or failure of infrastructure systems against a partial failure of parts of the system.

A special aspect of this sensitivity analysis approach is downward counterfactual analysis (Aspinall & Woo, 2019; Lin et al., 2020; Roese, 1997; Woo, 2019). A downward counterfactual is a thought about the past in which the outcome is worse than what actually happened. The concept of counterfactual thinking initially comes from cognitive psychology (Roese & Olson, 1995) and is used in natural hazard research to extend the range of possible hazardous events. The exploration of alternative (worse) realizations of past hazardous events is used to extend knowledge concerning the impact of rare extreme events. If applied in flood risk analysis, an observed extreme meteorological event is perturbed and alternative, physically plausible scenarios of the same event are modeled. The impacts of the counterfactual outcomes are analyzed and compared with the impacts of the observed event; for example, the consequences of a hypothetical deviation of the observed track of a storm or an increase in rainfall (Staffler et al., 2008). This allows you to answer questions such as "What if the rainfall event had been 10% more intense?" The downward counterfactual approach can also be adapted to map the sensitivities of river reaches to an increase of flood peak discharges in terms of flood impacts. Zischg and Bermúdez (2020) showed exemplarily how these sensitivities to climatic changes can be mapped globally, regionally, and locally. Also, for this approach, a coupled component modeling framework as described above was used.

The storyline approach

Similar to the approach of modeling downward counterfactual events by perturbating specific natural events, e.g., an extreme rainfall event over a delimited region, is the approach of developing storylines of events. Storylines are defined as "physically self-consistent unfolding of past events, or of plausible future events of pathways" (Shepherd et al., 2018). The concept of storylines was introduced to climate sciences with the aim of providing an alternative way of dealing with uncertainties in statements about future climate, especially about extreme climate events (Hazeleger et al., 2015). The approach of focusing on a selected specific event is an alternative to the probabilistic approach that focuses on ensembles of scenarios and simulations. In contrast to probabilistic methods, the focus is laid on understanding the driving factors of a specific event or a specific hazardous process and its consequences in terms of impacts. The concept of storylines is used in flood risk management to develop narratives of how a flood event may occur in space and time (deBruijn et al., 2016). These storylines of flood events are used for educating and training intervention forces. The focus lies on describing possible future events, the physical and socioeconomic factors that lead to them, and on analyzing how such factors are affected by the event. In this aspect, the storyline approach provides a comprehensible way of explaining the drivers of flood risk change.

An important aspect of storylines of extreme weather events and their impacts on society is the definition of the vulnerability of a certain system (economic system, housing, infrastructure, agriculture) to climate parameters. The system's vulnerabilities are most understood by system operators and thus a codevelopment or participatory modeling approach is used to define these vulnerabilities. A representative extreme weather event in which the intensity exceeds the threshold of system vulnerability is then selected to develop the event storyline. This opens possibilities for connecting top-down modeling approaches, such as the presented model chain from atmospheric processes to flood impacts, with bottom-up system sensitivity studies. The approach also allows for qualitative studies in flood impact analysis. This complements the probabilistic approach that is based merely on pure quantitative approaches. An example of a combined top-down and bottom-up approach is the study of Zischg, Felder, Weingartner, et al. (2018) that creates spatiotemporal patterns of extreme rainfall events with a probable maximum precipitation over a river basin with a Monte Carlo simulation approach and models the flood impacts of these events with a coupled hydrological-hydraulic model (Felder et al., 2017) and a flood impact model. As described above, this is a model experiment that uses many rainfall scenarios (model input) and holds the other modules (hydrology, hydrodynamics, flood impacts) constant. This study shows that the flood impacts in terms of monetary damage to houses vary by a factor of 4–6 depending on the rainfall pattern. Thus, the same amount of rainfall within a river basin produces a high variability of impacts. Out of the resulting set of flood impact scenarios, the scenario with the probable maximum loss can be selected and used in a storyline that describes the course and spatiotemporal pattern of a rainfall event that leads to the highest impact. This storyline of a flood event that leads to the highest damage can be used to identify the most critical points in a river system in terms of flood exposure and the preconditions of a catastrophic event.

The retromodel experiment approach

Another approach for modeling flood risk change is to hold a rainfall scenario or a flood scenario represented by a hydrograph as constant and vary the digital terrain model. One digital terrain model is representing the current state of the river morphology, and the other digital terrain model is representing a counterfactual situation. This could be, for example, a situation with newly planned river engineering measures designed for flood risk reduction. These planned situations can be virtually inserted into the current digital terrain model. This results in two different digital terrain models. These two situations are incorporated in a two-dimensional hydrodynamic model that simulates flood scenarios with and without flood protection measures. This results in different flood maps that can be fed into the flood impact model. The resulting differences in flood impacts can be used to analyze the effectivity of the planned flood protection measures in terms of reduced damage or flood risks. Moreover, the comparison of the two model results can be used for analyzing the ratio between the costs of the planned protection measure and its efficiency. In this case, the costs for implementing the protection measures must be known as well as the benefits in terms of the monetary value of the reduced damages. This ratio between the costs and benefits supports the choice between multiple alternatives. For each alternative of flood protection measure, both the implementation costs and the reduction of impacts must be known quantitatively if the protection measure with the best ratio between costs and benefits must be selected from a set of alternatives.

However, in some cases, the effects of existing flood protection measures need to be known. This is the case when assessing the need for restoration or maintenance of old flood protection measures. Lateral levees of rivers can become unstable over decades and must be rebuilt or reinforced after a while. River reaches with a large length of lateral levees therefore require huge investments in analyzing the levee stability and in restoring levee stability. Large investments in existing infrastructures, e.g., in flood protection levees that help to keep risk levels constant, are publicly less well appreciated than investments in new structures that reduce existing flood risks (Zischg, 2016). Thus, the investments have to be defended by arguments and often quantitatively. This is done by showing the effects of collapsing lateral levees during flood events and their impacts. In this case, the current digital terrain model is modified by changing the lateral levees, either by deleting parts of the levees for simulating dam breaks or by totally eliminating the existing flood protection structures and reconstructing the original river morphology. The modified digital terrain model is then fed into a two-dimensional hydrodynamic model to simulate and compare the extent of flood events with or without flood protection measures. The difference between the flood impacts of these scenarios is used to calculate the net benefits of the existing flood protection structures.

In this case where the current state of the river morphology (with lateral levees) is the starting point and the counterfactual situation (without lateral levees) is a historical state, the model experiment is called a retromodel experiment. This concept of retromodeling was first defined by Remo and Pinter (2007) and Remo et al. (2009). Retromodeling is defined here after (Remo et al., 2009) as "the use of archival hydrologic and geospatial data in state-of-the-art hydraulic models to assess historical conditions." The comparison between modeled historical and modern conditions can be used to assess overall flood risk change and to assess a specific driver of change, namely the change in fluvial morphology.

In contrast to purely synthetic model experiments, a retromodel consists of several historically documented (observed) spatiotemporal stages of river morphology and settlement structure. The main advantage of this approach is that the overall long-term dynamics are known and do not have to be assumed as it is the case for simulating future changes. Zischg, Hofer, Mosimann, et al. (2018) used this model setup to analyze the temporal evolution of the flood risk in a floodplain in Switzerland, for an ex-post evaluation of the key drivers of change, and for analyzing possible counterfactual pathways for flood risk evolution under different governance settings. In the following, examples of model experiments will be shown.

Quantifying the isolated effects of drivers of change with model experiments

In this chapter, I present selected case studies that exemplarily show the application of model experiments for analyzing the effects of changing flood risk drivers on overall flood risk evolution. The first example describes the intended and unintended consequences of an early river engineering measure, i.e., the Kander river deviation of 1714 AD, on the hydrologic regime and on flood risks in the Aare river basin, upstream of Bern in Switzerland. This case study is both a real experiment on anthropogenic interventions in natural river systems and a model experiment that allows us to study the cascade of consequences. This example also allows us to explain the general approach of quantifying one isolated effect of changes in the river system on the overall flood risk change. The second example quantifies the cumulative effects of river training measures on flood hydrology and flood impacts at a river basin scale. The third example assesses changes in flood hazard maps due to climatic changes. Other examples describe the effect of climatic changes on sediment balance and riverbed erosion or aggradation and their impacts on flood risk change. An overview on the spatial variability of the riverbed erosion and aggradation process in the main river network of Switzerland is given. Model experiments analyze the feedback mechanisms of these processes to flood impacts and flood risks. Moreover, examples show the effects of changing exposure and changing vulnerability on flood risks.

The Kander river deviation of 1714 AD and its effects on hydrological change

The Kander river is a tributary of the Aare river in Switzerland. The Aare river basin has a catchment area of 17,709 km^2 and thus drains a relevant part of

Northern Switzerland, mostly Alpine areas and Alpine forelands. The Aare river confluences near Koblenz with the Rhine river. The river basin has a mean altitude of 1061 m a.s.l. with an altitudinal range between 305 and 4130 m a.s.l (www.hydromaps.ch). 2% of the area is covered by glaciers and 30% by forests. The Kander river is an Alpine torrential river that nowadays discharges into Lake Thun. The catchment area is 1094 km^2, and the mean altitude is 1686 m a.s.l., ranging from 558 to 3675 m a.s.l. Glaciers cover 4% of the catchment and forests cover 23%. The Aare river basin is a representative example of Alpine river systems that have been anthropogenically modified since the early 19th century. The river basin has over 50 hydropower facilities, eight reservoirs, and the main lakes are regulated for different purposes. The river morphology is highly artificially shaped by river engineering measures. One of the earliest anthropogenic interventions to the river system is the Kander river deviation in 1714 AD. To look at the effects of this early human intervention on the hydrological regime of the Aare river basin, I focus on the river basin of the Aare river upstream of Bern, Switzerland (see Fig. 3.2).

The Aare river basin has an area of approximately 3000 km^2 and thus is defined as a mesoscale catchment. The river network consists of 26 tributaries (subcatchments) that confluence into the floodplains of the main valley. The main valley is divided into seven floodplains. These are the floodplains of the river reaches "Hasliaare river," between Meiringen and Lake Brienz; the

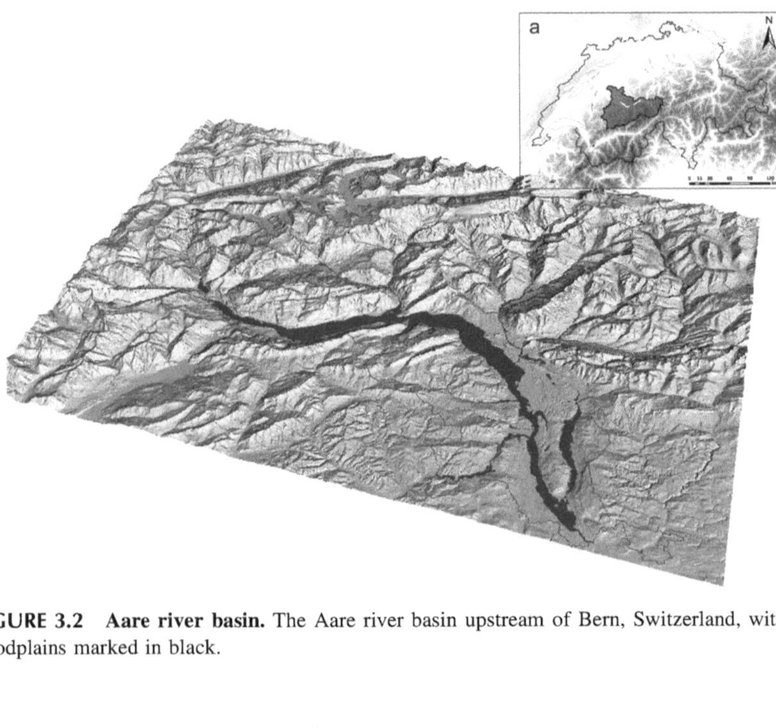

FIGURE 3.2 Aare river basin. The Aare river basin upstream of Bern, Switzerland, with the floodplains marked in black.

coastal areas of "Lake Brienz"; the floodplain of the Lütschine river in the city area of "Interlaken"; the coastal areas of "Lake Thun"; the floodplain of the city "Thun"; the floodplain of the "Aare river" between Thun and Bern; and the floodplain of the tributary "Gürbe river" between Burgistein and Belp. The flood processes in the Aare river basin are dominated by both riverine and lake flooding. The Hasliaare river floodplain is dominated mainly by riverine flooding. However, the delta of the Hasliaare river floodplain is also affected by flooding from Lake Brienz. The lateral shorelines of Lake Brienz and Lake Thun are exposed to lake flooding. In contrast, the city of Interlaken is exposed to four different flooding processes. In the eastern part, this floodplain is exposed to lake flooding from Lake Brienz. The western part is exposed to lake flooding from Lake Thun. A high water level in Lake Brienz leads to a high discharge in the Aare river, between the two lakes. Consequently, the central part of the city of Interlaken is exposed to riverine flooding. From the southern boundary of the floodplain, the tributary Lütschine river flows into Lake Brienz with possible riverine flooding. The city of Thun is exposed to both lake flooding from Lake Thun and riverine flooding from the Aare river when lake level is high. The floodplain of the Aare river between Thun and Bern and the floodplain of the Gürbe river are exposed to riverine flooding. The discharge of the Aare river downstream of Thun is dominated mainly by the outflow from Lake Thun and secondarily by its tributaries.

History of the Kander river deviation

The deviation of the Kander river into Lake Thun was a pioneer project in the early 18th century. Before 1714 AD, the Kander river flowed into the Aare river at a location that confluences with the Zulg river (see Fig. 3.3). During floods, the Kander river and Zulg river transported sediment that subsequently dammed the Aare river at this confluence and produced backwater effects in the city of Thun (Grosjean, 1962). In particular, the Kander river has been well known for its large sediment yields (Vischer, 2003) that repeatedly shaped the alluvial fan near Thun. The river changed its river bed after major flood events and therefore the large fan was sparely vegetated and could not be used intensively for agriculture. The large area therefore remained common ground for a long time. Moreover, the sediment input from the Kander forced the Aare river to braid downstream of the confluence with the Kander river (Vischer & Fankhauser, 1990) and blocked the confluence of the Zulg river with the Aare river. The Kander river was also known for producing high peak flows that repeatedly affected the city of Bern. Thus, the coevals of the beginning of the 18th century faced several threats: the flooding of the cities of Thun and Bern and the villages between the two cities; the riverbed aggradation of the Aare river between Thun and Bern due to the sediment supply of the Kader river that led to repeated changing of the river course; the aggravation of the ship transport due to the river braiding; the degradation of the land in the

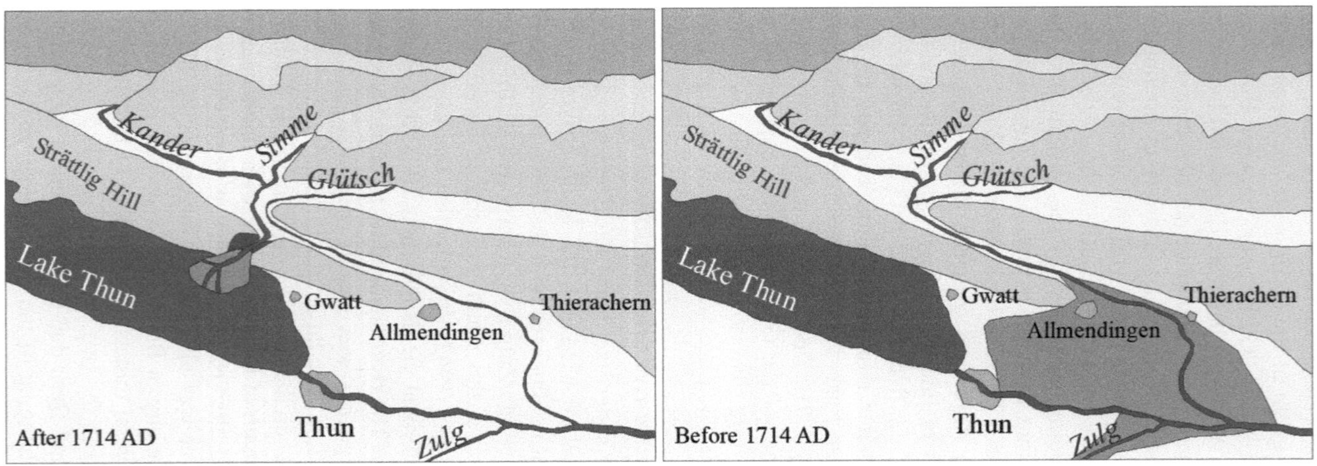

FIGURE 3.3 The Kander river deviation of 1714 AD. *Modified from Egli, H. R., Hasler, M., & Probst, M. (2019). Geographie. Wissen und verstehen (fifth ed.). hep verlag.*

floodplains; the repeated destruction of hydropower facilities such as mills; as well as the increased need for maintenance of the minimal river shoreline stabilization measures. The swampification of the floodplains increased the spread of the malaria disease (Bögli, 2018). Thus, the riparian communities requested the help of the government of the Republic of Bern for solutions to solve these problems. From 1680 AD, the first ideas popped up to deviate the Kander river into Lake Thun.

Only after the flood event in 1693 AD, with widespread damage to houses and infrastructure, the idea of deviating the Kander river into the Lake Thun was tangibly discussed. The discussions resulted in a concrete request from the local communities to the government of Bern to deviate the Kander river into Lake Thun (Bögli, 2018). The government founded a commission that evaluated different variants. The commission concluded in 1699 AD that the deviation was in principle feasible. One variant was the deviation of the Kander river from Allmendingen to the lake, by deviating the river at the right side of its alluvial fan. In rare cases during exceptional floods, the river flowed in this direction when river sediments blocked the original riverbed. The other variant was to cross the Strättligen hill. This hill is a middle moraine between the Kander glacier and the Aare glacier of the last ice age.

The main goal of both project variants was to make use of the retention capacity of Lake Thun for attenuating the flood events coming from the Kander and Simme catchments. The lake was assumed to dampen the peak discharge of the Aare river downstream of Thun. With this retention effect, the city of Bern would gain from reduced peak discharges of the Aare river that are highly influenced by the Kander river.

Soon, both project variants encountered heavy opposition. The opponents in the communities along the shores of Lake Thun feared a worsening of water quality. The water of the Kander river was full of dissolved sediment and turbid material, it was cold because of the high glaciation in the catchment. They feared that the turbid and cold water would cause negative effects on the fishery. The city of Thun used the lake water as drinking water, and the citizens feared a decrease in drinking water quality. Moreover, the city of Thun feared a rise in lake levels because of the additional contributing watershed of the Kander river. A lake level rise would negatively affect all infrastructure along the shorelines and along the lake outflow, namely the Aare river. In Strättligen, where the morainic hill would be cut through for the river deviation, the citizens feared the interruption of the road from Thun to Spiez and the Bernese Oberland. The commission that planned the project encountered these opponents with promises that the potential unintended consequences could be faced with structural measures. The higher lake level and the higher outflow from the lake could be managed with additional structural measures for protecting the riverbanks along the Aare river in Thun. They stated that in case of a lake level rise, a second relief canal could be constructed that would increase the lake outflow capacity.

After a positive evaluation by the commission, the government of Bern declared the project 1708 AD as a priority investment. Finally, the commission preferred the project variant that planned the cutting through the Strättligen hill to deviate the Kander river into Lake Thun over the other variant. The commission expected the following benefits: The communities downstream of Thun would not be inundated anymore, the need for maintenance of the flood protection measures along the Aare river would decrease, the whole riverbed of the Kander river downstream of the deviation could be transformed into agricultural land, and the shipping along the Aare river would gain in safety and regularity. The contras were the high costs, the potential negative consequences for the city of Thun because of increased lake flooding, and other unknown negative consequences.

In 1710 AD, the commission gave three experts the task of planning the project in detail; lieutenant of artillery Samuel Bodmer and the engineers Emanuel Gross from Bern and Pietro Morettini from Locarno. Samuel Bodmer (1652–1724) possessed several mills in Amsoldingen and was a lieutenant in the artillery of the Republic of Berne. He was a surveying engineer and cartographer for the government and led the project. Emanuel Gross (1681–1742) was a mathematician and chief engineer of the government of the Republic of Berne. Pietro Morettini (1660–1737) was a leading military engineer and had experience with tunnel construction and the construction of fortifications. The engineers expressed their doubts about the project. In particular, they expressed their concerns about the assumption of the unlimited retention capacity of Lake Thun. They suggested increasing the lake outflow capacity of Lake Thun in parallel with the Kander river deviation to avoid inundation of the city of Thun due to the lake level rise. The parliament of the Republic of Bern decided on February 11th, 1711 AD, to deviate the Kander river into Lake Thun. The parliament decided to postpone the additional measures for confronting lake level rise in Thun that had been proposed by the engineers and decided to implement these measures only in case of evident need (Bögli, 2018). The responsibility for construction works was given to a board of directors, the so-called Kander board. The board also had to confront opposition and technical problems. With this decision, the first geoengineering project of Switzerland was started.

The project plan was elaborated by Samuel Bodmer ("Plan und Grundriss von dem so genannten Cander Wasser" (Bodmer, n.d.)). Fig. 3.4 shows the original plan. It shows the original course of the Kander river and the planned deviation. Moreover, the plan shows a cross section of the cross-cut through the Strättligen hill. The cross section shows the steep slope at the lake side of the Strättligen hill. The plan also shows how the lake outflow in Thun should be enlarged by opening the castle mouth as a second river course. Downstream of Thun, the plan outlines the straightening of the Aare river that also aims at increasing the lake outflow capacity. In addition, the plan was accompanied with an explaining text. Bodmer also discussed the additional measures needed

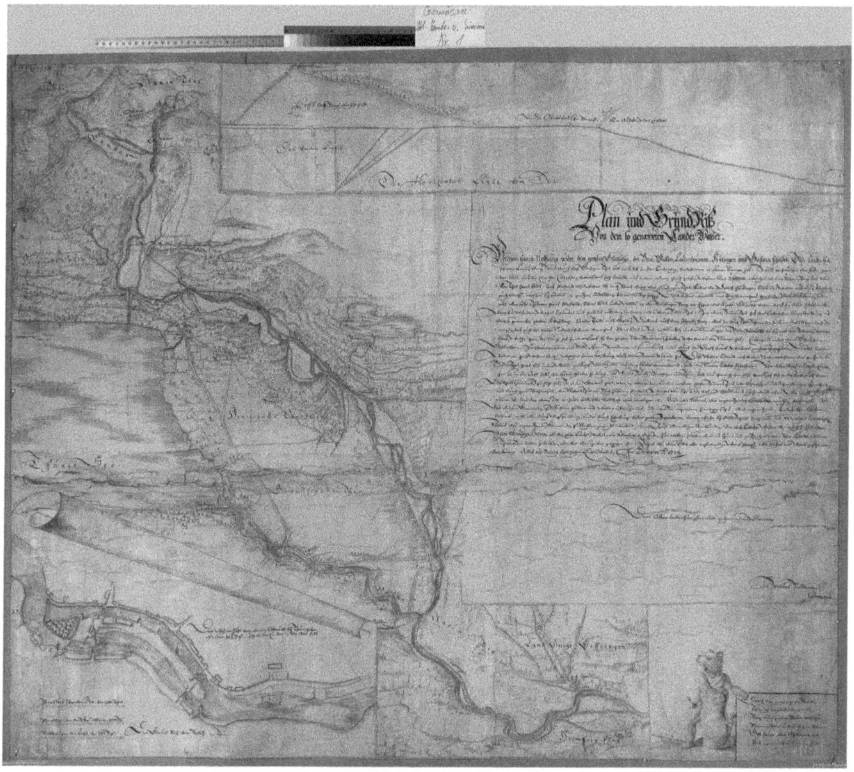

FIGURE 3.4 Project plan for the Kander river deviation into lake Thun. *From Bodmer, S. (n.d.). Plan und Grundriss von dem so genannten Cander Wasser. Plan und Grundriss von dem so genannten Cander Wasser.*

to attenuate the negative consequences of the lake outflow, i.e., the redesign of the infrastructure along the shorelines and the adaptation of the existing weirs. The river deviation itself should be implemented by excavating a channel through the morainic hill of Strättligen with opencast mining methods (Fig. 3.5). The morainic material was assessed to be stable to enable steep channel banks (slope of 120%) and thus to minimize excavation volume. The plan lacked a fixation of the steep slope (20%) at the downstream end of the cannel through the morainic hill.

The works started in April 1st, 1711 AD with up to 400 workers that manually excavated the channel through the morainic material. The works were interrupted from May to August 1712 because of the Second War of Villmergen. After the war, the works did not continue immediately. Members of the Kander board doubted the efficiency of the plan of Samuel Bodmer. A member of the board, Samuel Jenner, proposed digging a tunnel through the Strättligen hill instead of an open channel. In a second attempt, in spring 1713 AD, a tunnel

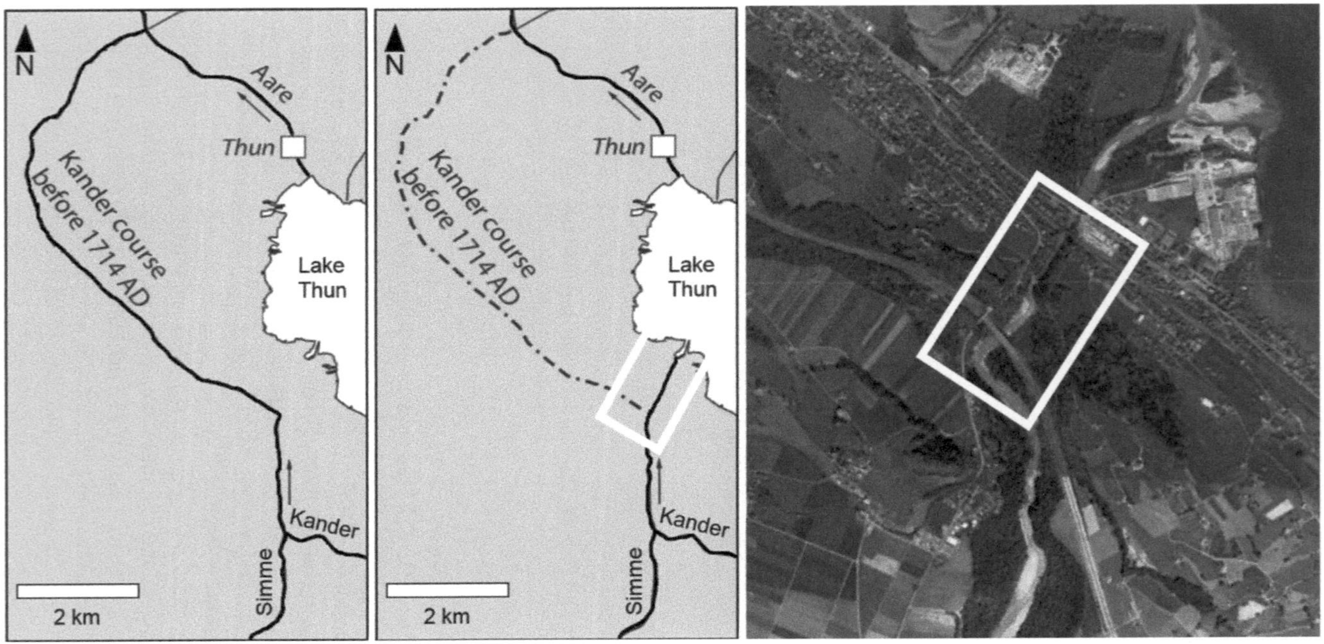

FIGURE 3.5 Topographical situation of the Kander river deviation and the related areas around Thun. *Modified from Vischer, D. (2003). Die Geschichte des Hochwasserschutzes in der Schweiz. Berichte Des BWG, Serie Wasser, 5. Background orthofoto: Swisstopo.*

was therefore drilled through the Strättligen hill with a length of 250 m, a width of 12 m, and a height of 4.5 m (Fig. 3.6). The aim of this tunnel was to deviate only flood waters from the Kander river into the Lake Thun to make use of the lake's retention capacity to reduce flood peaks. The proposed board member, Samuel Jenner, took over the responsibility for continuing the project. The tunnel was finished in December 1713 AD, and its functionality was tested by letting some water from the Kander river flow through the tunnel (Bögli, 2018). In spring 1714, the former engineer of the first attempt, Emanuel Gross, again expressed his concerns regarding the need to increase the lake outflow capacity before opening the tunnel for the Kander river. He suggested to stop the project. On May 18th, 1714, the Kander board decided to deviate the Kander river through the excavated tunnel. Only a small share of the residual water should flow through the old riverbed after the river deviation to guarantee the water provision for the animal stock, irrigation, and the hydropower utilities and production facilities downstream of the river deviation.

A cascade of unintended consequences

Beginning in July 1714 AD, the river was partially deviated into the tunnel. However, the bottom of the tunnel unexpectedly eroded quickly and within several days all the Kander's flow was conveyed through the tunnel. Two months later, the tunnel collapsed due to fast incision and lateral erosion and the resulting river deviation, called the Kander correction, was implemented. In this period, the backward erosion propagated from the lake level up to the river bed far upstream from the point of the river deviation. This erosion and incision quickly made human intervention irreversible. In the ensuing years, the new channel in the morainic and underlying older fluvial material proved to be geomorphically unstable and large amounts of erosion occurred within and upstream of the deviated river reach. For example, historical documents indicate that the Kander correction eroded ~27 m of the riverbed in the first 2 years since its inception (Koch, 1826; Wirth et al., 2011). In the lower part of the Kander river, the Kander correction shortened the river course by 8 km and

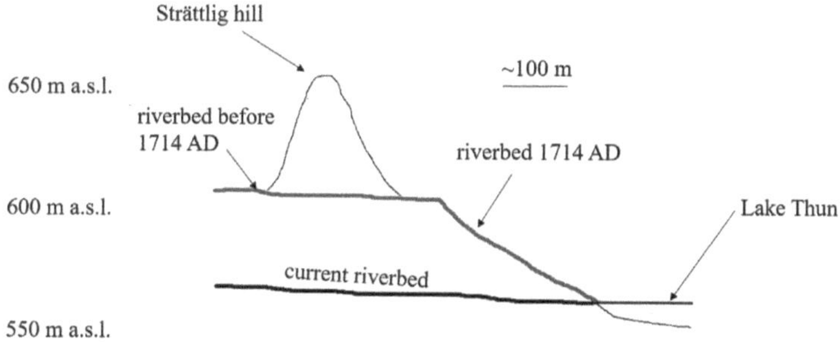

FIGURE 3.6 Cross section of the Kander river deviation through the Strättligen hill.

steepened the river gradient from 0.3 degrees to 3.7 degrees (Koch, 1826) (Fig. 3.6). This steepening of the channel increased stream power and bed shear, which contributed to the rapid channel erosion (Ramirez et al., 2020). The former plan to maintain a certain minimum flow in the old riverbed downstream of the deviation to provide the downstream villages with water quickly was made impossible to achieve. The downstream villages, between the river deviation and the former confluence with the Aare river, lost their water resources from 1 day to the other. The remaining water springs that were originally fed by the Kander river groundwater slowly run dry. The road from Thun to Spiez, the main access road from Thun to its hinterland, soon collapsed. The transport had to be shifted to ship transport over Lake Thun.

This human intervention also triggered a cascade of other unintended effects on the hydrological regime and the water management of the Aare river basin. The settlements located in the downstream part of the old river course lost their water supply for agriculture. The contribution area of Lake Thun almost doubled from 1300 to 2400 km^2 from 1 day to another. Hence, within 24 h after the Kander correction, the lake level rose by 30 cm (Bögli, 2018). Thereafter, the city of Thun was flooded in the years 1714, 1715, 1718, 1720, 1721 (Bögli, 2018) by high lake levels and high discharges at the lake's outlet. Due to the higher lake level of Lake Thun, the outflow and flow velocity increased remarkably. Consequently, all hydropower facilities along the Aare river in Thun, which were the backbone of the local industry (mills, sawmills etc.), collapsed as a result of high flow velocities and lateral erosion. The lake outflow capacity had to be improved in the following years by excavating and opening a second river course within the city of Thun and stabilizing the shoreline. The management of the unintended consequences lasted until recently; e.g., in the year 2009, nearly 300 years after the initial human intervention. A bypass tunnel was installed at the outlet of the lake that enables the preventive lowering of the lake level before a predicted large rainfall event in the Aare river basin upstream of the lake leads to high inflows into the lake.

Anthropogenic interventions following the Kander river deviation

Soon after the opening of the tunnel, the situation became irreversible. The first anthropogenic intervention following the river deviation was the restoration of the **water supply** for the villages downstream of the river deviation. The mills and production facilities lost their energy supply and agriculture lost water for the livestock. Soon, also the water springs fed by the groundwater of the former riverbed dried up. The solution to this problem was another river deviation. The small torrent Glütschbach originally confluenced with the Kander river a few hundred meters upstream of the deviation point. The affected communities and the Kander commission soon planned to deviate the Glütschbach torrent into the old riverbed of the Kander river. The Glütschbach now does not fully confluence with the Kander anymore but partially flows through the Glütschbach valley along the old riverbed of the Kander river. Downstream of the river deviation, the construction of small water channels was required to deviate water from the main channel toward the production facilities. Thus, a network of small water channels emerged on the former alluvial fan of the Kander river 1 year after the main river deviation.

The massive erosion of the Kander river after the deviation affected the **transportation network**. The collapse of the tunnel immediately affected the road from Thun to Spiez and Frutigen. The collapse of the tunnel eroded the road. The riverbed erosion process was very dynamic and accompanied by lateral erosion. Therefore, it took a while until it was possible to construct a bridge over the new river course. At this time, the transport of persons and goods had to be reorganized. The transport had to be shifted to ships from Thun to Spiez. Thus, the transport in the Bernese Oberland needed to be reloaded twice. Only in 1716, 2 years after the river deviation, the erosion process slowed down and a new bridge over the Kander river could be constructed. This bridge collapsed again in 1746 (Bögli, 2018). The erosion process shifted upward, and slowly the bridges upstream of the river deviation collapsed due to the river incision and following lateral erosion. They had to be reconstructed several times in the decades following the river deviation.

The massive erosion of the Kander led to sediment delivery into Lake Thun. Investigations of the lake sediment by Wirth et al. (2011) showed that the Kander river deposited 182.5 mio. m^3 of sediment into the lake from 1714 to 1852. This led to the **forming of a delta**. Two years after the river deviation, the delta occupied an area of 220′800 m^2. Until the beginning of the 20th century, the delta grew up to an area of 805′000 m^2. The lake in the area of the delta is shallow and therefore the government feared the sedimentation of the lake. Thus, sediment mining started in 1913 (Bögli, 2018) to avoid lake sedimentation. Around 80′000 m^3 of sediment is mined every year.

The city of Thun experienced the most negative consequences. Before the river deviation, parts of Thun and the surrounding villages were recurrently flooded by the Kander river at the **former alluvial fan**. These floods with high sediment transport led to riverbed aggradation and relocation of the riverbed during flood events. Sometimes the river deposited sediment at the confluence with the Aare river. The sediment deposited in the confluence area led to backwater effects of the Aare river that eventually flooded the city of Thun from downstream (Vischer, 2003; Vischer & Fankhauser, 1990). However, these flood events lasted for a few days only due to the torrential character and the water disappeared fast. The lake level only rarely raised to flood levels.

After the river deviation, these floods from the Kander river originating from the alluvial fan south of Thun or from the confluence downstream of Thun vanished. However, the massive increase in the contributing area of the lake led to a fast and sustained increase of the lake level. Also, the variability of the lake level increased remarkably. The marsh and reed zones along the lake south of Thun became wetter than before and became unuseable for agriculture. The additional flood volumes led to flooding of the villages along the **lake shorelines** and of the **city of Thun**. Underground floors of the houses in the city were regularly flooded and the above-ground floors annually. The water levels and flow velocities along the Aare river through the city of Thun, the **lake outflow**, increased and with it the erosion capacity. The riverbed incised, and due to lateral erosion, infrastructure along the river collapsed suddenly. The production facilities lost their energy provision from the hydropower along the river or were directly damage. The major part of the productivity of the city was lost. The **lake-river system** obviously was in an imbalance and had to adapt to the new hydrologic regime. Thus, the human system had to adapt too. The first adaptation measures were the removing of all hydraulic obstacles in the river channel. This aimed at increasing the lake outflow capacity. Downstream of Thun, the riverbed of the **Aare river** was widened and an additional channel with a length of 900 m was excavated to increase the discharge capacity of the Aare. However, these adaptation measures were found to not be efficient. In 1718, houses near the riverfront collapsed and a ship transporting persons shattered to the bridge pillars in the city due to the strong current. The accident led to 10 fatalities (Bögli, 2018). The collapse of the infrastructure along the Inner Aare and the accident finally made the need for an increased outflow capacity evident.

At the time of the river deviation, only one outflow river course through the city of Thun existed. Nowadays, there are two river courses, the **Inner Aare and the Outer Aare** rivers. The Inner Aare was the original river course. The Outer Aare river course was originally the town moat. It was closed to the Inner Aare by an earth dam. The earth dam was used to deviate water toward the mills and production facilities. The original plan for the Kander river deviation mentioned the opening of the castle moat for the Aare river to increase the lake outflow capacity. However, this accompanying measure had not been implemented together with the river deviation because of the high costs. After the need for increasing the outflow capacity became evident, the government of the state of Bern decided to finance this measure. Therefore, the town moat was deepened and widened in 1720 and opened for the flowing water by removing the dam between the Inner Aare and the town moat. However, the dam also guaranteed a minimum lake level. This was required for the ships coming from the lake that wanted to reach the mooring places in the city. The removing of the dam led to lower lake levels during periods without precipitation. The lake-river system was in imbalance again. While the frequency of high lake levels decreased a little because of the opening of a second lake outflow, the frequency of low lake levels increased. A low lake level was related with low river levels downstream of Thun. This frequently inhibited the shipping along the Aare river between Thun and Bern and thus impeded the transportation of persons and goods. Thus, in the year 1718 AD, the lake-river system in Thun was characterized by alternating high lake levels and high flow velocities during rainy weather conditions and low flows during dry weather conditions. The quiet, low-variability behavior of the system prior to the first human intervention was exchanged with a highly variable system behavior. This made another measure necessary, the **construction of weirs** for **regulating the lake level**. At the lake outflow, where the removed dam was located, a weir was constructed in 1720. The weir aimed at keeping the lake level at a minimum level during the dry periods and thus to allow ships to reach the inner city from the lake. For downstream shipping, the weir regulated the water volume and the operation of the weir allowed it to keep water back in the lake and open the weir on specific working days. The artificially increased lake outflow allowed shipping with the flood wave from Thun to Bern. The weir regulation was announced in anticipation of the official information system of the city to let all ship operators plan their transportation logistics. In 1724, a second weir was constructed in the Inner Aare. The weirs in Thun were the first regulable weirs in Switzerland. They were steadily optimized during the following decades. The lake level of Lake Thun has been regulated since those times.

The Aare river downstream of Thun lost part of its sediment supply. After the Kander river deviation, the sediment coming from the Kander and Simme river basins was deposited in Lake Thun. This sediment volume was missing for the river sediment balance of the Aare river downstream of Thun. This soon led to increased riverbed incision, especially in combination with the flood events following the river deviation. The flood events led to large inundations in the flatter areas of the floodplain between Thun and Bern and to high incision rates of up to 1 centimeter per year in the steeper river reach near Thun. Both the flood events and the fluvial erosion required high maintenance efforts by the local communities to stabilize the riverbanks. The footpaths along the riverbanks that supported the upstream pulling of the ships after unloading the transported goods in Bern were repeatedly destroyed. The shipping routes changed continuously. Moreover, the maintenance of the riverbank stabilization measures required a lot of wood and manpower. Thus, at the beginning of the 19th century, the local communities requested support from the state government. In these times, the engineering sciences became relevant and thus the techniques to "correct" the natural rivers, i.e., to reduce flooding and enable shipping, became widely known and accepted. In 1810 AD, engineers began to study the river system between Thun and Bern to propose river correction measures. These analyses led to **river corrections** in the following years. First, the **confluence of the Zulg** torrent was adapted. This tributary brings large amounts of sediment that is often deposited in the confluence area without being transported away from the main river. The confluence with the Aare river was newly shaped in a way that the sediment was brought to the Aare river in a way that it could be further transported downstream and did not block the Aare river anymore. Between 1824 and 1859, the river reach between Münsingen and Belp was straightened to a normal trapezoid profile with a riverbed width of 40 m (see Fig. 3.7). The construction of the railway in 1859 required to stabilize the river course in the upper part, especially in the area where the railway crossed the Aare river downstream of Thun. Consequently, the Aare was straightened between Thun and Uttigen between 1871 and 1892. The straightening and river corrections were accompanied with the construction of lateral levees in the form of earth dams.

These human interventions to the river systems led to other unintended consequences, namely a marked increase of the **fluvial incision** rate. The riverbed incision was expected to be in a certain range, and the erosive power was considered in the construction works to naturally support the implementation of the river corrections. However, the incision rate was much higher than expected. A few kilometers downstream of Thun, the Aare incised by 6.5 m, and soon the backward erosion reached Thun. To avoid river incisions in the city area, a block ramp had to be installed in 1877. The fluvial erosion is lasting until today. At the end of the 20th century, the old lateral levees and riverbank stabilization constructions were at risk of collapse because of continuing riverbed incision. Moreover, the groundwater wells for drinking water provision for the villages between Thun and Bern and for the city of Bern had to be rebuilt because of the sinking of the groundwater table that is related to the riverbed incision. The Aare river between Thun and Bern furthermore had a low ecological quality because of the massive **anthropogenically modified river morphology** and the related low ecomorphological diversity. Since the beginning of the 21st century, an integrated project to restore parts of the river reaches is being implemented. The riverbed is being widened to stop erosion and to laterally connect the remaining riparian forests with the flowing water. The old river engineering measures are being removed where possible before their expected sudden collapse during a flood event.

In 1999 and 2005, extreme flood events occurred in the Aare river basin. These events led to a hitherto unknown rising of the lake level of Lake Thun and to subsequent flooding in Thun. Fig. 3.8 shows how sensitive the flood exposure is against rising lake levels in Lake Thun. The number of flooded houses increases remarkably with every 10 cm of lake level rise. Moreover, the floodplains between Thun and Bern as well as the city of Bern were flooded during these events. These two flood events after a long period without floods during the 20th century raised awareness in Bern and Thun about the flood risk. It was acknowledged that the flooding in Thun was related to the early river engineering project, the Kander river deviation of 1714 AD. Thus, a project started to assess possibilities to finally solve this unintended consequence of lake flooding. The technological advancement of the **weather forecasting systems and of hydrologic prediction** capabilities allowed now to better regulate the lake. The lake level in principle could be lowered before

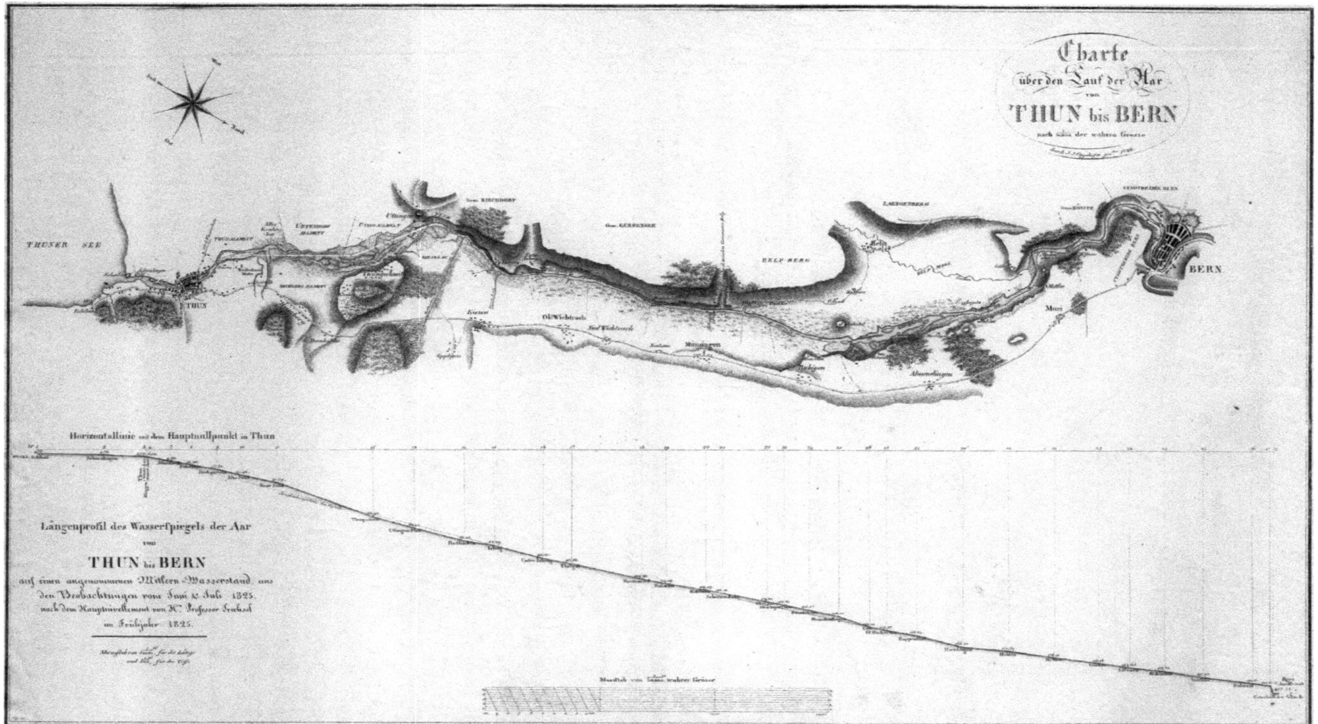

FIGURE 3.7 Map of the old river course of the Aare downstream of Thun and the planned straightening. *From Oppikofer, J. J. (n.d.). AA V 341 Aare. In Karte über den Lauf der Aare von Thun bis Bern, Staatsarchiv des Kantons Bern.*

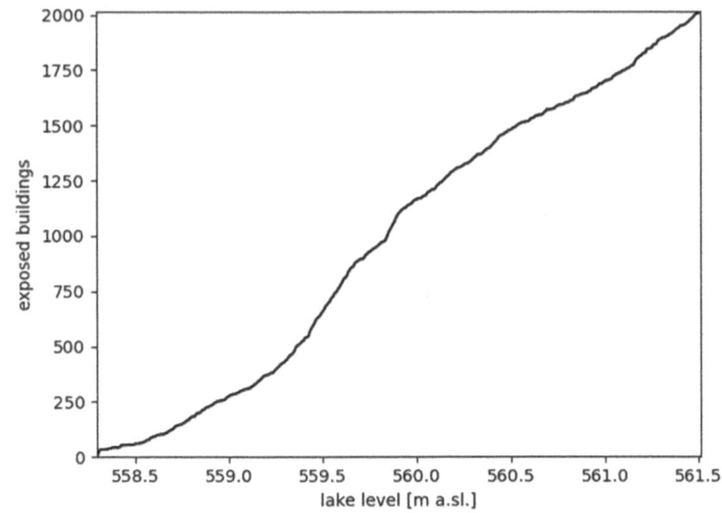

FIGURE 3.8 **Relationship between lake level and flooded buildings.** Schematic representation of the relationship between lake level and flooded buildings in Thun and surroundings. The y-axis shows the number of flooded buildings in case of flooding depending on the lake level of Lake Thun.

an extreme rainfall event to increase the retention capacity for the flood volume. This would increase safety for both Thun and Bern. However, the implementation required another adaptation to the lake-river system in Thun. The minimum base level of the lake outflow at the intersection between the lake and the outflow river is too high to lower the lake level within a reasonable period, i.e., in the period of reliable weather forecasts. The relatively high base level of the outflow allows only a limited outflow discharge when the lake level is low. Thus, the outflow capacity had to be increased technically. A lowering of the base level of the outflow riverbed was not possible because of environmental concerns and because of the configuration of the installed weirs. Therefore, a **relief tunnel** was planned. The entering of the relief tunnel was constructed below the base level of the outflow riverbed, and therefore the lake outflow capacity could be increased by 100 m^3/s. This relief tunnel can be regulated and is managed by the cantonal authorities. The lake regulation rules are coordinated with all requirements and stakeholders. In case of a predicted heavy precipitation event in the upstream river basin, the lake level of Lake Thun is now lowered as much as needed to be able to withhold the forecasted inflows and thus to flatten the flood peak downstream of Thun. This solution is now for the benefit of Thun and the downstream villages.

It is the final puzzle piece of the adaptation of the coupled human–environment system to new conditions that had been set by human intervention

300 years ago. It is also the last human intervention to the lake-river system in a cascade of interventions needed to encounter the cascade of unintended consequences that followed the first human intervention.

The correction of the other river reaches in the Aare river basin followed soon. The **Gürbe river** was corrected and trained around 1850. The **Hasliaare river** (the river reach of the Aare river between Meiringen and Lake Brienz) was corrected in 1866, the **Aare river in Interlaken** was corrected in the period 1852–62 from Lake Brienz to Interlaken, and in the period 1890–94 from Interlaken to Lake Thun. The **Lütschine river** at the alluvial fan in Interlaken was corrected and trained in the mid of the 19th century. All main river reaches were trained with a uniform trapezoidal cross section and lateral levees.

In the 20th century, many reservoirs for hydropower generation were built. Over 50 medium-sized to large hydropower facilities are currently operating in the river basin. Eight reservoirs for seasonal storage of water for hydropower generation are operative. They can store up to 195 mio. m^3 water. The operation of these facilities alters the hydrologic regime in addition to the previously mentioned interventions into the river system.

Model experiment setup for quantifying the cumulative effects of human interventions to river systems on flood risk change

To summarize the history of all human interventions to the river system of the Aare river basin upstream of Bern, we can state that this river system is an anthropogenically influenced system. Only a very few river reaches are left in its natural state. The ecomorphological quality is degraded. Only recently, revitalization measures are being implemented to restore ecosystem's functionality and quality. Human interventions influence nature and the former are influenced by the natural systems. Therefore, we can define the river system in the Aare river basin as a coupled human and natural system (Liu, Dietz, Carpenter, Alberti, et al., 2007; Liu, Dietz, Carpenter, Folke, et al., 2007) or as a social-ecological system (Biggs et al., 2021).

The human influence on the natural system can be studied from many different perspectives, ranging from an ecological perspective to a socio-technical perspective. From the perspective of this book, the flood risk change and complexity perspective, the Kander river case study is interesting to study the human influence on flood risk and its change. Thus, here we used this case study to analyze the cumulated effects of all human interventions to the river system on flood risk change. Therefore, a model experiment is set up according to the general principles described in the previous chapter. We developed a coupled hydrologic and hydraulic model that is able to simulate the discharge from the upstream catchments and the routing of the water through the main floodplains. In this model setup, the meteorological input, i.e., rainfall and temperature, and the hydrological parameters, i.e., land cover and land use, are kept constant. In contrast, the hydraulic model is extended. One hydraulic model represents the current state of the river system, i.e., the current digital terrain model with all flood protection measures and all river corrections considered. The counterfactual hydraulic model represents a historical state of the river system prior to any hydraulically relevant anthropogenic intervention, i.e., the river system before the Kander river deviation and the subsequent river corrections.

The rainfall in the mountain catchments and the discharge calculated by the hydrologic model are then routed through the floodplains of the main river valleys and the lakes in both systems representations. The discharge at the outlet of the river basin differs when comparing both states of the river system. The difference between the resulting hydrographs provides the basis for quantifying the change in the hydrologic regime. Moreover, the proposed model setup can also be used to calculate a number of extreme rainfall scenarios and to calculate inundation maps for each rainfall scenario. The resulting flood maps are overlaid with the houses that are considered as the values at risk. Thus, the number of exposed houses or the damage to houses can be compared when using two different system states (Fig. 3.9). The changes in the damage can be used for quantifying flood risk changes on the river basin scale. In the following, the single modules of the coupled model and the model experiment will be described in detail.

The coupled hydrological-hydraulic model

For the rainfall-runoff modeling, we applied the hydrological model PREVAH (Viviroli, Zappa, Gurtz, & Weingartner, 2009) to model the hydrological processes in the Aare river basin upstream of Bern, Switzerland. The study area covers an area of about 3000 km^2 and has a mean elevation of about 1600 m a.s.l., ranging from 500 to 4200 m a.s.l. The river basin can be roughly divided into an upper section with a mountainous character and a lower section with a pre-Alpine character. The upper section of the catchment consists of a steep mountainous and partly glaciated landscape. Several alpine peaks within this area exceed 4000 m a.s.l., and parts of the catchment area are glaciated (8% of the total catchment area). The subcatchments in this mountainous area drain directly into the two connected lakes, Lake Brienz and Lake Thun, that cover an area of 30 and 49 km^2, respectively. The main valley of the Aare River basin consists of a relatively flat floodplain with two lakes, where widespread flooding can occur. The lakes are natural but artificially managed and are oriented along an approximately east—west axis in the lowland part of the catchment. The outflow of Lake Thun flows over into the lower part of the catchment and its floodplains along the main valley floor. The mean annual rainfall in the catchment amounts to 1500 mm, of which 500 mm evaporates and 1000 mm is discharged (Felder & Weingartner, 2017). The discharge regime is influenced by the presence of glaciers that cover about 8% of the

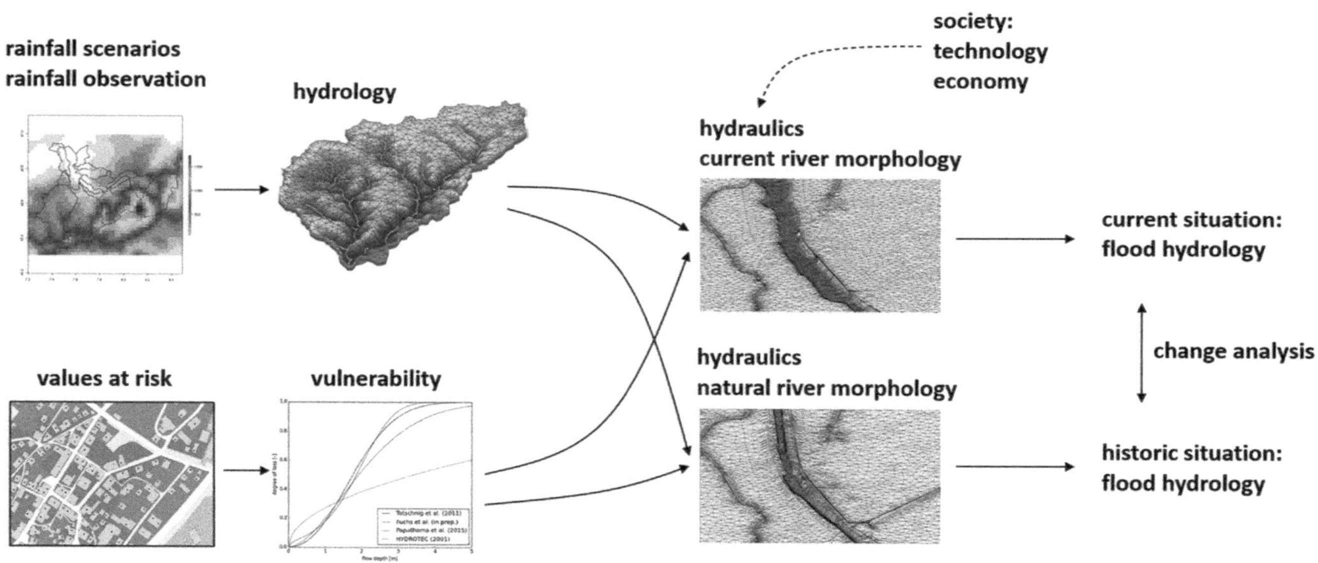

FIGURE 3.9 **Model experiment.** Model experiment setup to analyze the cumulative effects of all human interventions to the river system on flood risk change in the Aare river basin. The meteorological input and the values at risk are set as constant, while the digital terrain model was varied, representing two system states, the status of the historical natural and the status of the anthropogenically modified system.

total area, meaning that the average discharge at the river basin outlet in Bern is relatively low (70 m^3/s) in winter and relatively high (180 m^3/s) in summer.

The following main river reaches are considered in the model chain:

1. Hasliaare river, from Meiringen to Lake Brienz (floodplain: 15 km^2; contributing area: 451 km^2);
2. Lake Brienz, static inundation model (lake area: 31 km^2; contributing area: 1138 km^2);
3. Interlaken, area between Lake Brienz and Lake Thun, and the fan of the Lütschine River (floodplain: 28 km^2);
4. Lake Thun, static inundation model (lake area: 49 km^2; contributing area: 2450 km^2);
5. Thun (floodplain: 8 km^2);
6. Aare river reach between Thun and Bern (floodplain: 42 km^2);
7. Gürbe river reach between Burgistein and Belp (floodplain: 15 km^2; contributing area: 116 km^2).

The PREVAH model uses hydrological response units that are directly routed to the (sub-)catchment outlet. For the Aare river basin, the hydrological response units were built based on characteristics (altitude zone, slope, aspect, land use, soil type, and glaciation) of the subcatchments at a 2 km resolution. Twelve parameters were calibrated (14 in the case of glaciated catchments). Fifteen subcatchments were independently modeled; their location is shown in Fig. 3.10. For nine out of these 15 subcatchments, calibration was performed using hourly resolved discharge data, leading to NSE skill-scores between 0.70 and 0.92 for the calibration period (2001–10) and between 0.65 and 0.88 for the validation period (2011–14). The other six subcatchments were parametrized using the regionalization approach proposed by Viviroli, Zappa, Schwanbeck, et al. (2009).

The hydrodynamic model aims at simulating the water fluxes coming from the subcatchments through the main valleys and through the lakes. This model is able to consider the effects of the weirs on lake levels and on the lake outflows. Moreover, the hydrodynamic model is able to consider the retention effects of the lakes and of the floodplains and is therefore more reliable than a simple hydrological model without specifically adapted and calibrated routing modules (Felder et al., 2017; Grimaldi et al., 2018; Hoch et al., 2017, 2019). In this case study, we used the 1D hydrodynamic model BASEMENT (Vetsch et al., 2017). The model is based on the continuity equation and solves the Saint–Venant equations for unsteady one-dimensional flow. In order to consider floodplains and storage outside the riverbed, the river cross sections were expanded to cover the whole floodplain. Cross sections of the riverbed and the directly adjacent levees were provided by the Swiss Federal Office of Environment FOEN (FOEN, 2018a). These cross sections were expanded to potential flood-prone areas beyond the levees using data from a digital elevation model with 0.5 m resolution and a vertical accuracy of 0.2 m

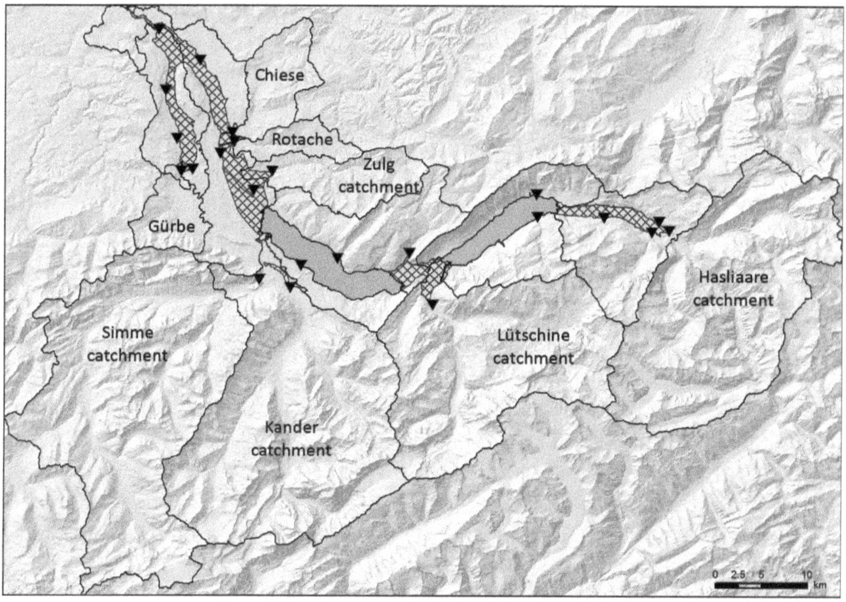

FIGURE 3.10 **Aare river basin.** Spatial setup of the interface between hydrologic and hydrodynamic models. The black lines show the catchment delimitations. The calibrated subcatchments are indicated by labels. The black triangles indicate where the output of the hydrologic model will be used as input for the hydrodynamic model.

(KAWA, 2011). The cross sections were set straight and perpendicular to the flow direction with an average cross section spacing of approximately 150 m, as recommended in studies of other catchments (Castellarin et al., 2009). The hydrodynamic model was calibrated by empirically adjusting the roughness coefficient (Manning–Strickler). The values were adjusted by reconstructing the water surface elevation of observed flood events. Particular attention was given to calibrating the roughness coefficient for the peak discharge. Additionally, hydrodynamic parameters that define the characteristics of weirs (factor *m* of the Poleni equation) and pipes (contraction factor) were adjusted (Felder et al., 2017). The weirs configuration was set in a way that the discharge out of the lakes was maximized, as this is likely to be the case during flood events. The hydrodynamic model was then able to reconstruct the rating curve at the catchment outlet with an error of ± 2 cm in water level for observed flood events.

The outputs of the hydrologic model are fed into the hydraulic model as upper boundary conditions or as lateral inflows. The model coupling is external, which means that there is no direct interaction between the hydrologic and the hydrodynamic models. The range of the hydrodynamic model was set to incorporate all significant flood-prone areas and potential retention areas. It was assumed that minor retention areas inside the subcatchments have a negligible effect on the peak flow at the catchment outlet. The coupling interfaces between the hydrological and the hydrodynamic models are shown in Fig. 3.10. The hydrological model was not applied to the lakes because they directly receive the precipitation that falls upon them. In these cases, precipitation was directly fed into the hydrodynamic model.

Reconstruction of the historic natural river morphology

The coupled hydrologic-hydraulic model was set up to simulate the water fluxes through the main valleys and lakes in present-day status. However, the analysis of the hydrological change due to human interventions in the river system requires the reconstruction of the historical state of the river system, a state without any anthropogenic shaping of the river morphology, preferably the natural state of the river system to the latest point in time before the first human intervention took place. In our case, this is the river morphology shortly before the Kander river deviation in 1714 AD. The reconstruction of the natural river morphology allowed us to modify the hydraulic model and thus to simulate the water fluxes through the river system in its natural state. The comparison between the water fluxes of both river states allowed us to analyze and quantify the hydrological change.

However, the river morphology of the 18th century was not mapped and surveyed with earth observation measures as today. Fortunately, the engineers planning the early river corrections were mostly surveying engineers. It was professional sense in these times to study the system behavior of the rivers before planning any measures. Therefore, the planning phase of these early river engineering projects consisted also of topographic surveys. These surveys resulted in relatively exact topographic maps, also from a nowadays perspective of the technological advancement of earth observation.

The natural states of the river courses before the first relevant anthropogenic modifications took place were reconstructed on the basis of these historical topographic surveys and maps. The river corrections were planned on the basis of remarkably detailed topographic surveys by overlaying the planned new river channel on top of the old maps (e.g., Fig. 3.11). These project plans include plans and maps but also many mapped cross sections and water level measurements from different dates.

The historical maps were georeferenced on the basis of the hill shade of the current digital elevation model with a high spatial resolution (0.5 m). In this digital terrain model, the former river beds and channel geometries are still visible, especially in those areas that are covered now by farmland or alluvial forests and not massively shaped by land management and infrastructure. This evidence of former landscape patterns that are still visible in the terrain provides a useful base for verifying the accurateness of the historic maps and for the georeferencing. Specific signs in the historical maps, as for example, river braiding patterns and side channels, are still visible in the current digital

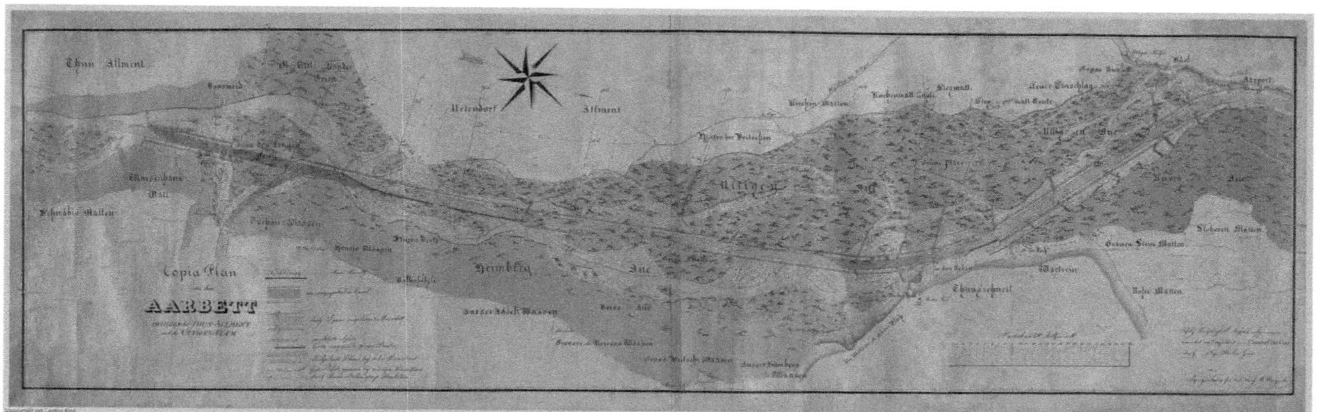

FIGURE 3.11 Plan of the Aare river correction. Example of a historic map that is showing the original river morphology and the planned river training on top. *From Baczynsky, J. (1834). AA V 64 Aare, Thun–Uttigen: zwischen Thun-Allmend und Uttigenfluh, mit bestehenden und projektierten Verbauungen, Staatsarchiv des Kantons Bern, Bern.*

elevation model and can be used as landmarks for georeferencing the historical maps on top of the current landscape. After georeferencing the historical maps, the old river channels and water surfaces were digitized. The current terrain can help to interpret and validate the mapped historical landscape. Then, the recent anthropogenic modifications of the terrain that are relevant obstacles to the hydraulic modeling were erased from the current digital elevation model, e.g., terrain cuts or landfills for highways or railway dams. These hydraulically relevant manmade features in the landscape also contain structural engineering works related to the river corrections, such as the incised riverbed of the corrected river channel and the lateral levees. This results in a digital elevation model with no data values in anthropogenically modified areas and contour line features with voids in those areas. These holes in the current digital terrain model were then filled by interpolating the terrain surface over the "holes" based on the information on the borders of the holes. In the case of larger areas where this filling of the holes in the digital terrain model based on neighborhood information was not reliable, we manually connected the contour lines ending at the left and right lateral borders of the void areas by conserving the adjacent fan slope and planar/longitudinal curvature. The digitized contour lines were then interpolated to a surface representing the natural terrain of the floodplain in the erased areas of the digital elevation model. This interpolated surface was merged with the present-day digital elevation model outside of the voids and thus the erased anthropogenically modified areas in the digital elevation model could be filled with the interpolated surface of the original geomorphology of the floodplain. The result of these modifications was a digital elevation model without any human landscape features but also without the historical river morphology. In the following step, the digitized areas of the historic riverbeds were incised into the cleaned terrain model. The incision depth was delineated from the historical topographic surveys and the historic cross-section geometries (see Fig. 3.12 for an example of historic cross sections). The historic riverbed of river reaches without available historic cross sections for assessing the incision depth were incised into the reconstructed digital elevation model with a depth corresponding to the river conveyance capability of a flood event occurring statistically every 2 years according to Bhuyian et al. (2017).

Fig. 3.13 shows the different steps of the procedure, and Fig. 3.14 shows the result of the procedure for the example of the Gürbe river.

The human dimension of hydrological change

The reconstructed terrain model provided the basis for the creation of the computational mesh for the hydrodynamic model. To consider the historical digital terrain model, the cross sections of the current digital terrain model were transformed to represent the reconstructed historical digital terrain model. This means that the locations of the cross sections remained the same,

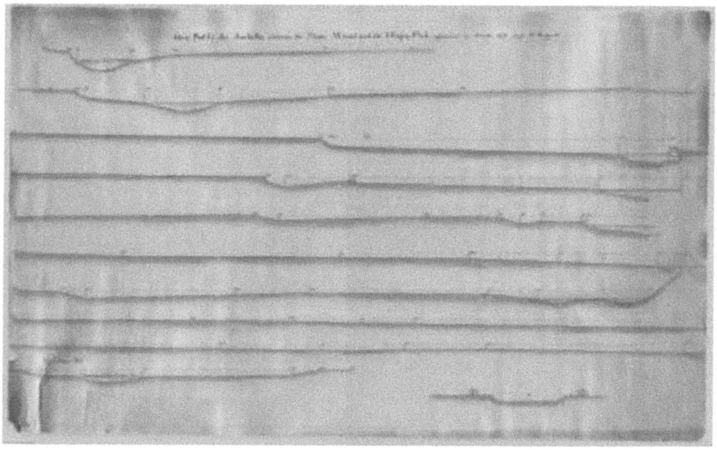

FIGURE 3.12 Historical cross sections. Example of historical cross section resulting from topographic surveys during the planning process of the Aare river corrections in the early 19th century. *From: Baczynsky, J. (1835). Aare, Thun - Uttigen: Querprofil des Aarebetts zwischen Thun-Allmend und Uttigenfluh, Staatsarchiv des Kantons Bern, Bern.*

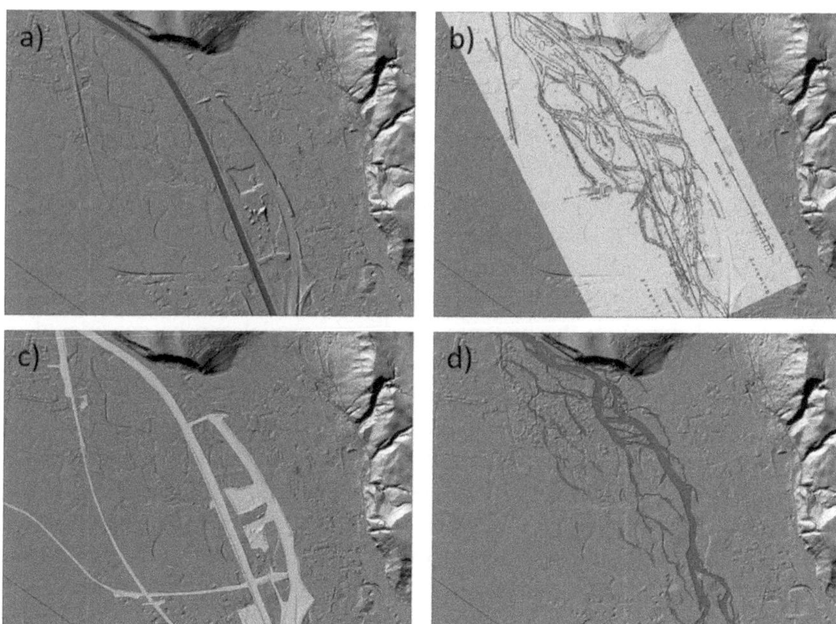

FIGURE 3.13 Elaboration of the historical digital elevation model. Example of the procedure for the reconstruction of the historical digital elevation model. Aare river reach between Thun and Bern. (A) current digital elevation model and riverbed, (B) georeferencing of the historical map with the help of terrain signatures, (C) mapping of hydraulically relevant anthropogenically modified terrain features, (D) historical digital elevation model of the Aare river before the river corrections in the early 19th century.

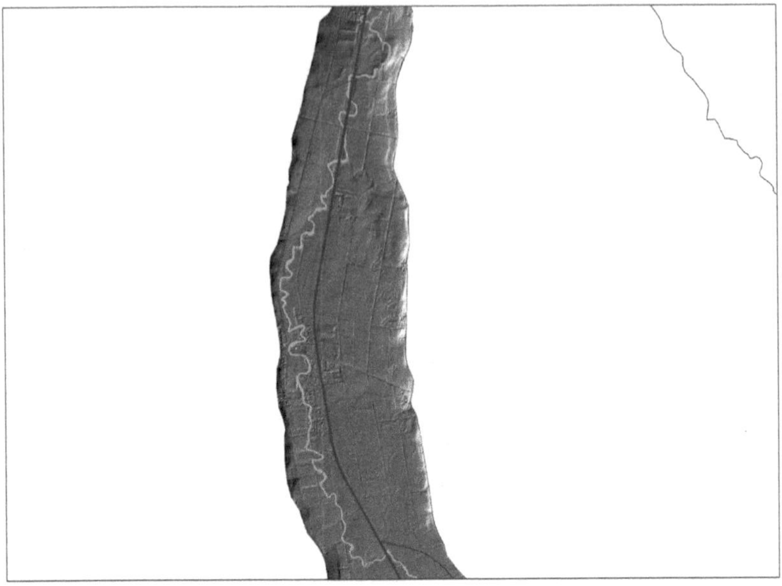

FIGURE 3.14 **Gürbe river.** Reconstructed riverbed of the Gürbe river before the correction works (black) and the current riverbed (white). *From Zischg, A. (2016). River corrections and long-term changes in flood risk in the Aare valley, Switzerland.* E3S Web of Conferences, 7, *11010. https://doi.org/10.1051/e3sconf/20160711010.*

but the z-coordinates of the survey points of the cross sections changed. The friction values assigned to the different roughness characteristics (main riverbed, riverbanks, and floodplains) were adapted to the historical situation to consider the lateral shifts of the river channels. The weirs that are implemented in the current hydraulic model were eliminated. Thus, the hydraulic model for the historic state of the river system has no weirs and river regulation measures, and therefore the historical hydraulic model can be considered as free from any human intervention in river geomorphology.

Now, two different hydraulic models can be used for simulating the water fluxes through the main valleys, the lakes, and the river system. If we keep rainfall constant, i.e., feeding the hydrologic models for the subcatchments with the same spatiotemporal rainfall data, we can assess the difference in the river routing by comparing the difference between the hydrographs of the basin outflow. This comparison leads to the assessment of the human dimension of hydrological change. Fig. 3.15 shows the two hydrographs, one hydrograph showing the basin outflow under current settings of the river system and the other showing the basin outflow without human interventions to the river system in the period 2005–16. We used this period because hourly spatiotemporal rainfall and temperature data are available for this period.

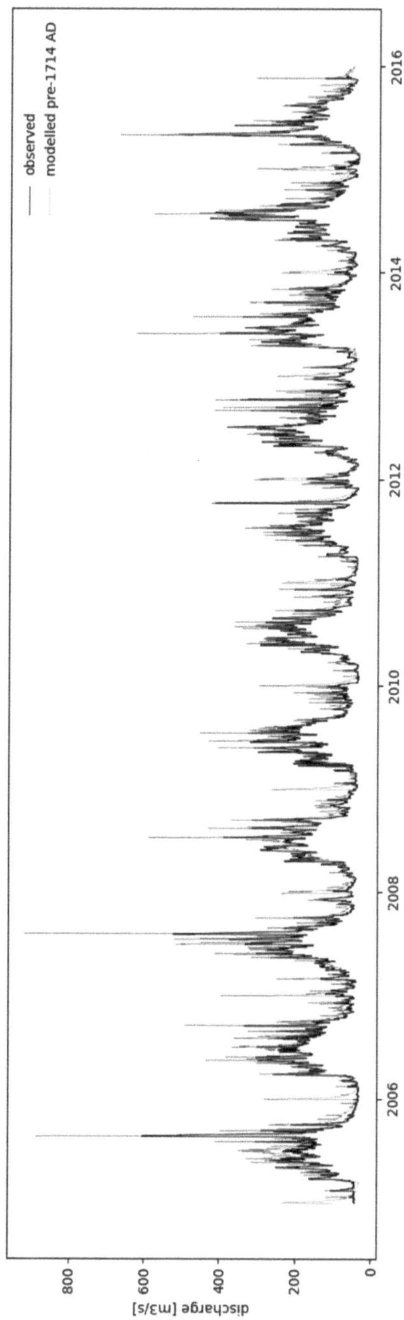

FIGURE 3.15 Human dimension of hydrological change. Observed hydrograph of the Aare river in Bern in the period 2005–16 (black line) and modeled hydrograph of river routing without human interventions to the river system (gray line).

The comparison between the two hydrographs shows that the peak discharges, especially in higher magnitudes, would be higher in a natural hydrologic regime. This makes the retention effect of Lake Thun evident that is gained by the Kander river deviation. The peak discharges would be on average 100 m^3/s higher in a natural state of the river system. A one in a 100 years event of the Aare river in Bern has a discharge of 551 m^3/s (FOEN, 2021). First damage occurs in Bern above a discharge of 430 m^3/s. A flood event greater than a nowadays 100-year event occurred once in the period 2005–16 but would have occurred six times in the counterfactual setting, i.e., the same rainfall over the river basin but routed through a river system in its natural state before any human intervention (Fig. 3.16).

The maximum discharge in the period 2005–16 is 605 m^3/s in the observation data and 920 m^3/s in the counterfactual simulation. The modeled hydrograph shows some peak flows where the observed hydrograph showed very low flows, especially in winter times. This means that a large number of flood events would have occurred. In the period 2005–16, nine flood events would have occurred in the counterfactual simulation that exceed the nowadays damage thresholds. The low flows are more frequent in the observation data and the high flows are more frequent in the counterfactual data (Fig. 3.17). The variability is also lower in the observation data because of the dampening of the flood waves by Lake Thun after the Kander river deviation and because of the lake regulation. The effects of the other river training measures are relatively low because the retention effect of Lake Thun dominates over the other factors. The river trainings have local effects on flood risks. These effects will be analyzed in the next sections.

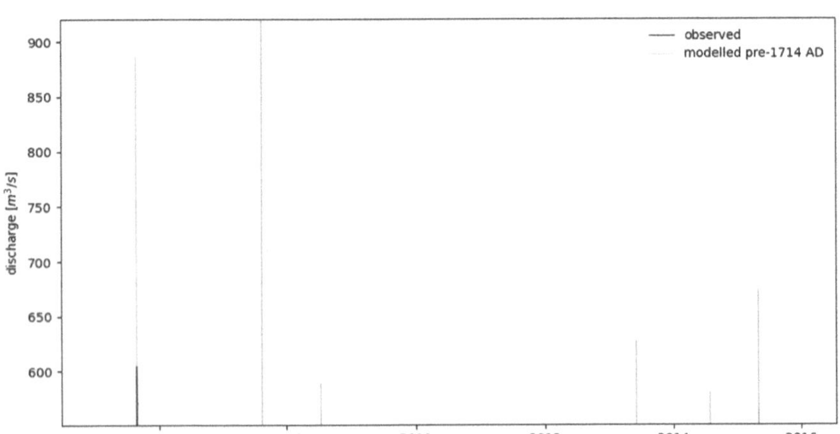

FIGURE 3.16 Flood events. Flood events with a magnitude greater than a 1-in-100 years event in the Aare river in Bern in the period 2005–16 (black) and in the counterfactual setting without human interventions to the river system (white).

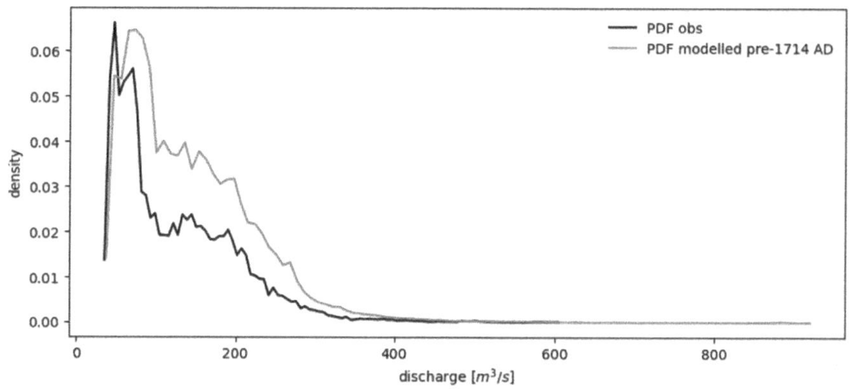

FIGURE 3.17 Observed and modeled hydrographs. Probability density of the observed hydrograph of the Aare river in Bern in the period 2005–16 (black line) and of the modeled hydrograph of river routing without human interventions to the river system (gray line).

Analyzing the cumulative effects of river training measures on flood impacts

The period 2005–16 we used for assessing the human dimension of hydrological change is too short for deriving extreme value statistics and thus for analyzing flood risk change. To assess the effects of human interventions on flood risks, we used different setups of the modeling chain. In a first setup, we used the observed flood hydrographs of a recent flood event in the river basin and used these hydrographs as input for the hydraulic models. In this model experiment, we used two different hydraulic models, one that represents the current state of the river morphology and the other that represents the counterfactual situation, i.e., the historical state of the river morphology. In this version of the model experiment, we compared the outcomes of the two hydraulic models in terms of changes in flooded area, flood depths, or exposed houses. If a detailed flood event documentation exists, as it is the case for the following example, the comparison between the actual and the counterfactual situation can also be made by comparing the event documentation and the counterfactual simulation. This simplified approach answers the question of what would have happened during the same flood event if the existing flood protection measures had not been implemented? The model experiment setup is schematized in Fig. 3.18.

In a more sophistic and complex model experiment, we changed rainfall input and simulated the flood processes with the whole model chain. Instead of using continuous spatiotemporal rainfall data as input for the hydrological model, we used a set of rainfall events as input data for the hydrological model. The hydraulic model then used the output of the hydrologic model to simulate the inundation processes in the floodplains and an impact model

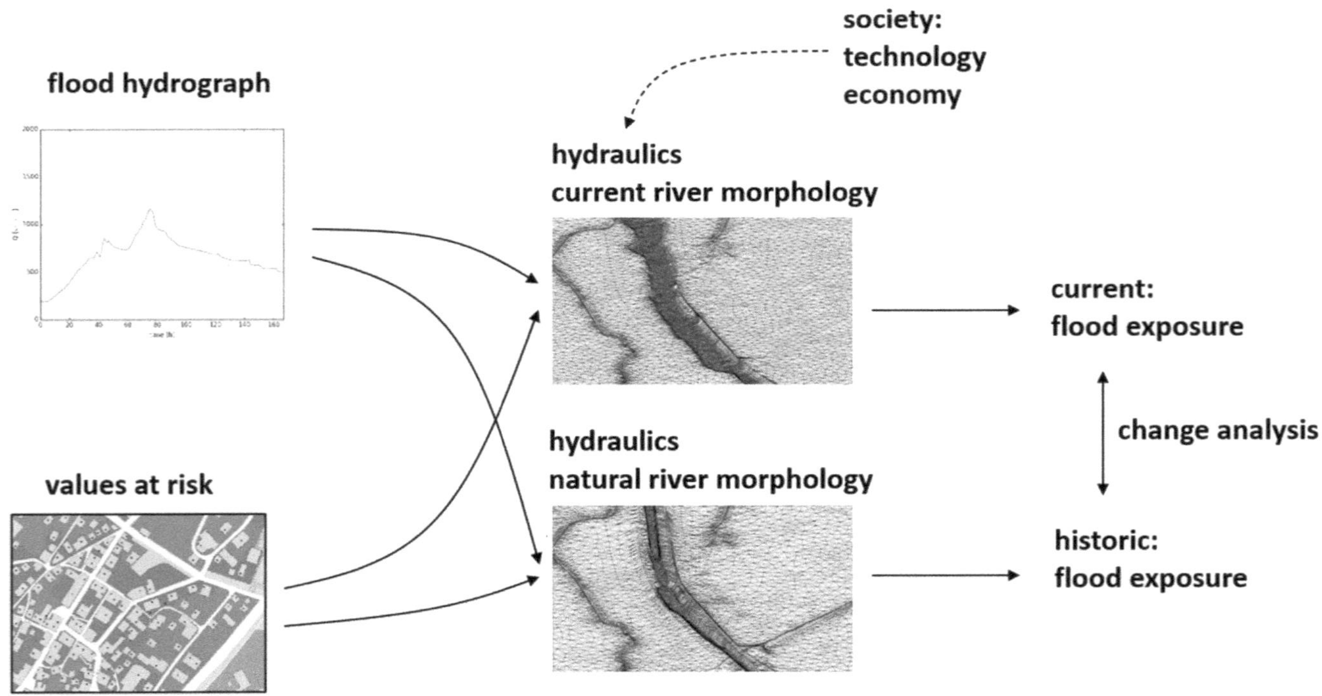

FIGURE 3.18 Model experiment. Setup of the model experiment aimed at comparing the impacts of a recent flood event with a counterfactual situation, i.e., the historic state of the river morphology.

assessed the impact of the floods. In the abovementioned model experiment to assess the effects of human interventions to the river system on hydrological change, we used a one-dimensional hydraulic model. This is capturing the retention effects of the lakes and the floodplains but is not calculating the flood depths and flood velocities in the inundated areas of the floodplains. However, a flood impact model needs the parameters of the flood magnitude in the floodplains for assessing the hazardous impact on the values at risk. This requires at least a two-dimensional model that computes the required parameters in space and time over the whole floodplain. The impact model then extracted the values over and around the objects at risk, and by means of a vulnerability function, the impact model quantified the damage to each inundated object. Hence, the setup of this model experiment was extended in contrast to the previous model experiment (Fig. 3.19). This model experiment setup differs from the former setup by different input data for the hydrological model, a 2D hydraulic model, and an impact model for describing the impacts of the simulated flood events.

Hydrodynamic model

In contrast to the previous model experiment, we used a two-dimensional hydrodynamic model for this model experiment. We used the flood inundation model BASEMENT in its 2D version (BASEplane v2.8) to represent the water fluxes through the river systems (Vetsch et al., 2017). It is a numerical model for solving the shallow water equations on the basis of an irregular mesh. We created a computational mesh for each of the floodplains, namely the floodplain of the Hasliaare river, the floodplain of Interlaken between the two lakes, the floodplain of the Gürbe river, the floodplain of Thun, and the floodplain of the Aare river between Thun and Bern. The computational meshes of the floodplains of Thun and Interlaken as well as the floodplain of the Hasliaare river were used for simulating river floods and lake floods. The inundation of the lateral floodplains of the lakes was simulated with a static lake flooding model, i.e., a bathtub approach using the lake level as boundary condition and the altitude levels of the building's ground floors.

The used 2D flood simulation model requires the generation of computational meshes with irregular triangles. A preprocessing module of BASEMENT generates an irregular mesh, and the z-levels of the mesh nodes were attributed from the digital elevation model of a spatial resolution of 0.5 m (KAWA, 2011). The main linear structures were considered in the mesh as break lines that were digitized by hand. The computational mesh was generated in two versions, one version is based on the current digital elevation model and thus on the current river morphology, and the second version represents the historical state of the river morphology. For this version, we used the reconstructed digital terrain models described in the previous section. The friction values were calibrated for each river reach independently by using

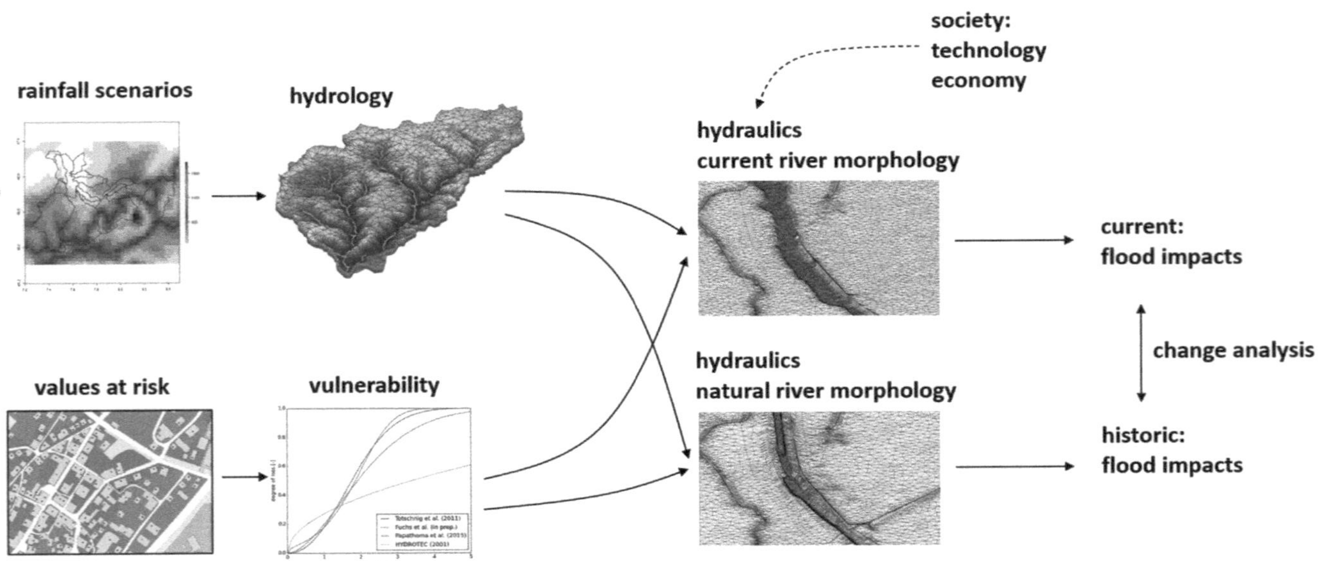

FIGURE 3.19 Model experiment. Setup of the model experiment for analyzing the cumulative effects of river training measures on flood impacts based on changed rainfall input.

the existing stage−discharge relationships of the gauging station in the respective river reach. We used different friction values for riverbed, riverbanks, floodplains, and forests in the floodplains. The friction values of the counterfactual computational mesh were classified with the same friction values but according to historical land use. The break lines were newly digitized for the historic terrain conditions. The average size of the mesh elements is approximately 10−30 m^2 in the river channel (depending on river size) and 100−300 m^2 in the floodplain. The inundation model was validated with insurance claim data and reproduced historical flood events well (Zischg, Mosimann, et al., 2018). Fig. 3.20 exemplarily shows an extract of the two computational meshes.

The effects of the river training measures on the surface area of water in the landscape were remarkable. The Aare river between Thun and Bern lost 7 km of flow length (−20%) and 64% of water surface area, the Hasliaare river reach lost 88% of the water surface, and the Gürbe river reach lost 22% of water surface area.

Flood inundation of recent flood events on a counterfactual river morphology

In this simulation, the observed hydrographs of the recent flood events in the Aare river basin were used as input for flood modeling. The hydrographs at each river reach were measured by the respective gauging stations in the river network. The observed hydrographs were used as an inflow at the upper boundary of the current and related historical computational meshes. For the Aare river reach between Thun and Bern, the hydrograph of the August 2005 event measured in Thun was used. The peak discharge was 543 m^3/s. For the Hasliaare river reach, the hydrograph of the October 2011 event with a peak discharge of 365 m^3/s was used. For the Gürbe river, the hydrograph of the July 1990 event with a peak discharge of 85 m^3/s was used for modeling the historic flood scenario. The simulated flooded areas of the historical states were compared with the event documentation dataset of the Canton of Bern (Kanton Bern, 2021). In terms of exposed buildings, the comparison between the two states of the river courses was done with the building stock of today. Obviously, an overlay of inundation areas of the old river course with today's buildings is not realistic. Without the river training measures of the 19th century, parts of the floodplains would not have been built-up because of the frequent flooding (Collenteur et al., 2015; Hutton et al., 2019; Montz & Tobin, 2008; Tobin, 1995). Therefore, we neglect here the feedback effect between the frequency of flooding and the adaptation of land use. However, this counterfactual analysis is appropriate to isolate the effects of the river training works on the extent of flooded areas.

The inundated areas of the Aare river reach during the August 2005 flood event would be eight times higher in the counterfactual simulation, i.e., a

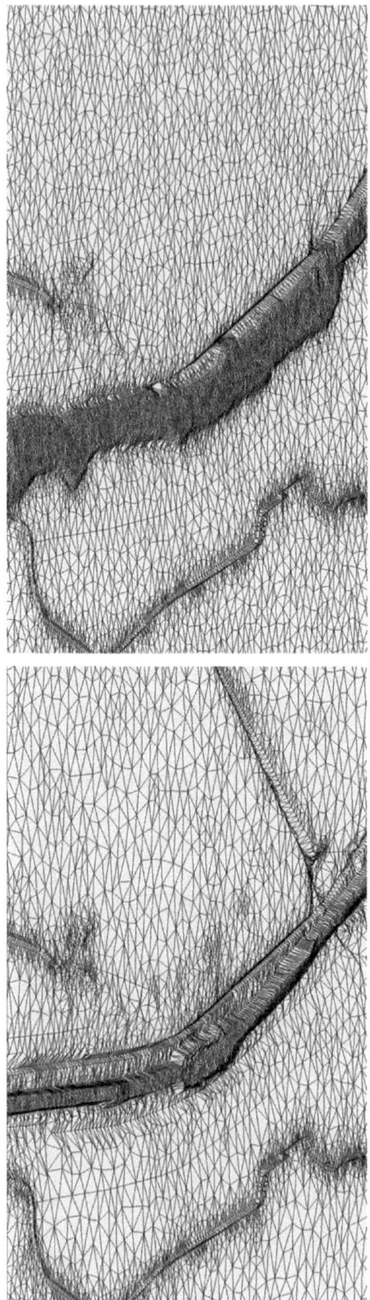

FIGURE 3.20 Computational mesh. Example of two computational meshes of the current (left) and historical river morphology (right).

status without any river corrections. The flood defense structures reduce the flooded area by about 88% (see Table 3.1). If the current building stock is assumed as constant over time in the historical flood scenario, then the number of residential buildings exposed to a flood similar to the August 2005 event but without any flood defense measures would be around 40 times higher (see Table 3.1). This means that the exposure of the current buildings is reduced by 97.5% by the river corrections in a flood event comparable to the August 2005 flood event (Figs. 3.21 and 3.22).

The comparison of the two states of the Hasliaare river reach shows a similar picture. Here, the October 2011 flood event was taken for comparison. The flooded area would have been nearly eight times higher without any flood defense measures. Therefore, the river corrections reduced the flooded areas in this scenario by 87%. If the building stock is assumed as constant over the whole period (building stock of 2012), then the number of exposed buildings would have been around 48 times higher in the historical scenario without the river corrections. This means that the exposure of the current building stock is reduced by 98% in a flood event comparable to the October 2011 flood event (Figs. 3.23 and 3.24).

In the Gürbe river, the results differ remarkably from the other cases. Here, the historical scenario affects fewer buildings than the actual scenario. From the building stock of 2012, 217 residential buildings are located within the documented flooded areas of the July 1990 flood event, whereas only 182 are exposed to the same flood event but on the historic terrain model (Fig. 3.25 and Table 3.1).

The flooded areas were reduced by the river training measures by 88% in the Aare river reach, by 87% in the Hasliaare river reach, and by 96% in the Gürbe river reach (Table 3.1). The number of exposed buildings was reduced by 97% and 98% in the Aare and Hasliaare river reaches, respectively. The situation of the Gürbe river differs from the other two river reaches. Here, the river correction led to a new location of the river and the settlements growth around the river. Nowadays, the Gürbe river crosses the settlements while it flowed mostly through the unpopulated floodplains in its original course.

The results enable a first attempt to assess and quantify the effects of the river corrections of the 19th century on today's flood exposure. The counterfactual simulation shows that the range in reducing the flood exposure is between 69% and 98%. The number of residential buildings exposed to the selected flood events would be up to 40 times higher in a situation without any flood defense structures. The special case of the July 1990 flood event in the Gürbe valley shows that the deviation of the river by the correction works is not reducing the exposure of buildings.

This model experiment setup allows us to answer the question of the actual value of the cumulative effort in flood risk reduction of the past decades and centuries. This question arises when it comes to decisions needed for the maintenance of the old infrastructure. In many times, the effort and the

TABLE 3.1 Comparison of flooded areas.
Sum of flooded areas and exposed buildings of the current situation and the counterfactual simulations.

	Flooded areas			Exposed buildings		
River reach	**Early-19th century [km^2]**	**Actual state [km^2]**	**Difference [%]**	**Early-19th century [km^2]**	**Actual state [km^2]**	**Difference [%]**
Aare Thun-BernAugust 2005 flood	20.4	2.5	−88	2076	51	−97
HasliaareOctober 2011 flood	8	1	−87	242	5	−98
GürbeJuly 1990 flood	4.6	0.2	−96	182	217	19

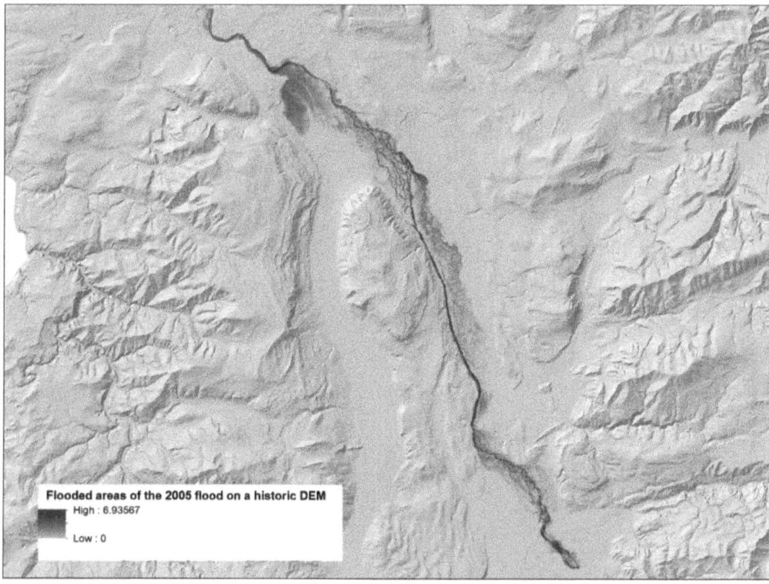

FIGURE 3.21 Flood simulation with historical river morphology – Aare river. Simulated flooded areas of the August 2005 flood event in the Aare river between Thun and Bern in a historic state of the river morphology.

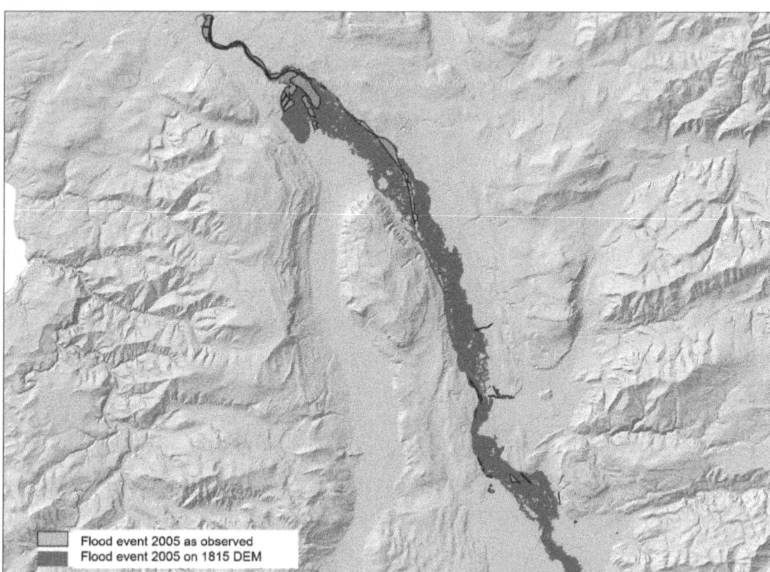

FIGURE 3.22 Effects of river morphology change on flood extent – Aare river. Comparison between the documented flooded areas of the August 2005 flood event (in white) in the Aare river between Thun and Bern and the hypothetic simulation of the same flood event in a historic state of the river morphology.

FIGURE 3.23 Flood simulation with historical river morphology – Hasliaare river. Simulated flooded areas of the October 2011 flood event in the Hasliaare river in a historic state of the river morphology.

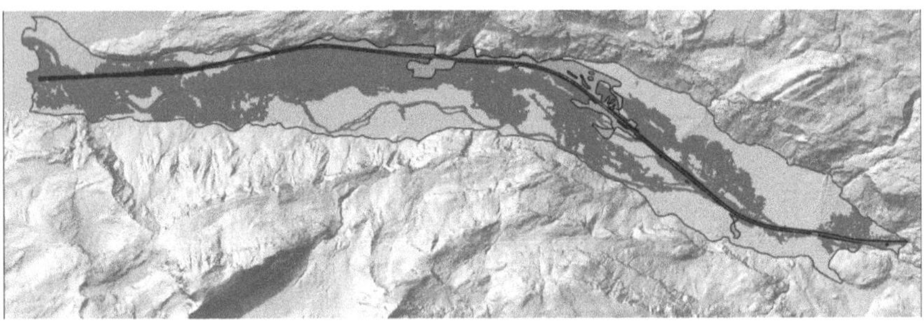

FIGURE 3.24 Effects of river morphology change on flood extent – Hasliaare river. Comparison between the documented flooded areas of the October 2011 flood event (in white) in the Hasliaare river and the hypothetical simulation of the same flood event in a historical state of the river morphology.

financial need for the maintenance of old infrastructure are very difficult to convey to the population (Thomi, 2010; Thomi et al., 2015). With this approach, arguments for the maintenance efforts could be formulated. Without continuous maintenance of the meanwhile 150-year-old flood defense structures, the damage will be remarkably higher in the case of a collapse. However, the example of the Gürbe river reach shows a differentiated view of this topic. Here, the flood defense structures increase the number of exposed buildings. The reasons are that the location of the current river course totally differs from the historical river course and that the retention effect of the floodplain is lost. Nevertheless, this example shows that the reallocation of the current river course to the former place could be an interesting option in flood risk management.

The limitation of the presented approach lies in the restricted selection of the flood scenarios. For each study area, only one selected flood event was chosen for the assessment of the effectivity of the existing flood defense

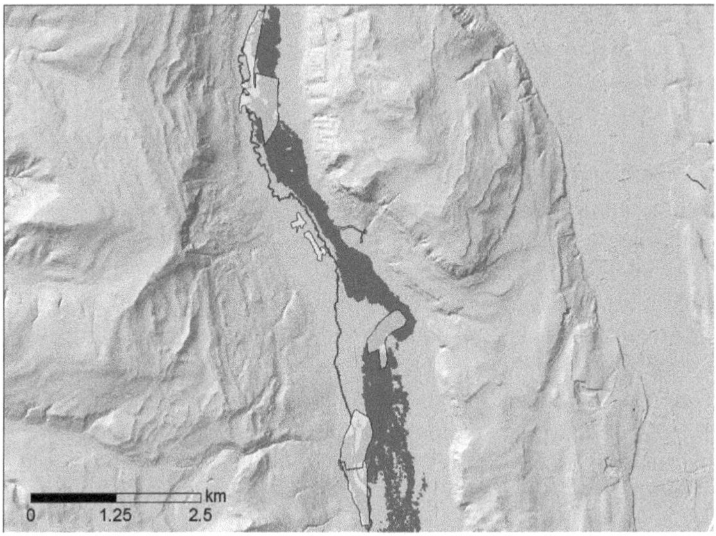

FIGURE 3.25 Effects of river morphology change on flood extent – Gürbe river. Comparison between the documented flooded areas of the July 1990 flood event (in white) in the Gürbe river and the hypothetical simulation of the same flood event in a historical state of the river morphology. *From Zischg, A. (2016). River corrections and long-term changes in flood risk in the Aare valley, Switzerland.* E3S Web of Conferences, 7, *11010. https://doi.org/10.1051/e3sconf/20160711010.*

structures. This may lead to a constrained explanatory power. In particular, discharges below the transport capacity of the current river reaches do not result in damage today. But, without any flood defense structures, these discharges would result in damage. Therefore, it is interesting to assess if the consideration of discharges with a medium frequency led to other conclusions than drawn here. With more flood scenarios, the effectivity of the existing flood defense structures could be assessed in more detail. Thus, in the following, I will show a more sophisticated model experiment.

Defining extreme rainfall scenarios

The previous section has shown the limitations of using only a very few flood events that occurred recently. An in-depth analysis of flood impacts, however, requires a larger set of flood scenarios. The higher the number of the set of scenarios, the more robust are the results of the risk analysis. Especially in mountainous river basins, the outcomes of impact analysis are sensitive to the spatiotemporal patterns of rainfall and the temporal pattern of peak flows in the river network (Zischg, Felder, Weingartner, et al., 2018). Peak discharge and the volume of a flood in each river reach within a river network depend on the intensity and the track of the triggering precipitation events, i.e., the spatiotemporal pattern of precipitation (Adams et al., 2012; Emmanuel et al., 2015; Paschalis et al., 2014). In addition to the storm track dynamics, the peak flow depends on the watershed characteristics (Singh, 1997). In mountainous catchments with high topographical complexity, the storm track and the precipitation pattern are influenced by the mountain ranges. Furthermore, the river network is influenced by geological and tectonic structures and is thus more complex in mountainous terrain than in low-lying areas. Thus, in upland areas, high variability in the spatiotemporal patterns of precipitation events and the resulting river flows has to be assumed.

In a complex river basin, the precipitation pattern, i.e., the spatiotemporal footprint of rainfall intensity, leads to a specific temporal pattern of the outflows from the subcatchments. Depending on the geometry and topology of the main river network, this timing of the outflows from the subcatchments influences peak discharge in the downstream river reaches. Hence, the relative timing of peak discharge arrivals in river confluences as a consequence of the spatiotemporal distribution of the rainfall pattern has to be addressed (Emmanuel et al., 2016; Nicótina et al., 2008; Nikolopoulos et al., 2014; Pattison et al., 2014; Zoccatelli et al., 2011). Thus, a sound analysis of extreme floods in a geographically complex river basin requires an assessment of the variability of chronological superimpositions of flood waves in tributaries and the effect of the resulting peak synchronization on the probability of inundation. Neal et al. (2013) highlight the importance of spatial dependence between tributaries in terms of inundation probability and magnitude. After Ochoa-Rodriguez et al. (2015), the temporal variation of rainfall inputs markedly influences the hydrodynamic modeling results. Emmanuel et al. (2015) showed that the spatiotemporal organization of rainfall plays an important role in the discharge at the outlet of the catchment and stated that a simulation approach is needed to study the effects of rainfall variability in river basins with a complex topology of the river network. Pattison et al. (2014) showed that tributary relative timing and synchronization are important in the determination of flood peak downstream. Depending on peak synchronization, the spatial distribution of flood damage can vary remarkably. Hence, some rainfall patterns can lead to high damage to infrastructure and houses while others do not. Zischg, Felder, Weingartner, et al. (2018) showed that the same sum of rainfall over a certain time period and over a river basin can lead to flood damage that varies by a factor of 3–6 depending on the rainfall pattern.

In conclusion, if the flood impacts in a river basin are highly dependent on the spatiotemporal rainfall pattern, then we have to assume that this effect is overlaid by the effect of the river training measures on the flood impacts. If the effects of the river training measures or all human interventions on flood impacts in a river basin should be studied, the confounding effect of rainfall variability on flood impacts must be excluded and the effect of river training measures must be isolated or disentangled. A data-driven approach is not possible because, first, the data series of river gauging are not long enough to go back in time of a natural state of the river system. Second, extreme flood events that are overflowing the riverbanks are, per definition, rare events. Thus, a timeseries of a 50-year period usually includes only a very few extreme flood events of a similar magnitude that occurred before and after the implementation of flood protection measures. Third, according to the high dependence of the spatial footprint of the flood events within the river basin on the rainfall pattern, only the comparison of two flood simulations that have the exact the same boundary conditions (i.e., the rainfall pattern) can lead to a robust assessment of the human dimension of changes in flood impacts. Thus, to systematically and quantitatively assess the effects of human interventions in the river system on flood impacts, we have to conduct a model experiment that keeps all boundary conditions as constant but varies the hydrodynamic model.

We based our model experiment on a set of generated extreme rainfall scenarios. Each of these rainfall scenarios is used as an input for both the hydrodynamic model that represents the historic natural state of the river network and the hydrodynamic model for the current state. The outputs of both hydrodynamic models are than overlaid with the same impact model. To generate the rainfall scenarios for the model experiment, we used a Monte Carlo simulation procedure. This procedure stochastically generates spatio-temporal patterns of a certain amount of rainfall sum over the river basin within a certain period. The rainfall is distributed in space and time with physical constraints, i.e., the procedure selects only physically plausible rainfall scenarios as for example scenarios where the rainfall falls over neighboring subcatchments and in temporal adjacency. In a first processing step, a random temporal distribution of the total precipitation for the predefined duration was generated. The variation of rainfall between one (hourly) time step and the following time step was limited to 20% at maximum. This avoids implausible temporal distributions and simulates a plausible increase and decrease of rainfall intensities. In a second step, the temporal pattern of the rainfall was distributed spatially in three meteorological regions and in the subcatchments within each meteorological region. The subcatchments and the meteorological regions were defined to consider the relatively independent behavior of specific parts of the catchment, e.g., lowlands and mountainous regions, in terms of precipitation amount and intensity (Felder & Weingartner, 2016). The randomly created precipitation pattern was checked against the spatial dependencies to fulfill a spatial consistency within neighboring catchments. Intensive precipitation must be concentrated in adjacent meteorological regions and adjacent subcatchments. The concentration of intense rainfall in meteorological regions and thus in affiliated subcatchments simulates the storm movement and the effects of the mountain crests in dividing rainfall patterns. If a rainfall pattern passes the plausibility check, it is added to the set of rainfall scenarios for flood simulation. The rainfall scenarios were assumed to occur during the summer season with a height of the freezing level above the maximal altitude. This means that snowfall is not considered. The procedure was developed by Felder and Weingartner (2016, 2017) and applied by Zischg, Felder, Weingartner, et al. (2018) and Zischg, Felder, Mosimann, et al. (2018) for assessing the effects of rainfall variability on flood damage during probable maximum precipitation events. In a first analysis, the cumulative effects of river training measures on hydrological change and on flood impacts will be shown by using this set of rainfall scenarios for the analysis.

The cumulative effects of human interventions to the river system on the hydrology of probable maximum floods

Before analyzing the effects of the river training measures and of the Kander river deviation on flood impacts, I want to explore how extreme precipitation events affect the hydrology, e.g., flood peaks, at the catchment outlet. The previous section discussed the effect of human interventions in the river system on hydrology by using a continuous rainfall-runoff simulation in the period 2005–11. Here, I analyzed the effects of the changed river system configuration on flood peaks in Bern by means of simulating a large set of very extreme rainfall scenarios as input in the hydrological model. The procedure is described in detail in Felder et al. (2017) for the current setting of the river network. To this analysis, here I added the comparison of the simulated flood hydrographs in Bern with the actual river morphology and simulated flood hydrographs in Bern with historical river morphology (counterfactual simulation without human interventions). Fig. 3.26 shows the setup of this model experiment.

The hydrological model used for this model experiment is the same as for the previous experiments, the PREVAH model. The outflows from the subcatchments were used as an input for the hydrodynamic model. We used the BASEMENT 1D model in two versions, one the model that represents the current situation and the other that represents the historic state of the river system prior the Kander river deviation and prior to all human interventions to the river system. As input for the coupled hydrologic-hydrodynamic model, we used a set of extreme rainfall scenarios with different spatiotemporal patterns but with the same amount of rainfall over the river basin within the same period. The rainfall sum over the river basin is the probable maximum precipitation. The value was set according to the guidelines of the World Meteorological Organization (WMO, 2009). The probable maximum precipitation is an estimation of the physically possible maximum of precipitation within a river basin of a specific size and within a certain period. The value for the probable maximum precipitation in the Aare river basin is 300 mm within 3 days. We generated 150 rainfall scenarios with this value that show different spatiotemporal patterns. The runoff from the subcatchments was then routed through the main river valleys and the lakes with the two different

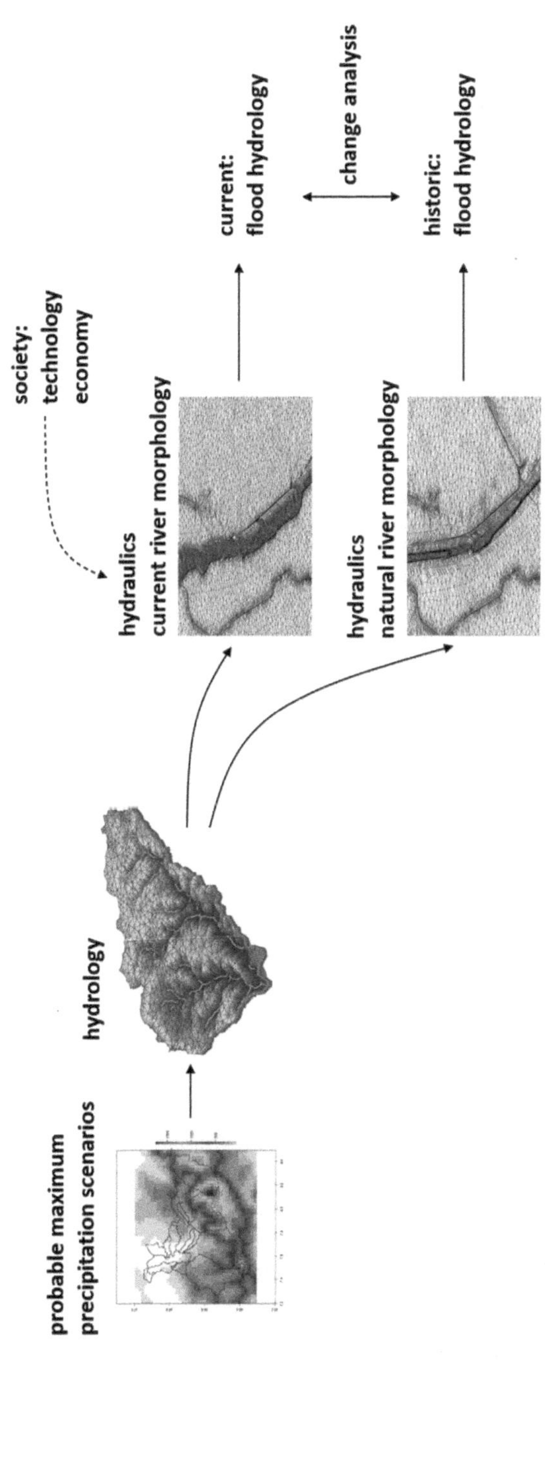

FIGURE 3.26 **Model experiment.** Model experiment setup for analyzing the cumulative effects of the human interventions to the river system on the hydrology of probable maximum flood events.

hydrodynamic models. Fig. 3.27 shows the differences between the flood hydrographs in Bern.

The simulations show that the Kander river has a remarkable influence on the peak discharge. While nowadays the probable maximum flood, i.e., the flood event with the highest peak discharge of the scenario set, has a peak discharge of around 1500 m^3/s. In contrast, a natural situation has peak discharges of slightly more than 2000 m^3/s. The comparison of the set of flood scenarios shows the dampening effect of Lake Thun. The flood peaks in Bern are more or less synchronized in the current situation. This is because Lake Thun has first to be filled, and once a certain lake level is reached, the outflow from the lake suddenly increases nonlinearly. The highest peak discharges in Bern are thus arising when high outflows from the lake are combined with the flood waves of the tributaries of the Aare river between Thun and Bern, especially the Zulg, Chiese, Rotache, and Gürber rivers. The counterfactual hydrographs that represent the natural situation show that the peak discharge of the Kander river, together with the peak discharge of the Simme river, leads to high flood peaks at Bern that are not dampened by the lake retention effect. The half area of the Aare river basin thus contributes directly to the discharge at Bern. Today, the ratio is only one-sixth of the total watershed area that is not influenced by lake retention effects. The time to peak at the river basin outlet is highly variable and direct. Thus, the dampening effect of Lake Thun and Lake Brienz is in the historical situation not as relevant as in the current situation. However, the highest peak discharges are also produced in the historical situation when a high lake outflow is superimposed with the peak discharge coming from the Kander and Simme catchments.

The effect of the Kander river deviation dominates over the other human interventions to the river system, namely river corrections. Fig. 3.28 shows the relationship between the peak discharges at Bern of the 150 extreme precipitation events in the current situation and in the counterfactual simulations. The figure on the left compares peak discharges at Bern of the river morphology in the Aare river basin before the Kander river deviation in 1714 AD. The generalization of the flood hydrographs in Fig. 3.28 to the peak discharges in Fig. 3.27 shows the remarkably higher peak discharges at Bern in a natural situation. All precipitation scenarios from the set of probable maximum precipitation scenarios have produced in the historical situation higher peak discharges than in the current situation. Thus, the city of Bern is benefitting from the cumulative effects of all human interventions to the upstream river system.

In contrast, the effects of all other human interventions together, namely the river corrections of the Hasliaare river, the river corrections and the weirs of the Aare river in Interlaken and in Thun, the river corrections of the Aare river between Thun and Bern, and the river correction of the Gürbe river, on the flood peaks at the river basin outlet are not as relevant as the Kander river deviation. In comparison to the modeled peak discharges of the current setting

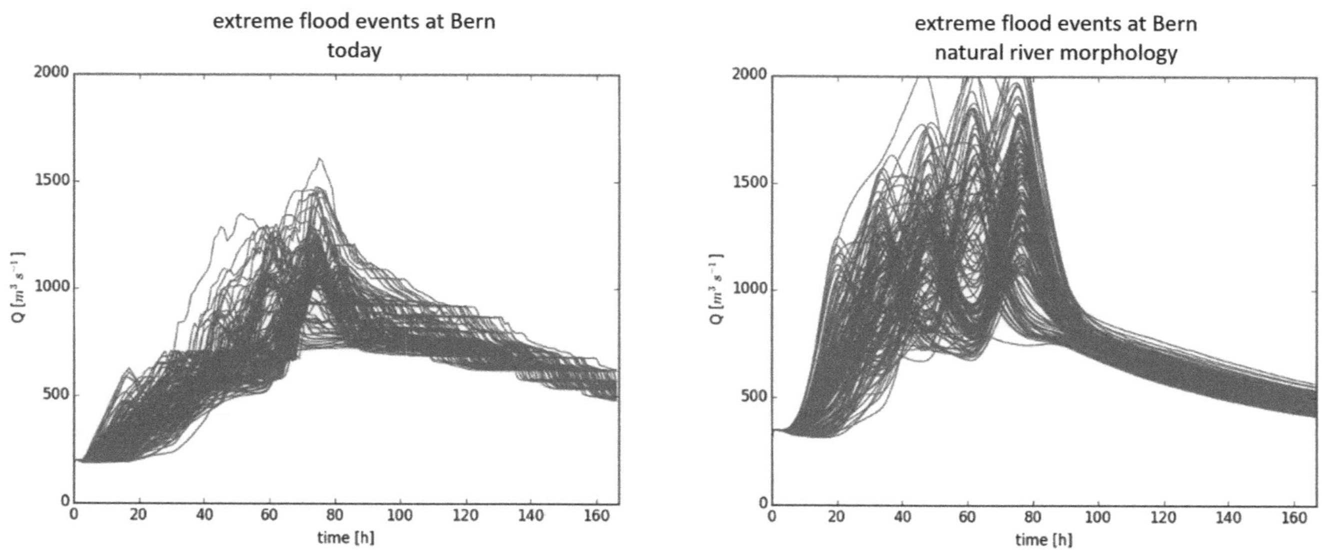

FIGURE 3.27 Flood hydrographs. Flood hydrographs at Bern of 150 probable maximum precipitation events with a hydrodynamic model of current river morphology (left) and historic (right) river morphology.

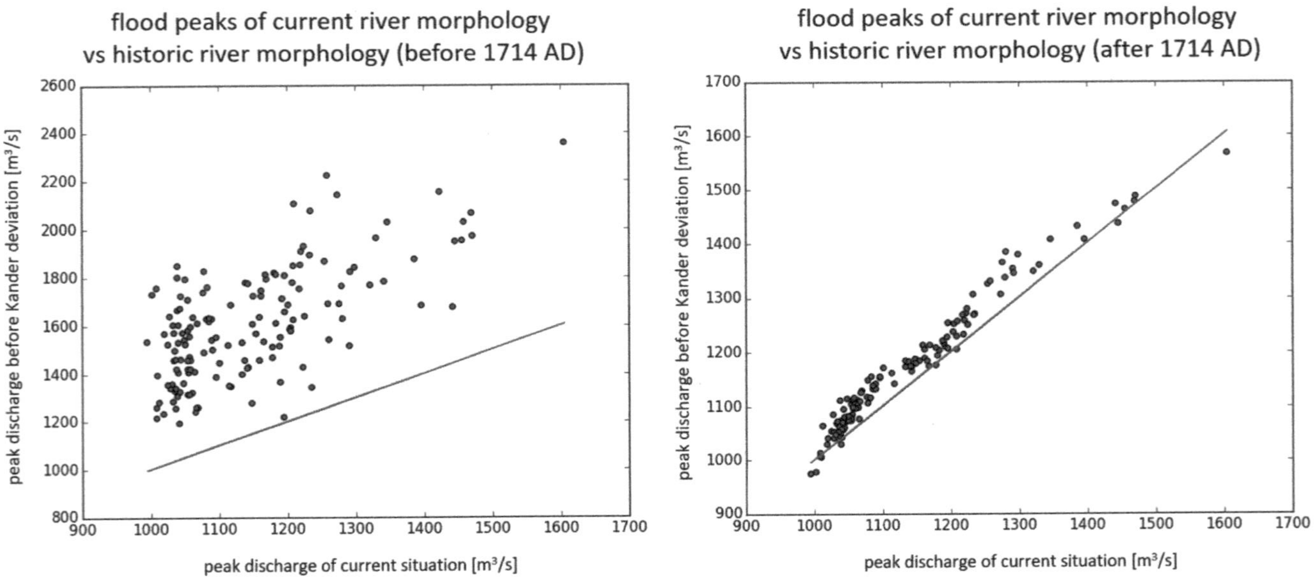

FIGURE 3.28 Flood peaks. Comparison of peak discharges at Bern of simulations of probable maximum precipitation scenarios in current and historic situation (left: before Kander river deviation in 1714 AD, right: after Kander river deviation in 1714 AD but before any other human interventions to the river morphology).

of the river network and river morphology, the peak discharges of the counterfactual simulation, the situation shortly after the Kander river deviation but before all other river corrections are only slightly higher than the peak discharges of the current situation. This is in contrast to the hypothesis that the construction of levees reduces the overflowing of water and thus reduces the retention effects of the floodplains. The construction of levees at the riverbanks reduces the likelihood of water flowing into the floodplain. The flow velocity in the floodplains is much lower than in the river channel and therefore the flood wave is expected to be dampened. River training and water embankments result in higher flood levels and thus increase flow velocity (Guerrin et al., 2014; Valentová et al., 2010). The flood wave is flowing through the river channel without attenuation or the flood peak downstream can be amplified by river engineering measures (Di Baldassarre et al., 2009; Munoz et al., 2018). The river system setting here is slightly different than previous studies in terms of floodplain retention. The natural floodplain of the Aare river between Thun and Bern was a large, braided river with only a few vegetated areas in the main riverbed. Thus, the lack of topographical sinks and storage areas in the floodplain limits the retention effect in the case of very extreme flood events. Instead, the flood wave in this river reach is amplified. In contrast to the natural river morphology, the current situation is influenced by other human modifications of the floodplain's topography. Examples of these human landscape structures are the construction of road networks. The highway through the main valley is mainly located in parallel to the river course. This means that the bridges that cross the highway and the bridges that cross the Aare river must be elevated. This in turn requires the construction of access roads to surpass the height difference, and these road segments must be elevated too. These access roads to the bridges across the highway and the Aare river cross the floodplain rectangularly. Thus, these road dams are hydraulically relevant as they function like retention dams. The unintended consequence of these anthropogenic landscape structures is the retention of flood water in the floodplain. This can lead to higher flow depths and backwater effects upstream and thus to the retention of flood water in the floodplain. Another retention effect of today's floodplains in contrast to natural floodplains is the effect of the lateral levees. In the case of the overflowing of water over the levees from the river channel into the floodplain, the water of the floodplain cannot flow back to the river channel further downstream unless the water level in the floodplain reaches the height of the levees. These effects lead to the attenuation of the flood peaks in the floodplain of the Aare river between Thun and Bern in the current situation in contrast to the historic situation. Thus, the interplay of processes in the floodplain that determine the retention capacity of the floodplains is complex.

The cumulative effects of human interventions to the river system on flood impacts of probable maximum floods

In contrast to analyzing the effects of river training measures on peak flow characteristics at the outlet of the river basin, the quantitative assessment of the impacts of a probable maximum flood event requires a different approach. The quantification of the impacts requires a 2D flood model. Therefore, we used the 2D hydrodynamic model BASEMENT as described above. The setting of the model experiment now is that a set of 150 probable maximum precipitation scenarios with different spatiotemporal patterns is used to model the discharges at the outlets of the subcatchments. In this model experiment setup, a new problem arises. The 2D flood models are computationally very expensive. Therefore, a 2D flood model covers each of the different floodplains and all flood models are then coupled together. This is important because running the flood model downstream first needs the flood model upstream still be finished with the simulation. A single model covering all floodplains limits the use of the model due to the high resolution needed for the impact analysis and the related high computational costs. Thus, the outflow of the upstream model(s) must be used as input for the downstream model. Only in this way, the consideration of the peak synchronization or the retention effects of the upstream floodplains on downstream flooding is enabled. However, the high number of flood scenarios to be simulated requires a different approach for tackling the computational burden. Therefore, we developed a nested approach. We first used the 1D hydrodynamic model to simulate the water fluxes through the floodplains. This allows quantifying the hydrographs at each location and time within the river network. Then, hydrographs were extracted from the 1D simulation at locations that define the upper boundary conditions of the 2D flood models. These hydrographs can also be shortened, i.e., extracting only those part of the hydrograph where the peak discharge exceeds the conveyance capacity of the river channel and thus only the parts where flooding will occur. This cutting of the hydrograph timeseries to the most relevant period saves computation time in addition to the nested approach. Finally, the extracted time series were fed into the 2D simulation models. This two-step procedure of the 1D model for the overall water fluxes and the 2D model to simulate the inundated areas is defined here as the nesting approach. The approach was exemplarily demonstrated by Bermúdez et al. (2017) to nest local-scale flood models into a regional-scale flood model. Thus, the model nesting approach allows us to analyze floods across scales. Felder et al. (2018) used this approach for downscaling extreme precipitation events from global climate model simulations to the local scale, and Zischg, Felder, Weingartner, et al. (2018) used the combined 1D and 2D models for analyzing the flood impacts of probable maximum precipitation events of the Aare river basin. In contrast to the present approach, these two studies used the flood model LISFLOOD-FP (Bates & De Roo, 2000).

The nesting approach allows us to save computational time. However, the number of scenarios and the number of floodplains together with the high spatial resolution of the flood simulations with BASEMENT were still not feasible to run within a reasonable time. Therefore, we created so-called surrogate models for each floodplain that describe the relationship between peak discharge and flood impacts. Synonyms for surrogate models are meta-models, model emulators, or response surface modeling (Ratto et al., 2012; Razavi et al., 2012). These computational approaches aim at reducing the complexity of models or model chains while preserving the dynamic nature of the original process-based model. The principal idea behind surrogate models is to emulate and to replace the complex simulation model requiring high computational resources with a simplified and fast-to-run model (Castelletti, Galelli, Ratto, et al., 2012). A surrogate model can be derived by simplifying the process-based model structure or by generalizing the studied system's behavior with a low-order approximation of a set of outcomes of a model experiment with the complex model (Castelletti, Galelli, Restelli, & Soncini-Sessa, 2012). Surrogate models are often used in applications that require a large number of model runs, e.g., in sensitivity analysis, in scenario analysis, and in optimization. In flood management applications, therefore Zischg, Felder, Mosimann, et al. (2018) developed surrogate models for each flood-plain in the Aare river basin. In contrast to a cascade of a hydrological model, a 1D hydrodynamic model, a 2D hydrodynamic model, and a flood impact model, the surrogate modeling approach eliminates the last two models and substitutes them with a simplified model. The 2D hydrodynamic model and the flood loss model are replaced by a 1D hydrodynamic model with a surrogate model for flood loss computation nested into the 1D hydrodynamic model (Fig. 3.29).

The surrogate model was built in three steps. First, for each floodplain, the range of system loads at the upper boundary condition was analyzed. For this, we used the simulations with the 1D hydrodynamic model and extracted the maximum peak discharge for each floodplain. In the second step, we elaborated typical flood hydrographs, so-called synthetic design hydrographs based on observed discharge timeseries. Synthetic hydrographs show the typical shape of a flood hydrograph. They are dependent on the river reach and its catchment characteristics. The theoretical background and the method for elaborating synthetic hydrographs are described in Serinaldi and Grimaldi (2011). For each river gauging station in the study area, observed hydrographs were normalized in terms of event duration and peak discharge. The resulting dimensionless normalized event hydrographs were superimposed and centered on the peak position. A two-parametric gamma distribution function was fitted to represent the typical shape of the event hydrographs, as described by Felder et al. (2017). This resulted in a synthetic unit hydrograph that represents a typical hydrograph shape of flood events in the corresponding catchment. The synthetic unit hydrographs were scaled to various peak discharges, whereas an

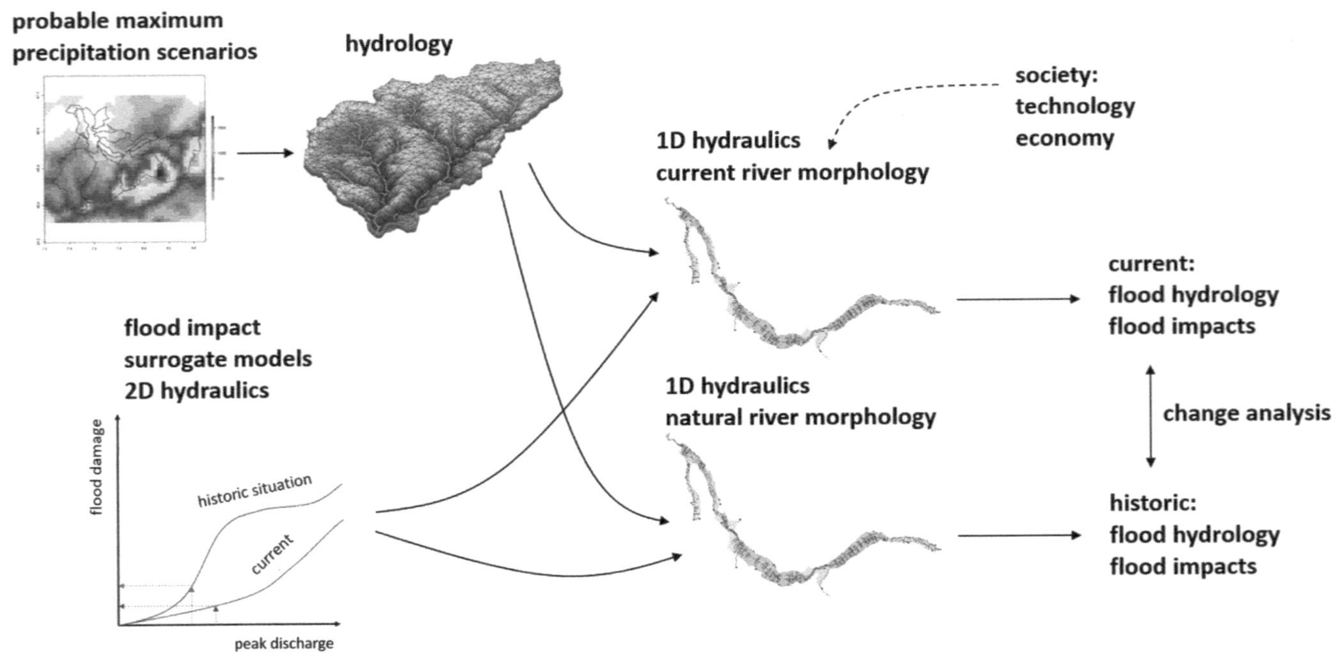

FIGURE 3.29 Model experiment. Model experiment setup for analyzing the cumulative effects of river training measures on flood impacts of probable maximum precipitation events with a surrogate modeling approach.

empirical peak–volume ratio was used to determine the corresponding event duration. The procedure was applied to generate synthetic design hydrographs for a continuous series of peak discharges for each of the floodplains affected by riverine flooding (Hasliaare river, Lütschine river, Aare river between Thun and Bern, Gürbe river). The synthetic design hydrographs were used as upper boundary conditions for the 2D inundation model of each floodplain.

In a third step, we developed the surrogate model of the flood impact module. The 2D hydrodynamic model BASEMENT provided the basis for the flood-loss model. Here, we considered structural damage to buildings (i.e., residential, public, and industrial buildings) as flood impacts. Damages to mobile assets, building content, movables, and infrastructure were not considered. The damage model consists of a dataset of buildings with attributes and a set of vulnerability functions. Each building is represented by a polygon and classified by type, functionality, construction period, volume, reconstruction costs, altitude level of ground floor, and number of residents. The dataset of the values at risk was elaborated on the basis of the Swiss-Buildings3D dataset of the Federal Office for Topography SWISSTOPO (Swisstopo, 2017). The reconstruction values of the buildings were calculated on the basis of the volume (derived by the Lidar surface and terrain models) and the mean prices per cubic meter and building function (SVKG-SEK/SVIT, 2012) accordingly to the practice in Switzerland and the method described in Röthlisberger et al. (2018). For each synthetic hydrograph, we simulated the flood depths with the 2D hydrodynamic model. The resulting flood intensity maps (flood depths) led to the calculation of the exposed buildings (objects at risk) and to the object-specific vulnerability of each building and therefore to the estimation of object-specific losses (Fuchs et al., 2012; Zischg et al., 2013). To assess the monetary value of the damage to the buildings, we used vulnerability functions that provide a degree of loss on the basis of the flood depth at the location of the building. The value ranges from 0 (no damage) to 1 (total loss). This degree of damage is subsequently multiplied by the specific reconstruction value of the building. Several vulnerability functions are considered in the damage calculation procedure to consider the uncertainty of this module. To compare the changes in flood impacts in terms of monetary damage, we used the vulnerability function of Jonkman et al. (2008). The multiplication of the reconstruction value of the house with the degree of damage leads to the flood damage of a specific exposed object (e.g., a single building). The sum of all damages in the floodplain enters into the "flood peak – flood loss" relationship of the floodplain. For each flood scenario and synthetic hydrograph, the damages were summarized. Furthermore, the number of exposed buildings and the number of exposed residents were summarized. With peak discharge describing the flood magnitude, the resulting flood damages and exposed buildings/residents increase. For each floodplain, the shape of this relationship between flood magnitude and flood damage is calculated. These floodplain-specific curves are the basis of the meta-model or surrogate model, for further analyses. The surrogate model can then be used to extend coupled hydrological-hydrodynamic model chains by nesting it into the 1D hydrodynamic model. The peak discharge of the hydrograph of the 1D hydrodynamic model was used for extracting the related damage from the surrogate model (Fig. 3.30).

The surrogate models in Fig. 3.31 show how variable the shape of flood exposure and flood damage curve over the range of flood magnitudes can be. For the floodplains of the Aare river, the Hasliaare river, and the Gürbe river, the relationships between peak discharge and exposed buildings or flood damage are shown. The current situation was calculated by overlaying the flood simulation results of the current digital elevation model with the current dataset of buildings. The counterfactual situation was calculated by overlaying the flood simulation results of the historical digital elevation model with the current dataset of buildings. These simulations represent a set of flood scenarios without any flood protection measures implemented and thus a natural river morphology. The floodplain of the Aare river between Thun and Bern shows a sharp increase in the number of exposed buildings around a peak discharge of about 800 m^3/s. In the counterfactual simulation, the gradient of exposed buildings in relation to the flood magnitude is less steep, but this shows that without the flood protection measures low magnitude flood events lead to a high exposure. However, the gradient of rising exposure with flood magnitude in the counterfactual simulation is relatively linear in contrast to the current situation. The damages of the current simulation are much less than the damages of the counterfactual simulation. The flood protection structures and the riverbed incision of the 150 years period since the river corrections reduce the damages remarkably. The damages show a nonlinear relationship between the flood magnitude and the flood damage in the current situation and a nearly linear relationship in the counterfactual simulation. This is merely the effect of the higher discharge capacity and the effect of a sharp increase of damages after the exceedance of the design event of the flood protection works. The Hasliaare river shows two different forms of the curves. In the current situation, the number of exposed buildings is increasing with a steep gradient between flood peaks of 400 and 500 m^3/s and with a flatter slope for the more extreme scenarios. In contrast, the counterfactual simulations have a higher slope right from the beginning of the first inundations. In terms of flood damage, the shape of the two damage curves of the Hasliaare river differs even more. For this floodplain, the effect of flood protection measures on flood risk reduction is evident. The exposure and damage curves of the Gürbe river are nearly parallel, but the counterfactual simulation shows a higher exposure and higher damage. These surrogate models provide the basis for estimating flood exposure and flood damages in the following model experiments.

The procedure for generating physically plausible rainfall scenarios was adapted from generating probable maximum precipitation events (300 mm within 3 days) to rainfall events of a certain magnitude. We analyzed the return

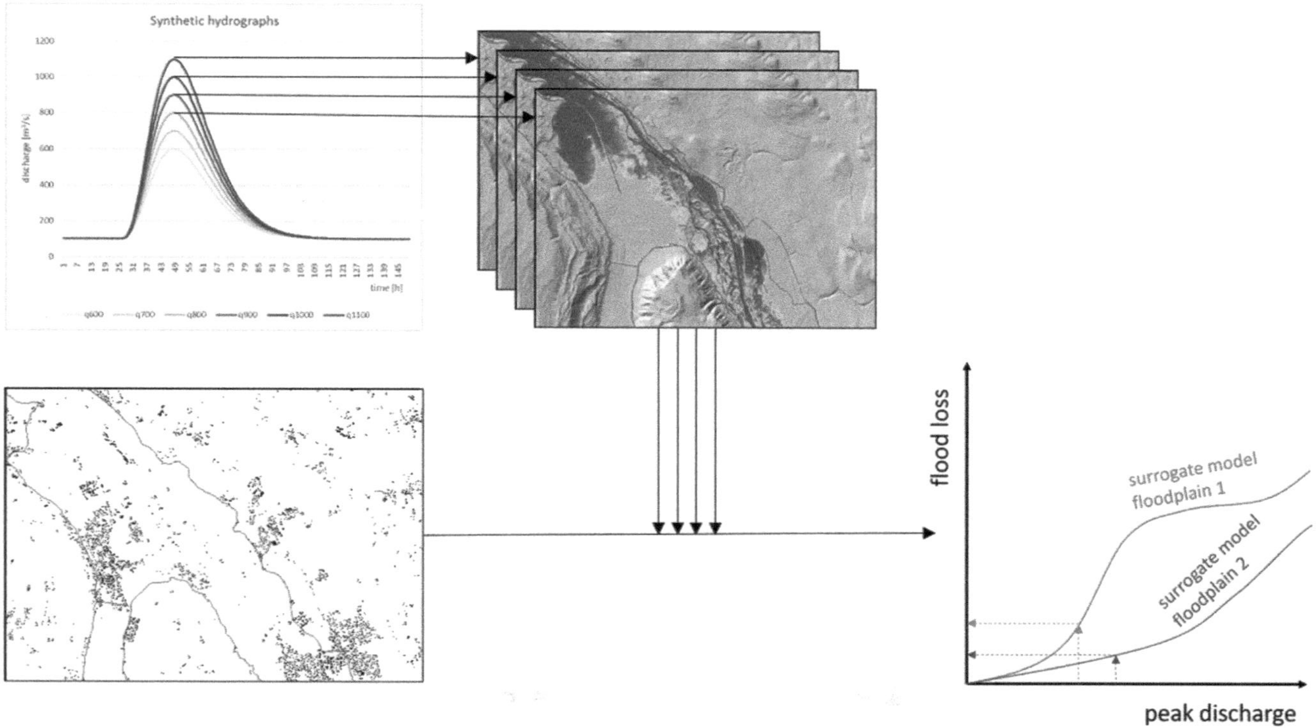

FIGURE 3.30 Surrogate modeling. Development of the surrogate models for the floodplains in the Aare river basin. *Modified from Zischg, A. P., Felder, G., Mosimann, M., Röthlisberger, V., & Weingartner, R. (2018). Extending coupled hydrological-hydraulic model chains with a surrogate model for the estimation of flood losses.* Environmental Modeling and Software, 108, *174–185. https://doi.org/10.1016/j.envsoft.2018.08.009.*

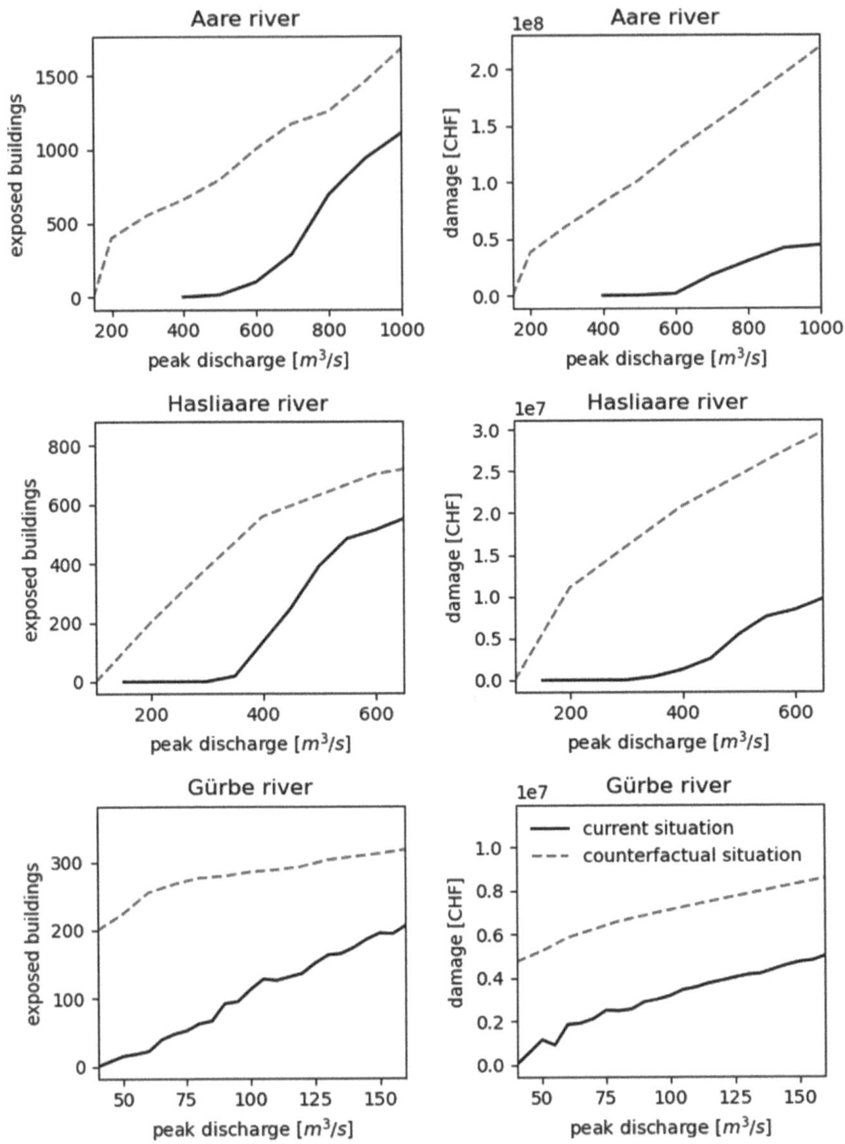

FIGURE 3.31 **Surrogate models.** Surrogate models for the floodplains in the Aare river basin under current conditions (black lines) and in the counterfactual simulations that represent a natural river morphology (gray dashed lines). The figures at the left show the number of exposed buildings and the figures at the right show the damage to buildings.

periods of precipitation sum over the Aare river basin upstream of Bern within 3 days. Then, we created 100 rainfall scenarios with different spatiotemporal patterns for the rainfall return levels of 10, 30, 50, 100, 150, 200, 250, and

300 years. In total, 800 rainfall scenarios of increasing magnitudes were created with the method of Felder and Weingartner (2016). This second analysis shows the cumulative effects of river training measures on flood impacts over a large range of flood magnitudes and not only for single flood events.

These rainfall scenarios were used as input for the hydrologic model. As for the previous model experiment, the hydrological model simulated the rainfall runoff for every subcatchment. In contrast to the previous simulation that was based on continuous rainfall and temperature data, here we used only 3-day rainfall data as input. Temperature and antecedent catchment conditions were set as constant as in the original study of Zischg, Felder, Mosimann, et al. (2018). The outflow from the subcatchments was used as input for the hydrodynamic model in two versions, the model that represents the current situation and the model that represents the historical situation.

The simulations show that human interventions into the river system had a remarkable effect on the hydrology of flood peaks. Fig. 3.32 shows that the flood peaks at Bern, i.e., the outlet of the Aare river basin upstream of Bern, would have been on average about 200–300 m^3/s higher in the counterfactual situation than in the current situation. As in the previous model experiments with the continuous simulation over the period 2005–11 or with the probable maximum precipitation scenarios, these simulations quantify the effect of the artificially created retention on the Kander river. The lake levels in Thun were remarkably higher over the whole range of flood magnitudes in the counterfactual simulation. The lake level would have been 20–40 cm higher if the Kander river had not been deviated into Lake Thun. The difference in the lake levels between the current and counterfactual simulation increases with the flood magnitude (Fig. 3.32).

Quantifying inner-basin flood risk transfer

This model setup of using precipitation events with a certain occurrence probability allows us to attribute probabilities to the flood impacts (Fig. 3.33). This in turn allows us to quantify the change in flood risk. Out of the 100 simulations for each of the eight return periods, a flood risk curve can be derived. For each return period, we calculated the average flood impact. The eight average values of flood impact then were used to calculate the area under the risk curve, i.e., the flood risk in terms of expected annual loss. We calculated the flood risk of Bern due to river flooding under the current and the counterfactual situation and the flood risk of Thun due to lake flooding under the current and the counterfactual situation. The comparison of the flood risks allows us to calculate flood risk change at river basin level. Flood risk due to lake flooding in Thun increased by 30% and the flood risk in Bern decreased by 78% when comparing the current and counterfactual simulations

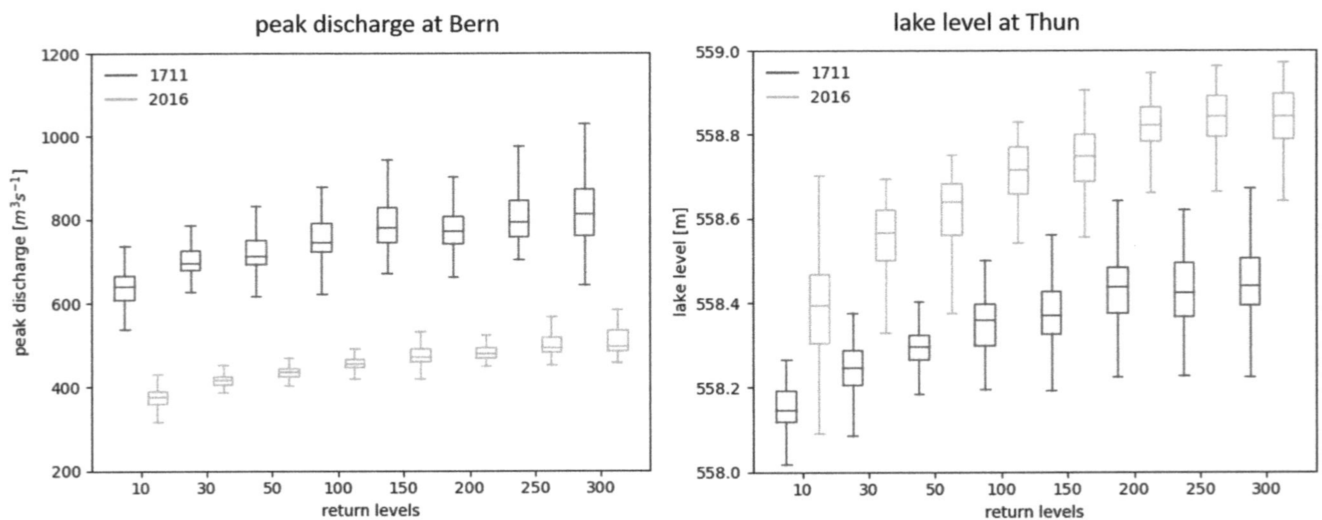

FIGURE 3.32 Cumulative effects. Cumulative effects of river training measures on hydrology for flood events with a range of magnitudes.

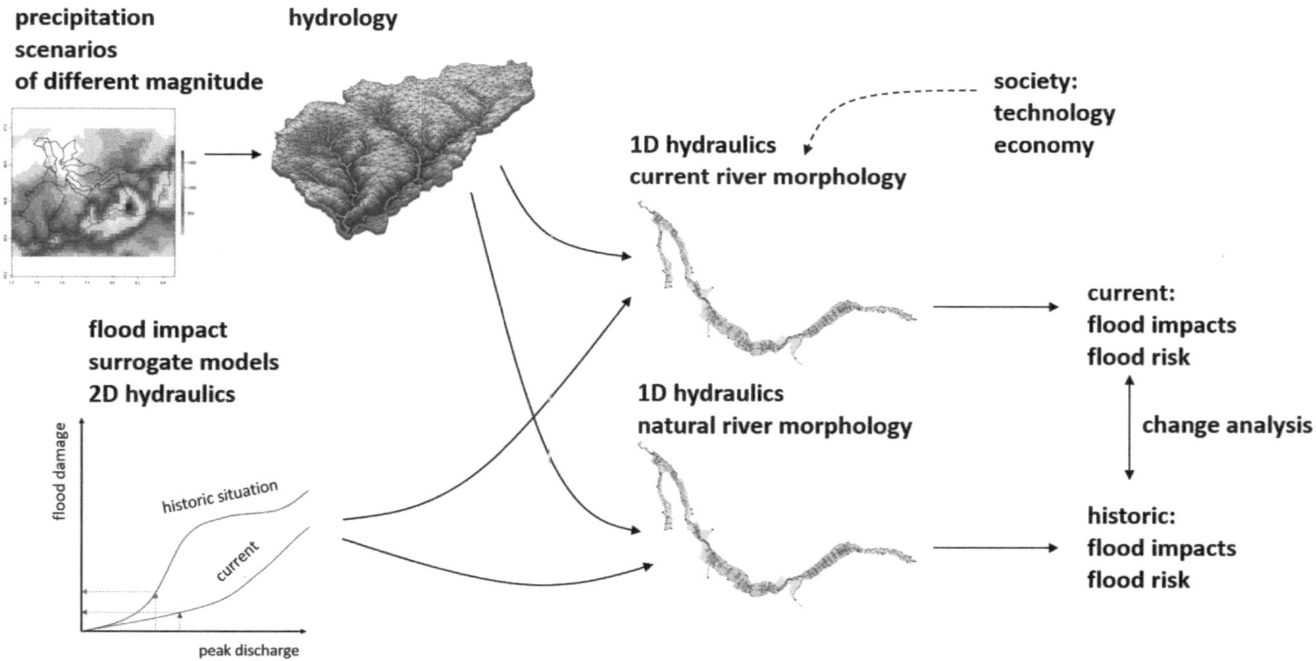

FIGURE 3.33 **Model experiment.** Model experiment setup for analyzing the cumulative effects of river training measures on flood impacts of events with a range of magnitudes.

(Fig. 3.34). In absolute values, the amount of flood risk reduced in Bern nearly equals the amount of flood risk increased in Thun.

This example clearly shows that flood risk can also be transferred from downstream to upstream. This complements the set of case studies that show how the flood risk was transferred from upstream to downstream, e.g., by heightening levees in upstream river reaches that reduce floodplain retention capacity and thus increase flood peaks in downstream river reaches (e.g. Di Baldassarre et al., 2009; Munoz et al., 2018).

Assessing changes to flood hazard maps due to climatic changes

Flood hazard maps classify the inundation processes in terms of process magnitude, i.e., the occurrence probability and flood parameters, as for example, flow depth and flow velocity. These maps are used in land use zoning and land use planning with the aim of preventing the construction of settlements in hazardous areas. Hazard maps are elaborated on the basis of current environmental conditions. They are periodically updated, especially when the environmental conditions change. The most relevant environmental condition is the frequency of flood events. This parameter is derived from discharge observation time series. From the extreme value statistics, a set of flood scenarios (hydrographs) is created and used as input for the hydrodynamic modeling. The resulting flood maps describe the flow conditions, and the overlay of all flood scenario maps leads to a combined hazard map.

In Switzerland, hazard maps are created according to national guidelines (BWW & BUWAL, 1997). Flood hazard is defined by a combination of intensity and probability of events (see Fig. 3.35) with thresholds in terms of

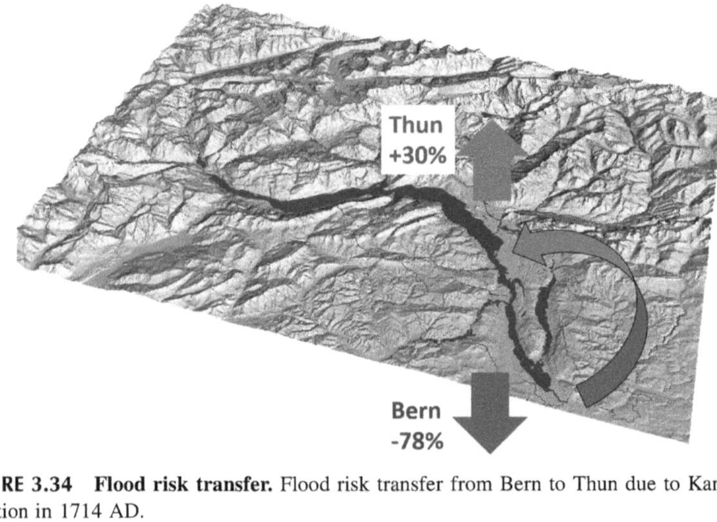

FIGURE 3.34 Flood risk transfer. Flood risk transfer from Bern to Thun due to Kander river deviation in 1714 AD.

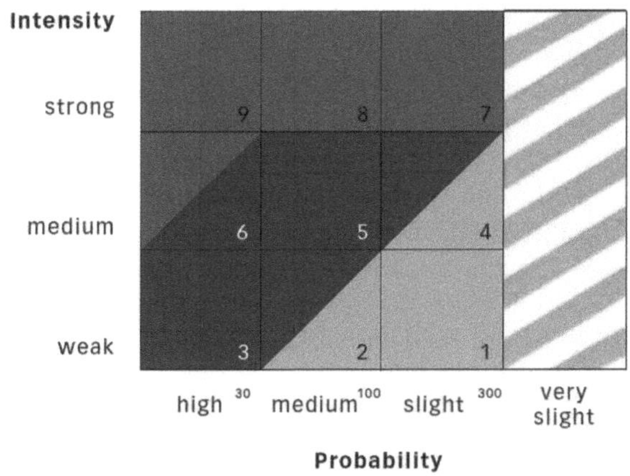

FIGURE 3.35 **Hazard classes.** Assessment matrix for the identification of hazard levels in Switzerland (black = high hazard; white = medium hazard; light gray = low hazard; dark gray stripped = residual hazard; black dash = no or negligible hazard). *Modified from BWW, B., & BUWAL. (1997). Berücksichtigung der Hochwassergefahren bei raumwirksamen Tätigkeiten.*

intensity at 0.5 m (water depth) or 0.5 m/s^2 (water depth × velocity), for the differentiation between "low" and "medium" intensity, and at 2 m or 2 m/s^2, respectively, to differentiate between "medium" and "strong" intensity. The probabilities of occurrence investigated correspond to return periods of 30 (high probability of occurrence), 100 (medium probability of occurrence), 300 (low probability of occurrence), and above 300 years (very low probability of occurrence), respectively. The spatial representation and combination of events of different return periods lead to the mapping of five hazard classes: "high hazard" (red colored areas), "medium hazard" (blue), "low hazard" (yellow), "residual hazard" (yellow-white striped), and "no or negligible hazard" (white) threat. In this paper, we focus on three classes—"high," "medium," and "low"—i.e., on areas endangered by flood events up to a return period of 300 years.

The combined hazard map is introduced in the communal land use regulation plans and is legally binding for the authorities and for land owners. After the implementation of the hazard map in the communal land use regulation plan, the map is named as the hazard zone map. In red hazard zones, the construction of new buildings is forbidden, and in blue hazard zones, the construction of new buildings is regulated. Buildings have to be constructed with object protection measures that provide local protection of the building against floods. Yellow hazard zones are notice areas that should raise awareness of home owners.

The main goal of the hazard zone maps is to avoid construction of buildings in red hazard zones. This is a long-term goal, and the principle works only if consistently followed for a long period. However, if the hazard zones change over time, the principle of long-term planning is hampered. If rainfall is expected to increase because of climate changes, then also the peak discharges must be expected to increase. As a consequence, the extreme value statistics and the set of flood scenarios for flood modeling will change. This in turn will affect the effectivity of the land use maps and will influence flood risk change. The problem could be if today's hazard map is valid for, let's say, the next 20 years. The map prohibits the construction of houses in red zones. Thus, the land available for construction tends to be assigned outside of the red hazard zones. However, since floodplains provide a feasible ground for construction due to the flatness and the hills surrounding the floodplains limit the constructible areas, buildings will still be constructed in floodplains. Instead of building in red zones, the blue or yellow zones will be preferred. So, if, for 20 years, the hazard zone maps shift potential construction from the hazard zones toward their outer borders, the hazard zone maps might create lock-in situations if the hazard zones will once be enlarged due to increasing flood frequency or magnitude in a changing climate. Therefore, the sensitivity of these hazard maps must be analyzed.

Thus, here we set up a model experiment to analyze the sensitivity of the method for elaborating flood hazard zone maps for land use regulation purposes. We set up a coupled hydrologic and hydrodynamic model that was fed with a set of rainfall scenarios that were derived from extreme value statistics. The resulting inundation maps were combined into a hazard map. In the model experiment, we tested if a potential increase in rainfall intensity affects the extent of the hazard maps. Finally, we overlaid both resulting hazard maps with the buildings dataset. The latter, as well as the flood protection standard (i.e., river morphology), was kept constant.

The focus of this case study lies not on the exact representation of the environmental systems by means of detailed process models and climate models but on testing the robustness of the methods and procedures for hazard mapping to changes in the needed input parameters. On the basis of a literature review, a possible increase of 20% of the precipitation intensity for each design event (reoccurrence interval 30, 100, 200 years) was assumed for this sensitivity analysis. The case study has been described by Staffler et al. (2008) and was conducted in the Rio Ridanna/Mareiter Bach river basin in South Tyrol, Italy. The river endangers parts of the city Vipiteno/Sterzing. The catchment area is 210 km^2. This study area is a representative example of an alpine river with hazard potential for settlements. Fig. 3.36 shows the model experiment setup for this analysis.

The results of the procedure described above were the hazard maps that describe the hazard situation on the basis of simplified assumptions representing present (named as "scenario 2000") and future (2050–2100) climate conditions (named as "scenario +20%"). For this scenario, we added 20% to the rainfall intensities derived from the extreme value statistics of a daily rainfall event. As a hydrological model, we used the rainfall-runoff model

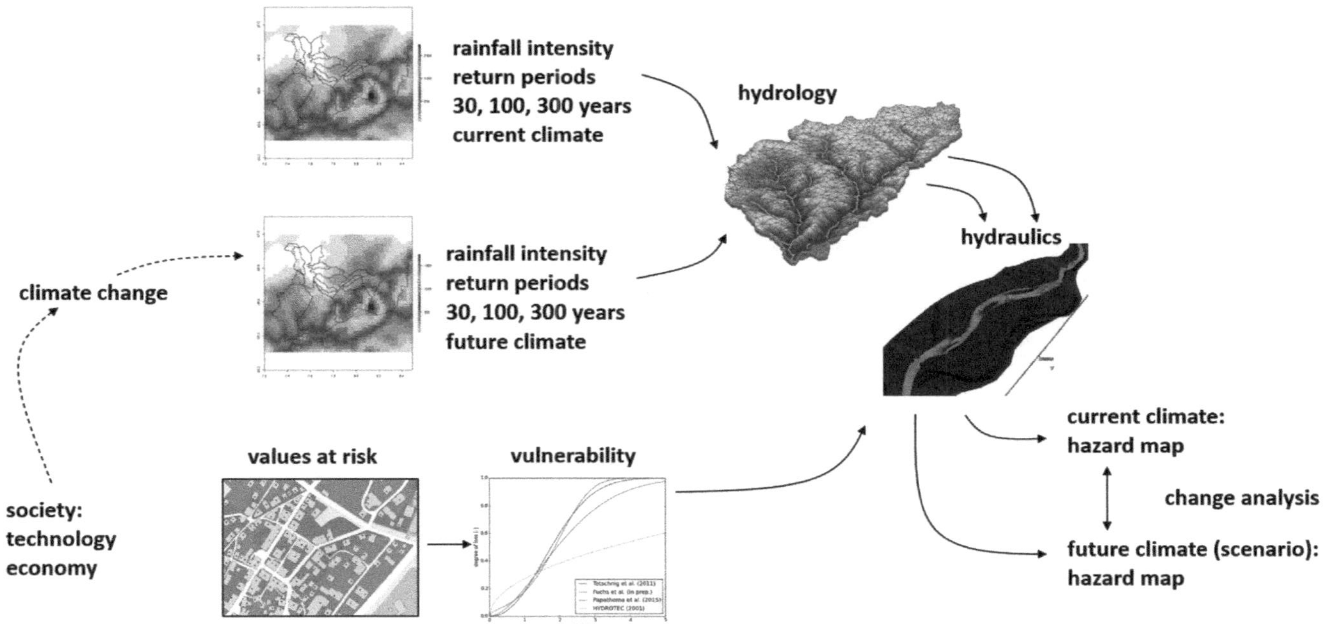

FIGURE 3.36 **Model experiment setup.** Model experiment setup for analyzing the sensitivity of flood hazard mapping procedure to precipitation increase.

Hec-HMS and the SCS approach (US Army Corps of Engineers, Hydrologic Engineering Center, 2012). For the simulation of the inundation process, we used the simulation model SOBEK of WL Delft Hydraulics (Deltares, 2012). The resulting hazard maps were overlaid with the buildings in the settlement areas. Thus, the changes in the hazard situation were demonstrated by the changes in the extent of the hazard zones and by the changes in the number of endangered buildings. Fig. 3.37 shows the changed extent of the flood hazard zones.

The increase of 20% of the precipitation scenarios resulted in an increase of peak discharge by 40% for the 30-year and 100-year scenarios, and 30% for the 200-year scenario. With the increased peak discharges, the extent of the hazard maps increased too. The flood hazard zones with a low intensity (yellow hazard zones) in future may be classified as blue hazard zones. This is especially crucial for the industrial buildings constructed during the last decade in the free spaces in the neighborhood of the rivers, avoided for settlement in the decades before. The potential shifts from blue hazard zones to red hazard zones do not show significant consequences for buildings. The expected damage of a flood event with a return period of 30 years ("scenario +20%") increased up to 1700% (in comparison to the "scenario 2000"). The expected damage of a flood event with a return period of 100 years increased up to 207% and up to 117% for an event with a return period of 200 years.

The model experiment showed possible consequences of climate changes to the hazardous situation of an alpine river. It showed that the assumed increase in the precipitation intensities potentially might have remarkable impacts on the natural hazard situation (process intensities). The calculated increase of discharge due to the assumed increase in rainfall intensity showed a significant accentuation of the already existing weak points in the flood protection structures and the resulting hazard situation. The modeling results show a remarkable increase in flooded areas and an increase in flow depths. Generally, the "yellow" hazard zones delineated on the basis of the current climate conditions tend to become "blue" hazard zones. This could lead to significant restrictions on land use. The study demonstrated that already known weak points in risk reduction systems as protection structures in future will become more important for risk management activities. This means that the stress–strain behavior of these weak points in cases of discharges exceeding the channel capacity must be studied. The knowledge of the behavior of protection structures in the overload case could provide a decision base for the elaboration and training of crisis management plans.

Changes in sediment supply for torrential floods

In mountainous areas, climate change will affect flood processes not only by increasing precipitation intensity but also by slowly changing the

FIGURE 3.37 Sensitivity of hazard maps to increasing rainfall intensity. Results of inundation modeling reclassified following the guidelines for hazard zone mapping. (A) Flood hazard map representing «scenario 2000». (B) Flood hazard map representing «scenario +20%». *Modified from Staffler, H., Pollinger, R., Zischg, A., & Mani, P. (2008). Spatial variability and potential impacts of climate change on flood and debris flow hazard zone mapping and implications for risk management.* Natural Hazards and Earth System Science, *8(3), 539–558. https://doi.org/10.5194/nhess-8-539-2008.*

environmental conditions in the catchments. Besides increasing rainfall intensities, the altitude level of the rainfall/snowfall border will increase with temperature rise. Therefore, larger amounts of rain than snow will fall on mountainous catchments in a warmer climate. As snow has a retarding effect on flood peak generation, the decrease in snowfall during extreme precipitation events and the related increase of liquid precipitation will lead to an increase of peak discharges at the catchment outflows.

In areas of the Northern Alps below 1500 m a.s.l., an increase in flood peaks is expected in winter due to higher soil water content, the rising of the rainfall/snowfall limit level, and increased liquid precipitation (KOHS, 2007, pp. 55–60). In the pre-Alpine regions, the increase of precipitation in winter and the rising of the snowfall limit will have consequences for the activities of landslides in winter and spring. The increase in saturation leads to an increase in landslide activity. The expected increase in landslide activities in turn will contribute to higher bedload transport during extreme rainfall events in the following summer months.

Another indirect effect of the shifts in altitude levels due to rising temperatures is the rising of the lower boundary of permafrost zones. Due to the rising altitude level of glacier retreat and permafrost degradation, the sediment transport in the areas between approximately 2300 and 2800 m a.s.l. and with relevant bed load source areas at this altitude level is expected to increase. Since in these areas more precipitation will fall in liquid form, this trend is expected to be relevant for flood risk change.

To analyze the potential effects of these environmental changes on flood hazards, Staffler et al. (2008) elaborated a method for identifying mountainous catchments that are sensitive to these indirect effects of climatic changes. The first step of the procedure was to match the environmental parameters relevant for hazard assessment on the regional scale with the existing spatial data. Three types of catchments were distinguished: mountain torrents, torrential rivers, and alpine rivers with a flash flood character. The susceptibility of flood and debris flow processes in alpine torrents is mainly driven by the discharge, the sediment budget, and the sediment transport capacity. The sediment transport capacity is influenced either by short precipitation events (thunderstorms) or by longer precipitation events. The sensitivity analysis was conducted in a pilot area of South Tyrol, Italy. The study was based on the following assumptions for potential future climate conditions: The daily mean temperatures in summer and winter are increasing; the mean sum of precipitation in summer is decreasing or remains constant; the mean sum of precipitation in winter is increasing; and the intensity and frequency of short extreme rainfall events in summer and autumn are increasing.

It was assumed that the following factors are varying spatially and are relevant to the sensitivity of the torrent catchments to climate change:

- Share of areas located between 1000 and 2000 m a.s.l. to overall catchment area: It is expected that the snow cover in these areas will be reduced and the frequency of combined snowmelt/rainfall events will increase.

- Characteristics of bed load source areas: Bed load source areas can be divided into recent and older deposits. Recent deposits are fed by ongoing weathering and denudation processes. The quantity of mobilizable sediment storage in torrents eroding recent deposits is depending on the intensity of the sediment delivery processes and the period between extreme discharge events transporting the weathered material downstream. A long period between two extreme rainfall events leads to higher sediment amount by weathering processes, whereas a high frequency of extreme rainfall events leads to decreasing sediment availability because the weather material is transported downstream. Older deposits are composed of relict geomorphologic deposition processes (e.g., glacial moraines, Holocene alluvial sediments, landslides, or rockfall processes). The quantity of mobilizable sediment storage in torrents eroding older deposits is mainly unlimited. If the percentage of areas with older deposits to the total bed load source area exceeds a certain share of the total catchment area, the torrent catchments were classified as torrents eroding older deposits, otherwise as torrents eroding recent deposits. Landslides also influence the quantity of mobilizable sediment. An increase in precipitation may lead to an increase in the activity of landslides, especially in winter and spring. If the percentage of landslide areas with respective to the total bed load source area exceeds a threshold, the torrent catchments were classified as torrents mainly influenced by landslide activity.
- Share of available bed load source areas to the overall catchment area: The bed load sediment budget of alpine torrents depends on the quantity of sediment available for recruitment and the sediment transport capacity. The sensitivity of torrent activity against climate changes increases with a higher proportion of bed load source areas respective to the total catchment area.
- Permafrost degradation and glacier retreat areas: Permafrost influences the hydrology and stability of steep scree slopes, since ice-rich permafrost acts as a barrier to groundwater percolation and can imply local saturation within nonfrozen debris. Permafrost thawing in nonconsolidated material leads to an increase of pore water pressure and a loss of cohesion (Harris et al., 2001). The disappearance of ground ice bodies on scree slopes leaves caverns and destabilizes parts of these disintegrated slope areas. With accelerated permafrost thawing, the susceptibility of these sloped areas to landslides and debris flows and the triggered volumes is expected to rise.
- Share of areas with elevated surface runoff to the overall catchment area: Areas with reduced water storage capacities increase the surface runoff. The sensitivity of torrential rivers and rivers to climate changes increases with a higher proportion of areas with reduced water storage capacities respective to the total catchment area.

The delimitated torrent, torrential river, and river catchment areas were classified by the combination of these factors influencing the sensitivity of mountain torrents and rivers to climate changes.

The main results of the catchment sensitivity classification procedure showed a high spatial variability in the sensitivity of the torrent catchments to specific impacts of climatic changes. The results showed that the runoff of nearly all torrent catchments is expected to increase in summer. The runoff in winter is expected to increase only in torrent catchments having a high percentage of their total surface area below 2000 m a.s.l. The bed load transport in summer is expected to increase in high mountain areas and is expected to decrease in catchments at submontane levels. In some catchments eroding younger deposits (weathered material), a decrease in extreme events is highlighted after the available sediment is eroded. The replenishment of sediment that can be eroded by debris flow is too slow in comparison to the expected increase in heavy precipitation. In contrast, for catchments eroding older deposits, an increase in extreme debris flow events is pointed out. The bed load transport in winter increases in a small share of the mountain torrent catchments and does not change in most catchments. The frequency of small-scale debris flow and sediment transport processes is expected to increase in most of the torrent catchments.

The results of the approach to assess and classify the sensitivity of mountain torrent and torrential river catchments against the assumed climate changes showed where the future scenarios of natural hazards are expected to occur more likely and where not. The analyses pointed out that the impacts of climate changes to the hazard situation of torrential and river systems have a very high spatial variability. Thus, the catchment characteristics have to be considered when climate change impacts on flood risks in mountainous areas are to be analyzed. The procedure for the identification and localization of alpine torrent and torrential river catchments that are sensitive to climate changes provides an information basis for the identification of those sites, where the risk potential tends to increase.

Zischg et al. (2012) used the data of changing permafrost areas to identify the catchments that will be influenced by permafrost thawing and where an increase of unconsolidated debris material in the debris flow triggering areas is to be expected. They have found that in the province of South Tyrol, Italy, 20% of all torrent catchments where debris flow processes are likely to occur, are affected by permafrost areas. In these catchments, the sediment supply for the torrential processes will change in a warming future. 1.4% of the settlement area is exposed to these processes, i.e., debris flow runout areas of torrential catchments with parts of their catchment area being located in permafrost areas and quaternary deposits. This means that floods and debris flows coming from these catchments are relevant for the exposure of about 1000 buildings.

Analyzing spatiotemporal dynamics of fluvial erosion and riverbed aggradation

Bedload balance is one of the most important features of a watercourse and plays a central role in river morphology. River morphology or riverbed morphology generally refers to the appearance of the river or watercourse.

This includes channel form, channel structures, and bed substrate. To assess bedload balance, both sediment supply and transport capacity must be investigated for each single river section in the river network. There are three possible states of changing sediment balance:

- Bedload deficit: if the sediment supply is smaller than the minimum transport capacity in a river section, the river erodes and transports additional material from its bed. This may result in riverbed incision.
- Bedload equilibrium: the sediment supply equals the transport capacity of the river and the bedload is transported through the river section. In this case, the bedload balance is in equilibrium and the riverbed does not change in the long term. Periods with bedload deficit can alternate with periods of bedload surplus.
- Bedload surplus: The sediment supply is greater than the maximum transport capacity of the river. The bedload can thus not be transported further and is deposited in the river section; riverbed aggradation occurs.

The transport capacity is a complex interplay of different characteristics such as channel width, discharge rate or bed gradient, and roughness. It can be influenced by human interventions and environmental change. Human interventions in the river system often lead to a bedload deficit. As a consequence, bed erosion occurs in many river sections (Bressan et al., 2020). Hydropower plants and dams affect bedload continuity as well as sediment flow. Dams reduce the amount of sediment transported through the river network (Jordan et al., 2020; Zhou et al., 2020), which in turn can change the flow regime and the probability of flooding (Llena et al., 2020). The riverbed eventually becomes more incised due to reduced sediment supply and flood flow variability (Jordan et al., 2020). Sediment retention basins for flood risk protection and bedload transport barriers in the tributaries also have an impact on sediment flux in the main river, as sediment is retained there and consequently removed from the river system (Syvitski et al., 2005). Another factor that changes the sediment budget of a river is gravel mining (Brestolani et al., 2015; Martín-Vide et al., 2010). On the one hand, these activities serve as flood protection and, on the other hand, influence bedload transport. This is because the process removes sediment load directly from the watercourse, thus deepening the riverbed, and in the long term it leads to incision in the riverbed (Llena et al., 2020). As a result of the riverbed being lowered, the groundwater table changes accordingly (Jordan et al., 2020). Moreover, land-use changes can exert an influence on the river system, for example, by reducing erosion due to a newly denser forest cover in the upstream catchments and thus reducing sediment entering the river. Vice versa, when forest cover is cleared, the soil erodes more easily and thus more sediment is expected to reach the main river network (Llena et al., 2020). As a result, runoff changes, and, as a consequence, runoff capacity also changes (Slater et al., 2015). Furthermore, bedload balance is influenced by slope and embankment structures. These serve to stabilize the river banks and can, in part, disturb the balance between sediment availability and transport capacity, altering the river width (Llena et al., 2020). Similarly, artificial channelization can lead to an imbalance between sediment availability and river height, which in turn increases the risk of flooding (Munoz et al., 2018). Anthropogenic interventions and their effects on the river system operate at different temporal and spatial scales. However, they often lead to a change in the sediment budget and thus to an increased or decreased occurrence of floods.

From the flood risk management perspective, the change of riverbed, either riverbed incision or riverbed aggradation, is an important driver for changing river morphology. An example of a changing cross section is given in Fig. 3.38. A changed river morphology alters the discharge capacity, and eventually this change will alter the probability of inundation in the floodplain. Riverbed incision increases the discharge capacity and consequently, the risk of flooding is decreasing. In contrast, riverbed aggradation decreases the discharge capacity and increases the risk of flooding in the floodplain. The patterns of riverbed incision and riverbed aggradation are expected to vary remarkably in space and time. Therefore, we analyzed the spatial pattern of riverbed erosion rates in the main river network in Switzerland. To this end, we mapped (a) the river reaches with fixed riverbeds (e.g., river reaches with artificially fixed cross sections to avoid riverbed incision), (b) the obstacles for sediment transport within the river network, (c) the locations of sediment supply by lateral tributaries that are capable to deliver debris flows to the main river network, and finally (d) the rates of riverbed erosion. We collected and analyzed the existing engineering reports that analyzed river balances for specific river reaches, we geomorphologically mapped the fixed riverbeds, the

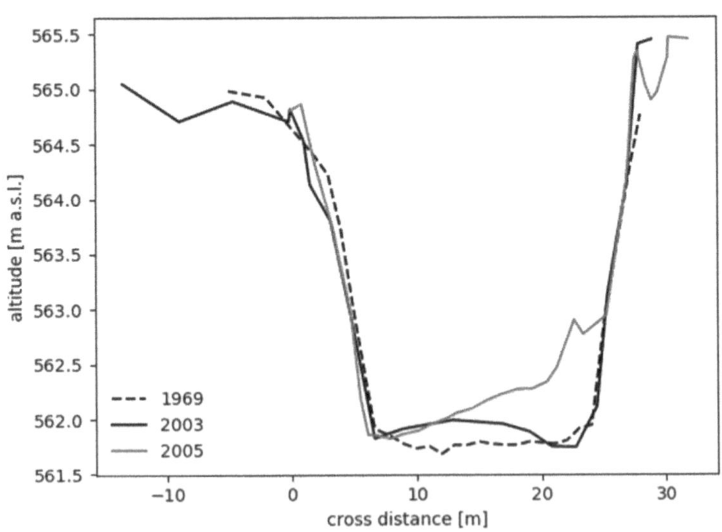

FIGURE 3.38 Cross section change. Changes in river cross section 252902 of the Hasliaare river from 1969 to 2005.

obstacles, and the sediment supply locations. To quantify the erosion rates, we analyzed the changes of cross sections. The main river network in Switzerland is regularly surveyed and the cross sections are repeatedly measured, e.g., every 10–15 years. This allows us to quantify the changes in the riverbed. The work therefore focused on the main river network of Switzerland, for which cross-sectional surveys or other data were available. Therefore, the resulting maps do not cover the entire river network but can give an insight into the spatial variability of the erosion rates.

The compilation of all data on the current development of river morphologies in Switzerland can be seen in Figs. 3.39–3.42. The map of potential sediment inflows from the lateral tributaries to the main rivers (Fig. 3.39) is an important basis for the analysis of small-scale local changes in river morphology that can be expected in the future. While the impact of climate change on flood risk in large river systems is still uncertain, for smaller catchments it can be assumed that bedload mobilization processes will become more frequent due to the observed and expected increase in short heavy precipitation.

Fig. 3.42 shows a high spatial variability of erosion rates. In many river sections, erosion and sedimentation alternate at a small scale. The rivers of the Central Plateau mainly have erosion rates between 1 and −1 cm/a. These rivers as well as the rivers in the foothills of the Alps have a high degree of fixed riverbeds (Fig. 3.40). Some river sections in the Alpine region show

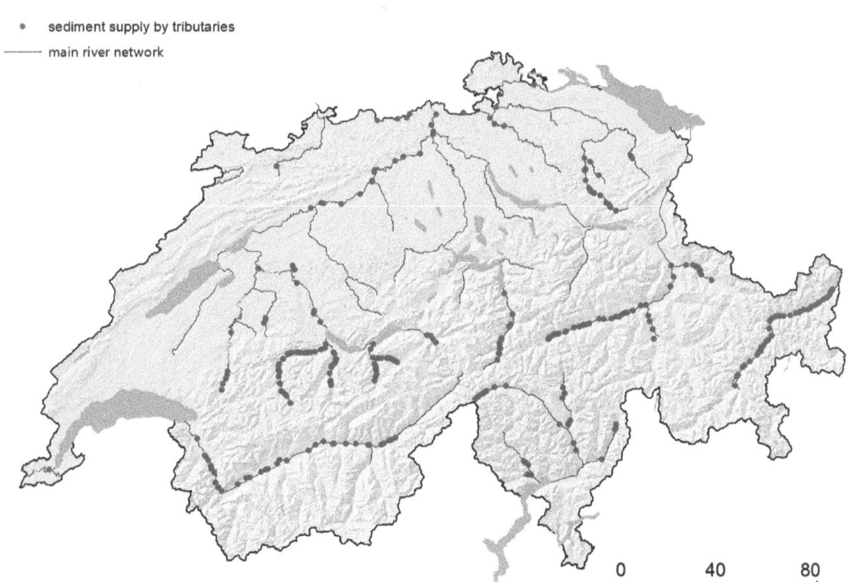

FIGURE 3.39 Sediment supply. Locations of potential sediment supply from the lateral tributaries to the main rivers.

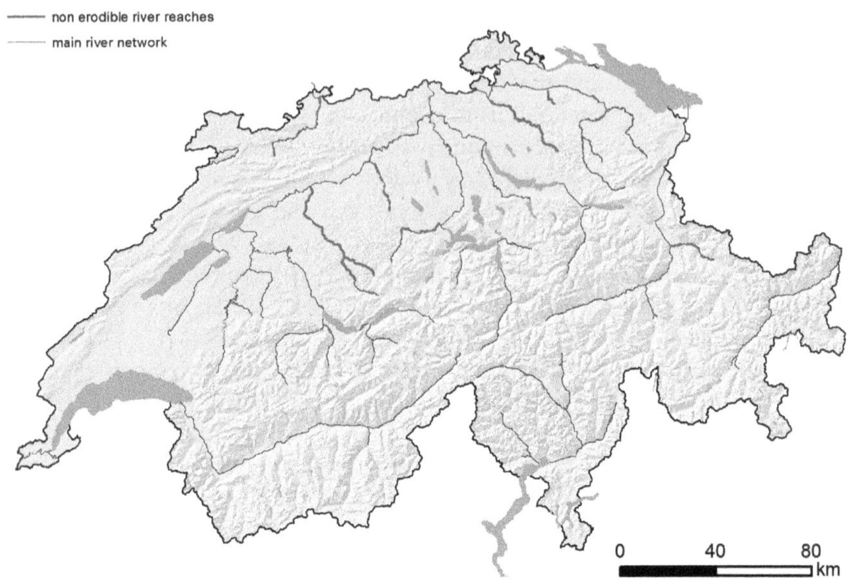

FIGURE 3.40 **Nonerodible riverbeds.** River reaches with nonerodible riverbeds.

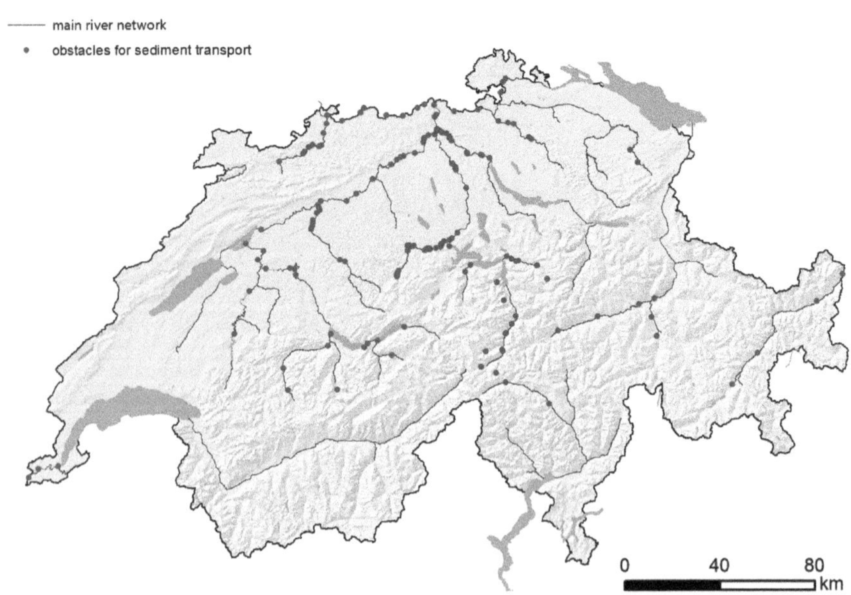

FIGURE 3.41 **Obstacles for sediment transport.** Obstacles for sediment transport in the Swiss river network.

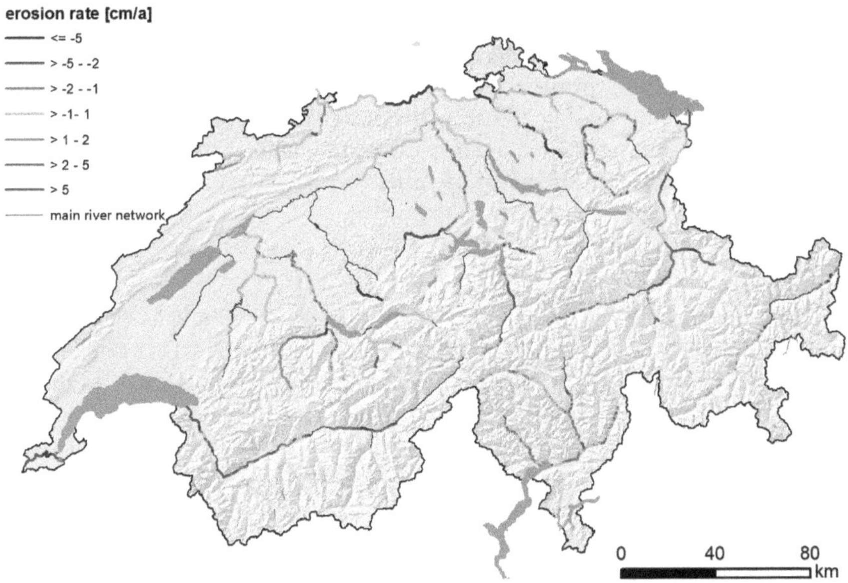

FIGURE 3.42 **Erosion rates.** Erosion rates in the main river network of Switzerland. Negative values show riverbed aggradation rates.

partly high and very high erosion rates, such as the river sections upstream of Bern or upstream of Lake Lucerne. In Valais, the middle course of the Rhone river shows a high erosion tendency, and in Ticino, the Maggia river. In the following, the main results are summarized for the main river systems in Switzerland.

Aare river

In the uppermost section of the Aare river (Hasliaare river) before the mouth of Lake Brienz, riverbed aggradation can be observed. The river reach between Thun and Lake Wohlen downstream of Bern is mainly characterized by erosion. The erosion rate has decreased slightly in recent years, which could be due to various watercourse rehabilitation and river widening projects. In the river reaches between Lake Wohlen and Lake Biel, the Aare riverbed again is aggrading. Between Lake Biel and the confluence of the Rhine, an aggradation tendency is predominant. The numerous hydropower plants reduce the flow velocity, which could be an explanation. Also partly responsible is the considerable sediment input from the tributaries Wigger, Reuss and Limmat. In general, most of the tributaries, as well as the Aare itself, show a high heterogeneity with respect to erosion and sedimentation tendencies. These alternate, in part, on a very small scale and in some places the erosion rates show extreme values.

Reuss river

For the Reuss river and its tributaries, a slight erosion trend is evident overall, with erosion occurring primarily at the Kleine Emme river and upstream of Lake Lucerne. The ongoing Reuss and Kleine Emme flood protection and renaturation projects will once again permanently alter part of the river section, which is why the effects of riverbed evolution on flood risk will be monitored.

Thur river

On many river reaches of the Thur and its tributaries, Sitter and Necker, small-scale sedimentation and erosion alternate, but overall erosion tendencies are evident. The riverbed of the Murg seems to be relatively stable, as its bed is strongly fixed with riverbed fixation measures. However, erosion tendencies seem to be a current issue, as various projects are being implemented or planned to stabilize the riverbed and for renaturation. Thus, the erosion is expected to stabilize in the near future.

Rhine river

In the upper reaches of the Rhine river, the Hinterrhein and Vorderrhein, erosion predominates. These river reaches should have a high positive sediment balance in a natural state, but the sediment budget is heavily reduced by human intervention. The Vorderrhein river shows more sediment input by tributaries than the Hinterrhein river. Nevertheless, erosion prevails. In the Hinterrhein river reach downstream of Thusis, bedload transport has been strongly reduced by river engineering measures, but nevertheless, there are more stretches with sedimentation in this section compared to the Hinterrhein. These inconsistencies in the results can be attributed to the different gradients of the two sections. The analysis of erosion rates shows that the Alpine Rhine and its tributaries are characterized by a bedload imbalance. Sedimentation and erosion alternate in these river sections, with erosion dominating overall. Numerous measures have already been taken to stabilize the riverbed, although sedimentation tendencies are frequently observed in these places. Downstream of Lake Constance, sedimentation dominates, whereby over large river reaches, the bedload balance is impaired to such an extent that bedload transport no longer takes place. This tendency applies in particular to the Töss river and partly to the Birs river.

Ticino river

In general, there is a high variability between sections showing erosion and those showing sedimentation, especially in the Brenno and Moesa rivers. Only in the Maggia and in the Magadino plain is there a clear signal of erosion. In the upper part of the Ticino river, slight aggradation tendencies can be observed. However, this is also an area with sediment-rich tributaries, which could possibly explain the aggradation rates. Another large area of riverbed aggradation is at the confluence of the Moesa with the Ticino.

Rhone river

The Rhone river showed a slight erosion trend between 1985 and 2001. Nevertheless, a high spatial variability of erosion and sedimentation can be observed. Most of the river section downstream of Lake Geneva to the border with France shows a remarkably high rate of sedimentation.

Inn river

The Inn river is basically an incising river. While relatively constant erosion rates were measured in the Upper Engadin, erosion rates vary more strongly in the Lower Engadin section. Despite the many tributaries that carry high sediment loads to the Inn river, the morphology of the Inn has changed little in recent decades. The reason for this is probably the reduction of bedload by various gravel extractions. The riverbed of the Inn is particularly incising in the Lower Engadine. This can be explained by the significantly steeper gradient in this section.

In conclusion, the spatial pattern of fluvial erosion is highly variable. If the current riverbed erosion or aggradation rates continue for the next decades, we have to expect remarkable changes of the discharge capacity and consequently of flood risk. The produced maps show where discharge capacity is expected to increase and thus flood risk will decrease.

Analyzing changes in flood exposure

In previous sections, we have seen how changes in the river morphology influence flood hydrology, flood exposure, and flood impacts. We conducted model experiments that kept the exposure and vulnerability as constant and looked at the impacts of human interventions to the river system that were implemented over centuries. Now it is time to look at another driver of flood risk change, namely the exposure of people, infrastructure, and buildings to flood hazards. Exposure is the presence of people, livelihoods; environmental services and resources; infrastructure; or economic, social, or cultural assets in places that could be adversely affected by hazardous processes. Exposure is a spatial indicator of population growth and economic growth. In regions with an increasing population or economic growth, a related increase in the spatial footprint of human settlements on the landscape is to be expected. Houses are built for residential and commercial purposes. Elmer et al. (2012) state that urban sprawl or settlement growth is the main driver of flood risk change in Eastern Germany. They must be connected together with infrastructure, e.g., roads, water and electricity grids, and communication grids.

It has been repeatedly claimed with respect to flood hazards in Europe that the main driver of increases in observed losses over the past decades is increased physical and economic exposure (Bouwer, 2013; Fuchs et al., 2015; Hallegatte et al., 2013; Jongman et al., 2014). These studies suggest that the spatial occurrence of losses is not so much dependent on the occurrence of specifically large events with high hazard magnitudes but more a result of an increased number of elements at risk in endangered areas (Fuchs et al., 2012). Long-term changes in flood risk, therefore, can be found to be a result of the significant increase in the numbers and values of properties endangered by natural hazard processes, and can be observed in both rural and urban mountain areas of Europe (Fuchs et al., 2015). Moreover, the robustness of natural hazard risk analyses depends on the availability of data on exposure and in particular accurate data on elements at risk, which also requires a temporal and spatial assessment of their dynamics.

The main driver of increasing exposure is thus population growth and economic growth. If we look at population growth since 1861 in Switzerland (BFS, 2021a), we see an increase from 2.5 million to 8.5 million residents (Fig. 3.43). This is an increase by a factor of 3.5 within approximately 150 years. The annual growth rates vary remarkably, with a range between −0.5% and 2.5%. The population growth rate peaked in 1961. The gross domestic product is increasing with population growth (Fig. 3.44). However, the growth rates have been decreasing since the 1970s (BFS, 2021b).

Model experiment setup for analyzing change in flood exposure

To analyze the role of settlement growth as a driver of flood risk change, we conducted a model experiment that varies from the ones described before. In contrast to the previous model experiments, here we kept the hazard constant over time and space and overlaid a spatiotemporal dataset of buildings with the hazard maps (see Fig. 3.45). The buildings in the dataset were attributed by the construction period. This overlay allowed us to elaborate the development pathway of exposed buildings. The analysis is described in (Röthlisberger et al., 2016).

For the determination of flood hazard areas, we used flood hazard maps provided by the Swiss cantons and elaborated according to national guidelines (BWW & BUWAL, 1997). As exposure data, we generated a homogenous data set of polygons of residential buildings and their period of construction for Switzerland by combining two datasets of (a) a topographic landscape model and (b) point data on residential buildings. From the topographic landscape model ((Swisstopo, 2021), version 1.1, 2012, TLM hereafter), we extracted the footprints of all present buildings in Switzerland. The information on residential use and period of construction as of the end of 2012 was taken from the federal register of residential buildings and dwellings (BFS, 2021c). This register was provided by the Swiss Federal Statistical Office and contains

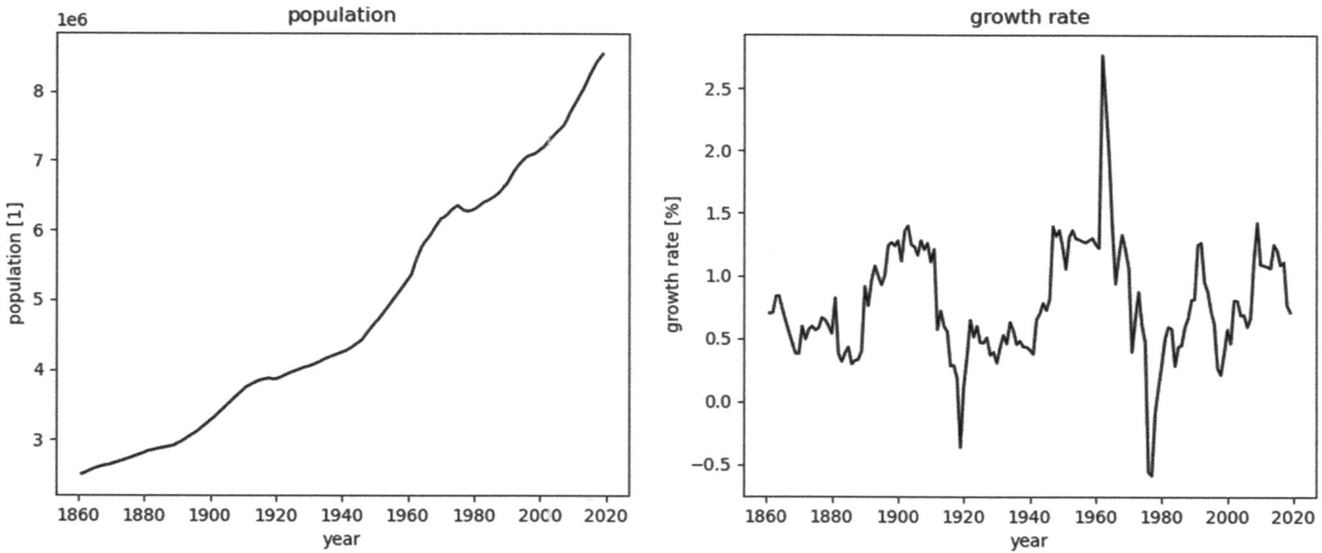

FIGURE 3.43 Population growth. Population growth in Switzerland 1861–2019. *Data from BFS. (2021). Bilanz der ständigen Wohnbevölkerung, 1861–2019. Bundesamt für Statistik. https://www.bfs.admin.ch/bfs/de/home/statistiken/bevoelkerung/stand-entwicklung/bevoelkerung.assetdetail.13707405.html.*

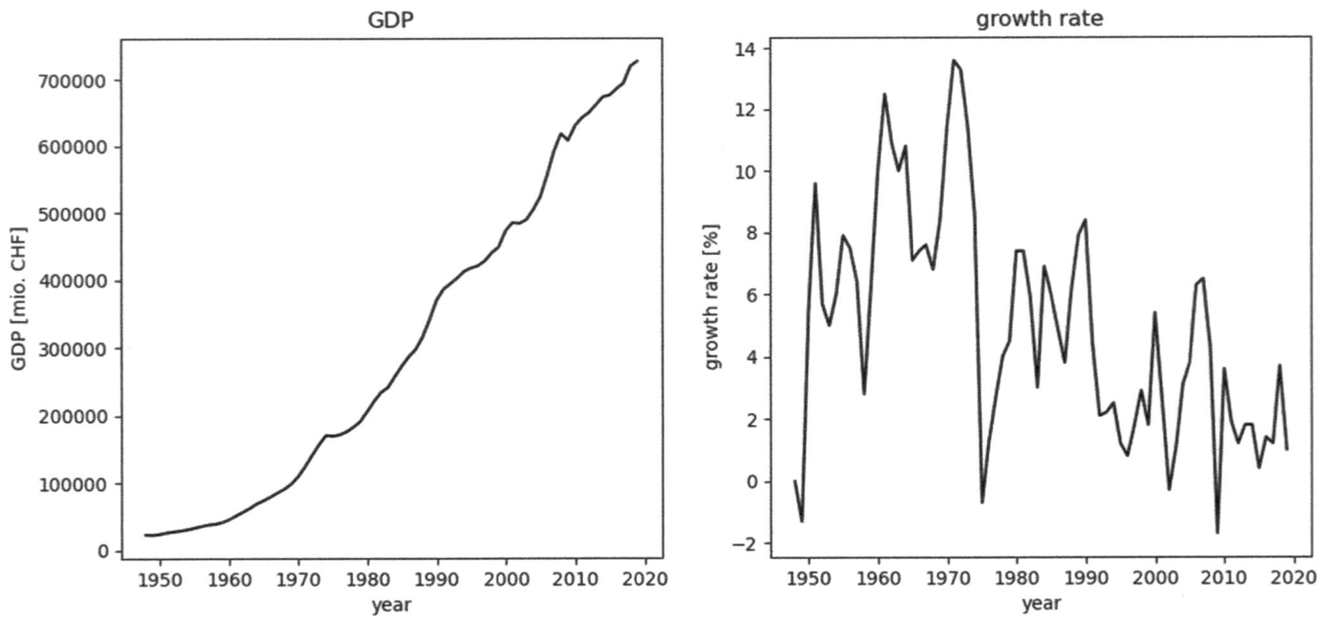

FIGURE 3.44 **Growth of GDP.** Growth of the gross domestic product in Switzerland 1861–2019. *Data from BFS. (2021). Bruttoinlandprodukt, lange Serie. Bundesamt für Statistik. https://www.bfs.admin.ch/bfs/de/home/statistiken/volkswirtschaft/volkswirtschaftliche-gesamtrechnung/bruttoinlandprodukt.assetdetail.14347493.html.*

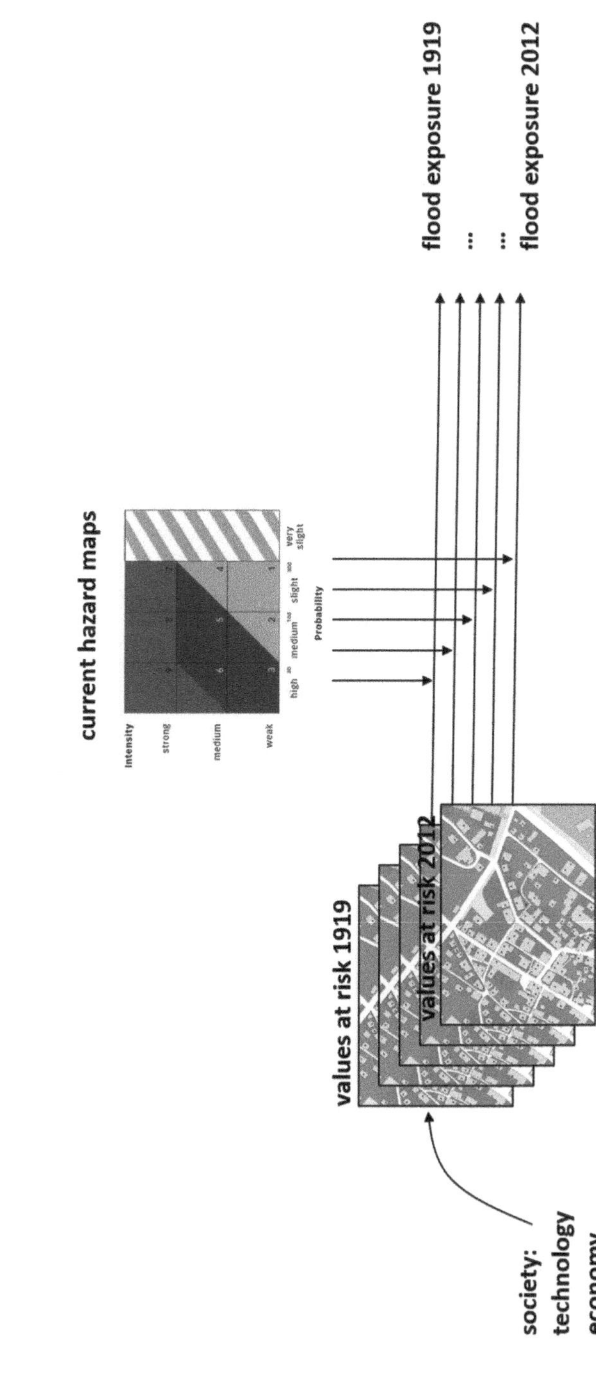

FIGURE 3.45 Model experiment. Model experiment setup for analyzing change in flood exposure in Switzerland from 1919 to 2012.

nearly 1.7 million point-referenced records of buildings (entrances) with residential use. The lengths of the 12 construction periods in the residential register differ considerably (from 2 to 27 years). In order to minimize these differences, we combined the seven shortest periods (of lengths between 2 and 5 years) to three (of lengths of 10 and 12 years), obtaining the following eight periods: (1) before 1919; (2) 1919–45; (3) 1946–60; four 10-year periods (4–7) between 1961 and 2000; and (8) 2001–12. The construction period is only available for residential buildings. Therefore, we focused the analysis of the exposed buildings on buildings with residential purposes only. We assigned the point data of residential registers to the building footprint polygons in order to assign the information on construction periods to the building footprints. We reduced the polygon dataset for buildings with residential utilization and with information on the construction period.

The building footprint polygons were intersected with the compiled and harmonized flood hazard maps. If a building intersects with more than one hazard class, it was classified at the highest hazard level it is intersecting with. For each of the eight time periods analyzed, the total number (and annual average) of the constructed buildings and the share of exposed buildings (at hazard levels "low," "medium," and "high") were calculated. We calculated the "share of exposed buildings" as the ratio of the number of "newly constructed buildings that are exposed to floods (at respective hazard level) to the number of 'total newly constructed buildings for which a flood hazard map exists." We computed the ratio separately for each time period. For the analyzed time period, we therefore assumed that no building has been demolished nor replaced, and that the flood hazardous areas remained unchanged.

Among the 2′053′539 building footprint polygons, 1′268′553 (61.7%) were classified as buildings with residential utilization. The deletion of the 50′270 polygons with inhomogeneous construction periods reduced the dataset further to 1′218′283 polygons. Finally, 239′619 polygons from the construction period 'before 1919' were removed, resulting in a dataset of 978′664 footprint polygons of buildings with residential utilization and homogenous construction periods between 1919 and 2012. These 978′644 building footprint polygons were used for the exposure analysis.

In Switzerland, a flood hazard map exists for 691′529 (70.7% of total 978′664) residential buildings constructed between 1919 and 2012. Therefore, these buildings were assessed with respect to flood exposure. Of the assessed buildings, 110′745 (16.01%) are exposed to floods. The exposure is mainly at hazard level "low hazard" (63′318 buildings or 9.16%). But some are also exposed considerably at level "medium hazard" (41′007 buildings or 5.93% respectively), whereas their exposure at hazard level "high hazard" is comparatively low, with just 6′420 buildings (0.93%) at that level.

The historical development of the settlements is the spatial manifestation of population growth. The number of buildings is growing in parallel with the population. However, the increase in buildings located in hazard zones is decoupled from the overall increase in buildings (Fig. 3.46). Nevertheless, the number of exposed buildings increased by a factor of 4. The overall buildings increased by a factor of 5.

In conclusion, this study shows a nation-wide analysis of flood exposure at a single building level. According to the definition of risk, i.e., risk is defined as the product of the probability of a hazardous event and its impact on values at risk, flood risk increased by a factor of 4 in the last century, if we assume the hazard component of the risk formula and the monetary values as well as the vulnerability of the buildings as constant.

Analyzing the effects of land use planning on flood exposure

Hazard mapping for floods in Swiss municipalities is almost complete, and the implementation of the hazard maps in spatial planning is underway. It took more than 20 years to cover all settlement areas with hazard maps. The introduction of hazard maps in the communal land use plans aims at regulating the construction of buildings in endangered areas. The hazard maps show the classified hazard in terms of magnitude and frequency. After the implementation of these maps in the communal land use regulation plan, these maps are called hazard zone maps. These maps "translate" the hazard classification into classes of land use regulation. Henceforth, the term "hazard map" is used

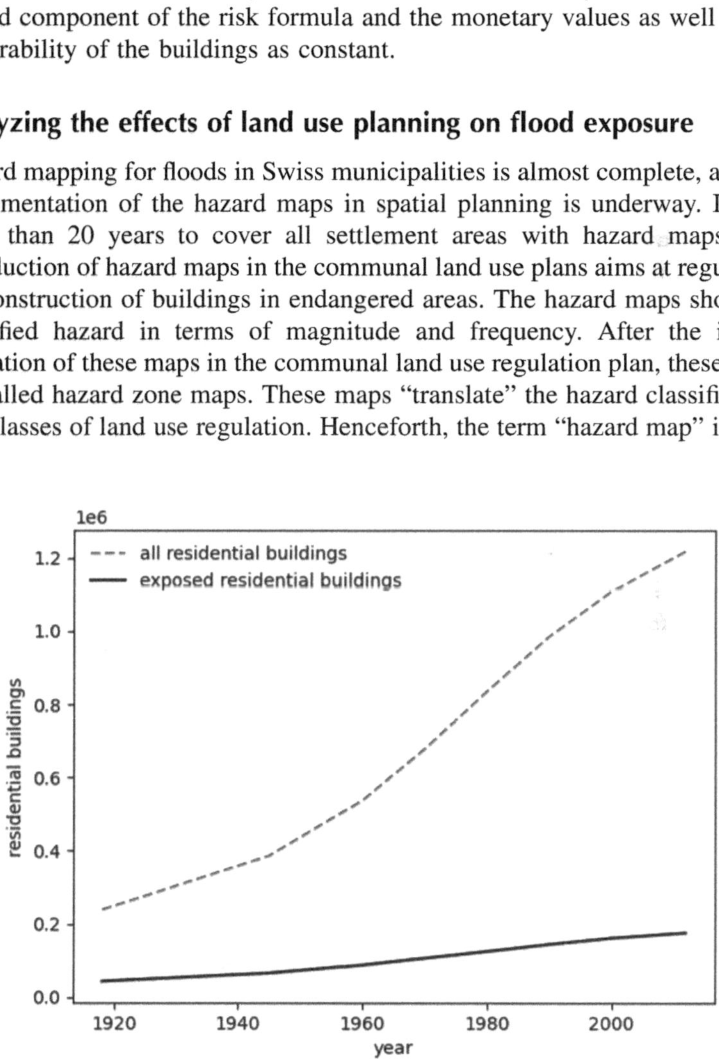

FIGURE 3.46 Growth of residential buildings. Growth of residential buildings in Switzerland (gray) and in hazard zones (black). *Modified from Röthlisberger, V., Zischg, A., & Keiler, M. (2016). Spatiotemporal aspects of flood exposure in Switzerland. In* E3S web of conferences *(Vol. 7). EDP Sciences. https://doi.org/10.1051/e3sconf/20160708008.*

for the information on hazards without legal implementation, and the term "hazard zone map" is used for the legally binding planning instrument.

Now, two decades after the introduction of this planning instrument in Switzerland, it is time to look back and to analyze the effectivity of the hazard maps and hazard zone maps, and the related land use planning procedure in terms of keeping flood exposure and therefore, flood risks constant. Therefore, we analyzed the effectiveness of these maps and investigated to what extent the hazard zone plans had an impact on construction activity in Swiss municipalities for the period 2010–15. The problem here is that the hazard maps have not been introduced everywhere at the same time. Instead, the elaboration of the hazard maps and their implementation into the communal land use regulation plans differed from municipality to municipality. Every communal authority implemented the hazard map at another date. To enable an analysis of the effectiveness of hazard maps in terms of avoiding construction in hazard zones, we investigated the year in which the hazard map had been finished and the year in which the hazard zone map had been officially implemented in the communal land use regulation plan. Only with this administrative procedure, the hazard map becomes legally binding for the authorities as well as for land owners.

The years of creation and implementation of the hazard maps of each Swiss municipality were compared with the building and housing statistics of the Swiss Federal Statistical Office (BFS, 2021c). This comparison showed that the implementation of the hazard maps for the period 2010–15 led to a slight increase in the number of buildings in the low and medium hazard zones. In the high hazard zones, the results are not statistically significant due to the small size of the study sample. However, it can be seen from the data, that, after the introduction of hazard zone map for floods, only a very few isolated buildings were built in the high (red) hazard zones. However, this effect was not seen in the Canton of Valais. It was found that Valais has a special legislation that allows construction under certain conditions in high hazard zones with slowly rising flood depths that can be predicted by hydrological forecasts. Other reasons for the construction of buildings in these areas can be mainly explained by special rules for construction of buildings outside of the land use regulation perimeters. Outside of the closer settlement area, i.e., the strict perimeter where a land use regulation plan must exist, the construction approval process is specifically regulated. In yellow hazard zones, widespread construction activities can still be observed after the completion of the hazard map. However, the construction rate is decreasing after the implementation of the hazard map in the land use regulation plan as a hazard zone map. A similar change in the spatial patterns can be observed in the construction rates in blue hazard zones. The number of communities that have a high construction rate in these hazard zones is less after the implementation of the hazard zone map in the communal land use regulation plan. A slightly different pattern is observable in the construction rate in red hazard zones. Construction in these zones is generally prohibited. However, there are still some communities that allowed the construction of buildings in red hazard zones. These are mostly given by exemption legislation of the Canton Valais. In this canton, a river correction project is planned that aims at reducing the risk of flooding in the main valley. Great parts of the main valley bottom are mapped as a red hazard zone. The communities allow the construction in these zones by anticipating the effects of the planned flood protection measures. Fig. 3.47 shows the spatial pattern of the construction rates of new buildings in the combined hazard zones (yellow, blue, and red hazard zones together). Statistical proportion tests show that the construction rates within and outside of hazard zones do significantly differ only in a very few communities (Bruchez, 2017). Overall, it can be stated that the hazard maps—except the red hazard zones—do not significantly reduce the exposure of buildings in hazard zones. However, as stated above, the construction of buildings in blue hazard zones is regulated, buildings can only be constructed in these zones if they implement object protection measures to reduce the vulnerability of the building to flood hazards. The data do not show if a building constructed in a blue hazard zone has object protection measures implemented or not. Moreover, the provision of information of the hazard (i.e., the hazard map) is less influencing the construction activity than the legally binding hazard zone map. Therefore, the

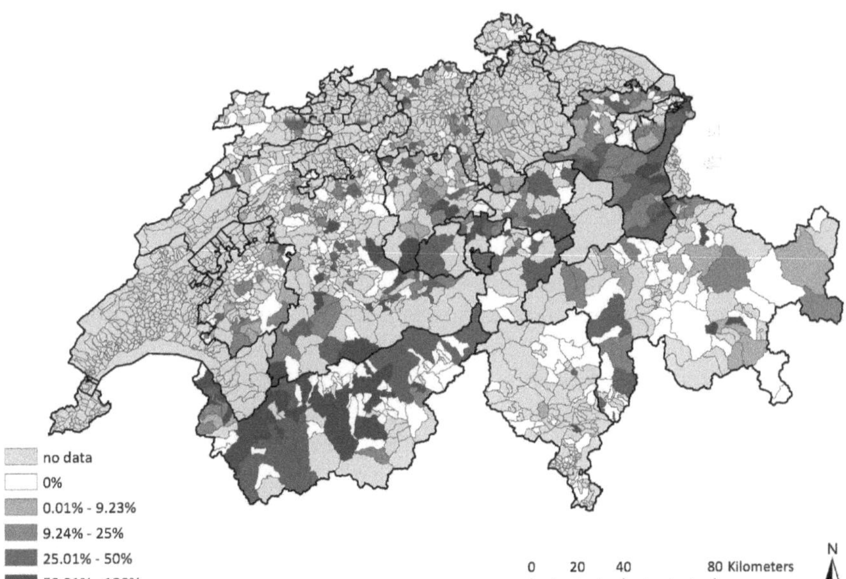

FIGURE 3.47 Construction of buildings in hazard zones. Share (%) of newly constructed buildings in hazard zones (low, medium, or high hazard) in the period 2010–15 after the elaboration of the hazard map. *From Bruchez, L. (2017). Effet des cartes de danger de crues sur la construction de nouveaux bâtiments d'habitation dans les communes de Suisse.*

implementation of the hazard map into the land use regulation plan is necessary.

Modeling the effects of changing vulnerability on flood risk change

In the previous sections, we looked at how changes in precipitation intensity and in river morphology influence flood hydrology and flood impacts. Moreover, we analyzed the effect of settlement growth on changes in flood exposure. The third main factor of the flood risk formula is missing yet, namely vulnerability. Thus, the next model experiment aims at testing the sensitivity of flood impacts to changes in vulnerability. Again, we focus on flood impacts on buildings. The vulnerability of buildings to floods can be positively or negatively changed by human intervention. A positive change of vulnerability will increase flood risk. An example of this kind of change is the introduction of new housing techniques that depend on complicated electricity and digital infrastructure. Servers, relays, sensors, and other electric equipment are installed in the basement or at the ground floor and if this equipment is affected by water entering the house, the whole house technique is damaged. Another example is the increasing value of sound or home-cinema devices. In contrast, if the house-owner implements object protection measures to avoid the entering of water into the building, the vulnerability of the house is decreasing. Object protection measures are structural property-level mitigation measures aimed at preventing or minimizing flood damage to persons or property. They can be applied to or directly next to existing buildings or new buildings. Object protection measures can be mobile or constructive measures. The first type of measure is only applied temporarily if a flood event is forecasted. Examples of these types of property-level protection measures are sandbags or dam beams that protect the building against the water. The second type of measure is a conceptual-structural solution that prevents water from reaching weak points in the building in the first place. An overview of property-level flood protection measures is given by Attems et al. (2019).

To analyze the effects of changing vulnerability, a model experiment was conducted. With this model experiment, we tested how flood risk in the Aare river basin upstream of Bern is influenced by the virtual implementation of object protection measures on the buildings in the floodplains of the Aare river and the lakes Thun and Brienz. To robustly assess these effects of property-level flood protection measures on flood risk change, we used the rainfall scenarios described above. We distributed the 3-day rainfall sum over the Aare river basin with a Monte Carlo simulation approach to 100 different spatio-temporal rainfall patterns for the return periods of 10, 30, 50, 100, 150, 200, 250, and 300 years. This results in a set of 800 rainfall scenarios that were fed into the coupled hydrological-hydrodynamic model. The hydrologic model was the PREVAH model and the hydrodynamic model was the LISFLOOD-FP model. The rainfall event generation process and the coupled hydrologic-hydrodynamic model are described in Zischg, Felder, Weingartner, et al. (2018). The 800 rainfall scenarios result in 800 flood maps with different spatiotemporal patterns of flood exposure. The dataset of the buildings was taken from the same study. The 800 flood maps were overlaid with the building dataset. This resulted in 800 datasets of flood depths at individual building scales. Thus, the buildings dataset with information on the building value (reconstruction value) was attributed to the flood depth for each of the scenarios. The flood depths at individual building levels were used to derive the degree of loss by means of a vulnerability function. The degree of loss describes the monetary damage to the house in terms of repairing costs in relationship to the monetary value of the building depending on the flood depth. The multiplication of the reconstruction value of the building with the object-specific and scenario-specific degree of loss resulted in the damage to each building per rainfall scenario. In contrast to the study of Zischg, Felder, Weingartner, et al. (2018), which used several different vulnerability functions, here we used only one vulnerability function. This simplifies the sensitivity analysis and makes it more feasible. However, it is remembered that the choice of the vulnerability function is one of the most relevant uncertainties in the overall risk assessment.

In the model experiment, we used the vulnerability function of Jonkman et al. (2008). This function describes the relationship between flood depth and the degree of damage. To test the effects of property-level flood protection measures on the overall risk, we assumed a threshold of a maximum flood depth under which no damage occurs. Once this threshold is exceeded, water enters into the building and the damage is as high as it would have been without object protection measures installed. Thus, it is assumed that no damage occurs if flood depths at the building level are under the assumed threshold. This is a rather simplified approach, but it represents a political choice of a minimum standard of building safety. In the model experiment, we assumed thresholds of 20 cm, 40 cm, 60 cm, 80 cm, and 100 cm for the standard safety level. For example, a threshold of 20 cm flood depth means that all buildings in the floodplain will be reinforced with temporary or structural measures that prevent damage from flooding up to a flood depth of this threshold. We assumed 100% efficacy. The thresholds were introduced in the vulnerability function (Fig. 3.48). The original vulnerability function is the equation proposed by Jonkman et al. (2008). The counterfactual vulnerability functions are the same function but with a degree of loss of 0 below the threshold for the water depth in the building. The setup of the model experiment to test the influence of the vulnerability function on flood risk change is schematically shown in Fig. 3.49.

In analyzing the results of the model experiment, we first looked at the number of flooded buildings and the flood depths at the individual building scale (Fig. 3.50). These two factors allow us to interpret the other results. As expected, the number of buildings located in the flooded areas is increasing

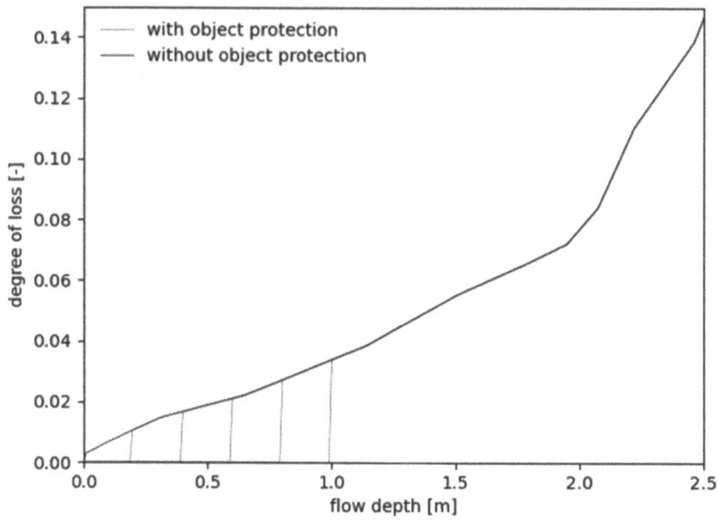

FIGURE 3.48 Vulnerability function. Vulnerability function with thresholds of 20 cm, 40 cm, 60 cm, 80 cm, and 100 cm for assumed standard safety levels.

with the magnitude of the rainfall scenarios. Generally speaking, the higher the rainfall sum, the higher is the number of flooded buildings. However, the increase is not linear. From the rainfall scenarios of return periods of 10–100 years, the number of exposed buildings is increasing at a higher rate than for the other rainfall scenarios. The figure shows a high variability. For example, there exist rainfall scenarios of a return period of 10 years that have a higher number of flooded buildings than some rainfall scenarios of a return period of 300 years. This can be explained by the spatial pattern of the flood processes. Depending on the city or village that is affected by the flood scenario, the number of exposed buildings can be much higher or lower than on average. The highest numbers of flooded buildings are produced by rainfall scenarios with a spatial footprint that leads to flooding in the city of Interlaken and the city of Thun. Interestingly, the median flood depths at all flooded buildings do not vary much. The average of all object-specific flood depths over all return periods is in the range of 0.49–0.52 m. The median object-specific flood depth ranges from 0.28 to 0.31 m. This shows that the increase in the number of exposed buildings due to the enlargement of the flooded areas is leveling the flood depths in this river basin. This is in line with the findings of Keller, Zischg, et al. (2019) and Zischg et al. (2021b).

The average damage sum over 100 rainfall scenarios for the rainfall return period of 10 years is 25.4 million Swiss Francs, and 26.7, 28.1, 30.2, 31.5, 33.8, 34.8, and 35.5 million for the return periods of 30, 50, 100, 150, 200, 250, and 300 years, respectively. Hence, the average flood damage of a

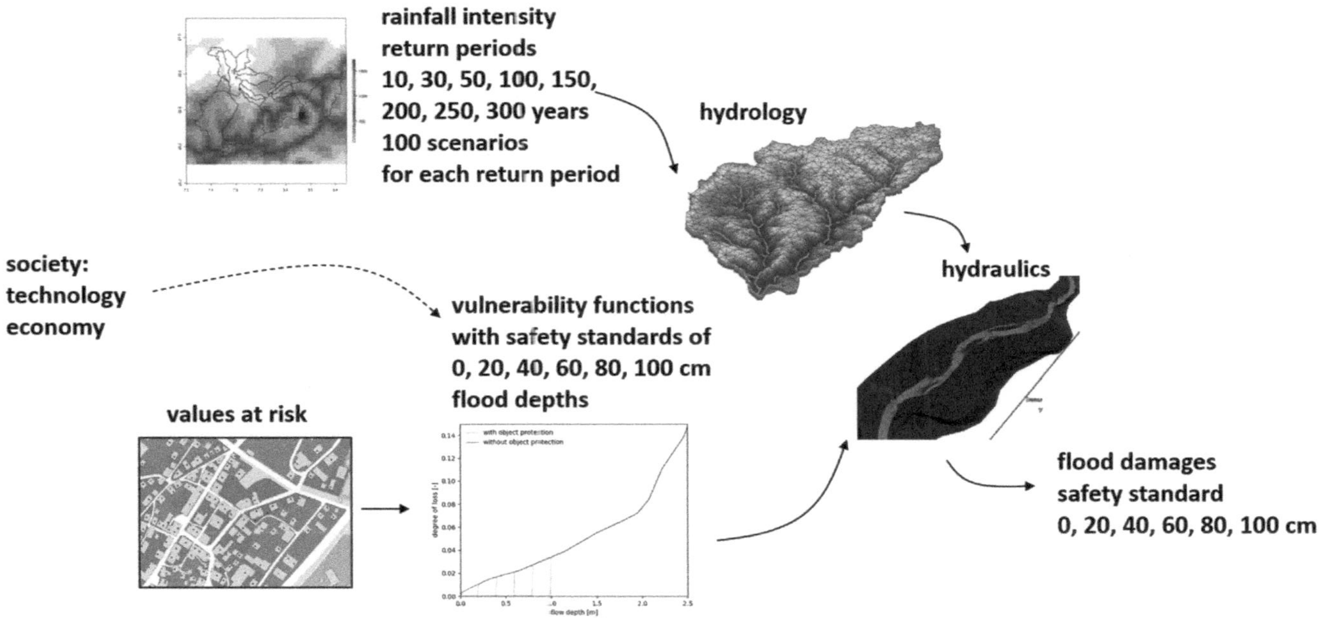

FIGURE 3.49 **Model experiment.** Setup of the model experiment to test the influence of property-level flood protection measures on flood risk change.

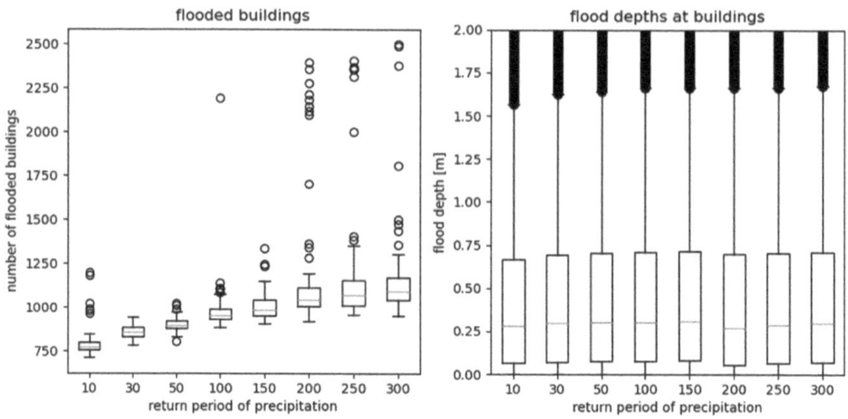

FIGURE 3.50 Flooded buildings. Number of flooded buildings and average flood depths at the buildings for rainfall scenarios with different spatiotemporal patterns and different return periods.

300-year rainfall event is by a factor of 1.4 higher than the average damage of a rainfall event of a 10-year period.

If we consider the effect of the property-level flood protection measures at individual building level, the average damage at a safety standard level of 100 cm is only about 47%–49% of the average damage without object protection implemented. Fig. 3.51 shows the effect of property-level protection measures on different standard safety levels. The average flood damage increased with the return period. Although a certain number of buildings were flooded because the flood depths are above the safety standard level, all scenarios of safety standard levels show markedly lower damage. The hypothetical introduction of a property-level flood protection with a safety level of 20 cm that prohibits the entering of water into the building from flood depths of up to 20 cm would reduce the average damage over the whole river basin by 10%. Safety levels of 40 cm, 60 cm, 80 cm, and 100 cm would reduce the overall flood damage by about 23%, 35%, 45%, and 52%, respectively. This shows the potentially highly relevant contribution of object protection for vulnerability reduction if implemented for all buildings in the flood zones. The damage reduction by property-level flood protection measures varies only a little over the return periods. The effect on damage reduction ranges for all return periods between 2% and 3%. There is no threshold in the return periods visible that could lead to the conclusion that the property-level flood protection measures would suddenly fail with increasing flood magnitudes.

In the last step of this model experiment, the flood risk change was quantified. To calculate the flood risk and its change, we calculated the expected annual damage per return period, i.e., the average sum of damage to buildings in over 100 scenarios of the selected return period multiplied by the probability of occurrence. Here, we calculated the occurrence probability by

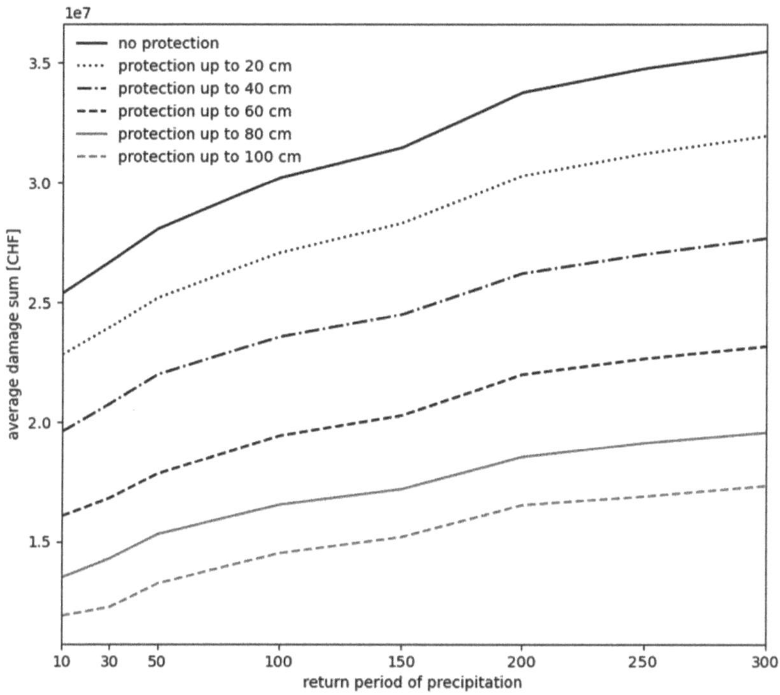

FIGURE 3.51 Property-level flood protection. Effects of property-level flood protection measures on flood risk at different thresholds for flood standard safety levels. The sum of damages over the whole river basin was averaged over 100 rainfall scenarios for each of the return periods of rainfall sum within 3 days.

dividing 1 by the return period. The averaged damage sum for each of the return periods (occurrence probability) resulted in the flood risk curve (risk footprint) of the whole Aare river basin. The expected annual damage over all return periods was integrated from the risk curve, i.e., by interpolation of the discrete flood scenarios used in this study (Eq. 3.1). The procedure is followed for each of the chosen thresholds for standard safety levels. The difference between the flood risk without property-level flood protection and the scenarios with safety standards introduced gave the flood risk reduction by the hypothetical implementation of the measures.

$$R \approx \sum_{i=1}^{n} \left(\frac{D_i + D_{i+1}}{2} \right) |P_i - P_{i+1}| \tag{3.1}$$

Where R is the flood risk in terms of expected damage in Swiss Francs per year, i is the scenario number, n is the number of scenarios, D_i is the damage scenario, and P_i is the probability scenario. The basic scenario is a scenario without property-level protection measures (100%).

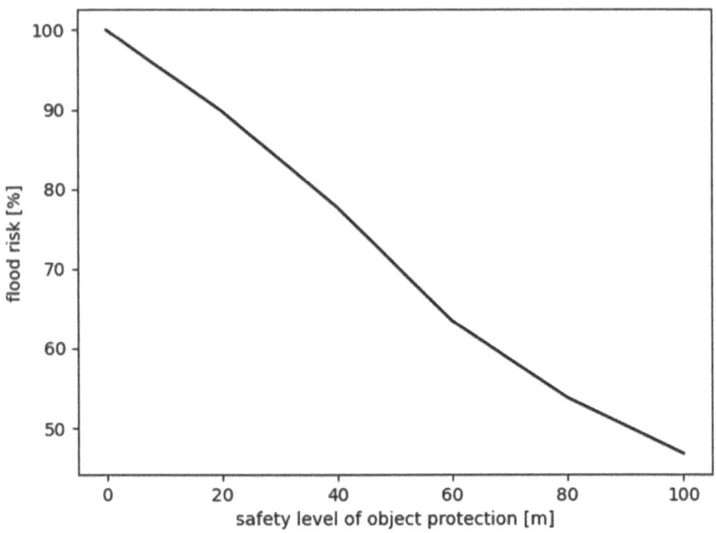

FIGURE 3.52 **Flood risk change.** Flood risk change due to property-level flood protection measures with different thresholds for flood safety.

The model experiment showed that the flood risk in terms of expected annual damage will be reduced remarkably if property-level flood protection was implemented. The implementation of a standard safety level of only about 20 cm reduced flood risk by 10%. As similarly observed for the damages, the risks were reduced by 12%, 34%, 46%, and 53% if flood safety standard levels of 40 cm, 60 cm, 80 cm, and 100 cm, respectively, were virtually introduced in the Aare river basin (Fig. 3.52).

However, the overall assumption of a generally valid safety standard for all buildings is very generalist. In practice, the object protection levels should be adapted for each building according to the locally possible flood depths. This allows a more targeted approach and avoids unnecessary regulations for buildings located in hazard zones with low flood depths. Nevertheless, the experiment shows that this kind of flood risk management measure could have a remarkable lever in flood risk management. These kinds of measures could complement river engineering solutions or even substitute structural flood protection measures in the river channel if there is not enough space for the implementation of river widening.

Analyzing the effects of feedback and time lags on flood risk change

Feedback effects of fluvial erosion and riverbed aggradation on flood risk change

Climate warming is very likely to result in an increase in heavy rainfall events in Switzerland (CH 2018 Project team, 2018). This trend can already be observed in several meteorological timeseries in Switzerland and will further intensify with increasing warming. The question is what impact an increase in heavy rainfall events will have on flood risk. It is assumed that an increase in the frequency and intensity of heavy precipitation events will also lead to an increase in the probability of occurrence and magnitude of flood events and thus to an increase in damage and risk.

An important research question in this field is how much and where flood risk will change. It is still an open question whether the change will be linear or nonlinear and what feedback processes can be expected in the "precipitation-runoff-flood-damage" process chain. For example, one such feedback process is the change in river morphology due to an expected future increase in flood event magnitudes. An increase in small-scale heavy rainfall events may lead to an increase in the frequency of debris flow and bedload transport processes in the smaller torrents. This may increase sediment delivery from the lateral tributaries to the main rivers, and thus it can potentially result in a riverbed aggradation of the main rivers. This leads to a reduction in discharge capacity and thus to an increase in the probability of flooding outside of the river channel. Conversely, in river sections with impeded sediment supply, an increase in erosion and thus a deepening of the river bottom can be expected. This would be an example of a negative feedback process where an increase in flood event magnitudes and frequency will increase the discharge capacity of the river channel and conclusively to a decrease of the likelihood of damage-related flooding.

In order to analyze these feedback processes, we conducted model experiments for (a) analyzing the sensitivity of discharge capacity to fluvial erosion rates, (b) the sensitivity of erosion rates to changes in peak flows, and finally (c) the sensitivity of flood risk to erosion rates.

Sensitivity of discharge capacity to fluvial erosion rates

The influence of a possible lowering or rising of the riverbed on the discharge capacity was estimated using various scenarios. For this purpose, a cross section of the Aare river near the gauging station of Bern was selected. We assumed a continuing riverbed incision and riverbed aggradation for the period 2020–2100. At the end of this period, a new dataset of cross sections was generated that represents the cross section after 80 years with the assumed erosion rate. With positive erosion rates, the riverbed will be at a lower level at the end of this period. In contrast, negative erosion rates mean riverbed aggradation. This will lead to a heightening of the riverbed. We simulated a slightly increasing hydrograph and determined the discharge where overflowing of the lateral dams begins. The discharge capacity corresponds to the maximum flow capacity without the water overflowing the riverbanks. We used the following erosion rates for the sensitivity analysis: 0, ± 0.5, ± 1, ± 1.5, ± 2, and ± 3 cm per year.

The results of this sensitivity study show the long-term effects of riverbed erosion and riverbed aggradation on discharge capacity (Fig. 3.53). After 80 years with an erosion rate of 1 cm/a, the discharge capacity nearly doubles. An erosion rate of 3 cm/a leads to a discharge capacity that is more than four

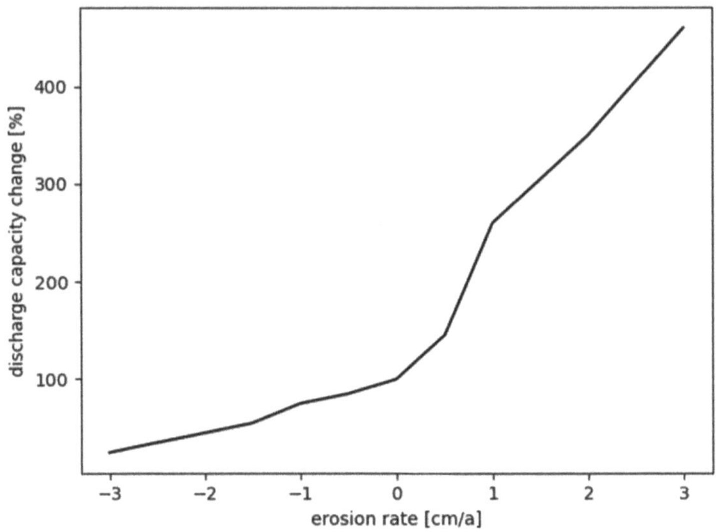

FIGURE 3.53 Discharge capacity. Discharge capacity in the Aare river after the end of the period 2020–2100 with different erosion rates. The erosion rate of 0 cm/a is the basis for the comparison (100%).

times higher than today. In contrast, an aggradation rate of 1 cm/a results in a reduced discharge capacity of 75%, and an aggradation rate of 2 cm/a results in a reduced discharge capacity of 45%. The sensitivity analysis showed a nonlinear behavior of the relationship between erosion rate and discharge capacity.

Sensitivity of erosion rates to changes in peak flow

Above, we have seen that the erosion rate markedly influences the discharge capacity in the long term. This is a geomorphological simplification based on assumptions. However, the question now is how the erosion rate is influenced by changed hydrology. Thus, we set up a model experiment for analyzing how the erosion rate changes with changing peak flows. We set up the hydromorphological model BASEMENT (Vetsch et al., 2017) in the 1D version for five cross sections of the Aare river downstream of Thun and calibrated the model in the observation period 2005–11. We used the observed discharge measurements of the gauging station of Thun (FOEN, 2018b) and calibrated the model to fit the mapped erosion rate as shown in the previous sections. The cross sections came from FOEN (2018a).

For the sensitivity analysis, we increased the magnitudes of peak flows in the river section by 5%, 10%, and 15%, respectively. Only peak flows exceeding the annual flood event were increased by these values. This step resulted in three counterfactual hydrographs. The aim of this model experiment was to determine how sensitive the riverbed erosion in this river section is to the increase in peak flows. Fig. 3.54 shows the changed hydrographs.

The model experiment resulted in a relationship between the assumed amount of peak flow increase and the related change in the erosion rate. The basic value of the modeled erosion rate was 5.7 cm in the period 2005−11, meaning an average erosion rate of 0.8 cm/a. If the peak flows (discharge above the annual maximum) were assumed to increase by 5%, the erosion rate nearly doubled. An assumed increase of the peak flows by 10% and 15% resulted in an increase of the erosion rate by 395% and 582%, respectively (Fig. 3.55). This means that a change in peak flow, e.g., caused by the effects of climate change, is a remarkable lever for changing river morphology and consequently, of flood risk. This can locally be a negative feedback effect.

Sensitivity of flood risk to erosion rates

The examples above show the sensitivity of a middle-sized main river reach, the Aare river downstream of Thun. The river reaches selected for this sensitivity analysis are not markedly affecting flood risk. Therefore, we selected an additional case study that represents a smaller river, a tributary of the main river Aare, and a river reach that is relevant for flood risk change. To analyze the sensitivity of flood risk to the erosion rate, we set up a model experiment for the Kander river at Reichenbach. The riverbed of this river reach has a width of 20 m and an average slope of 1.2%. The average erosion rate in this river reach is about 1 cm/a. However, the erosion rate varies remarkably and has a high local variability. In the upper part of the river reach, a soil fixation structure was implemented several decades ago to avoid riverbed incision.

In this model experiment, we tested the sensitivity of both the discharge capacity and the flood risk to the erosion rate. We assumed erosion rates of −1, −0.5, 0, 0.5, and 1 cm/a for the period 2020−2100. Positive values represent riverbed erosion, and negative values of erosion rates represent riverbed aggradation. We calculated the fluvial erosion after 80 years for each of the assumed erosion rates and adapted the cross sections by subtracting the fluvial erosion/deposition to the current cross sections.

As the inundation model, we used the BASEMENT model (Vetsch et al., 2017) in a coupled 1D-2D simulation setting. The cross sections of the 1D model were altered by considering the erosion rates and adapting the cross sections while the 2D model and the coupling was not altered. This resulted in one hydrodynamic model for the current river morphology and four counterfactual hydrodynamic models with the river morphology after a period of 80 years with different assumed erosion rates.

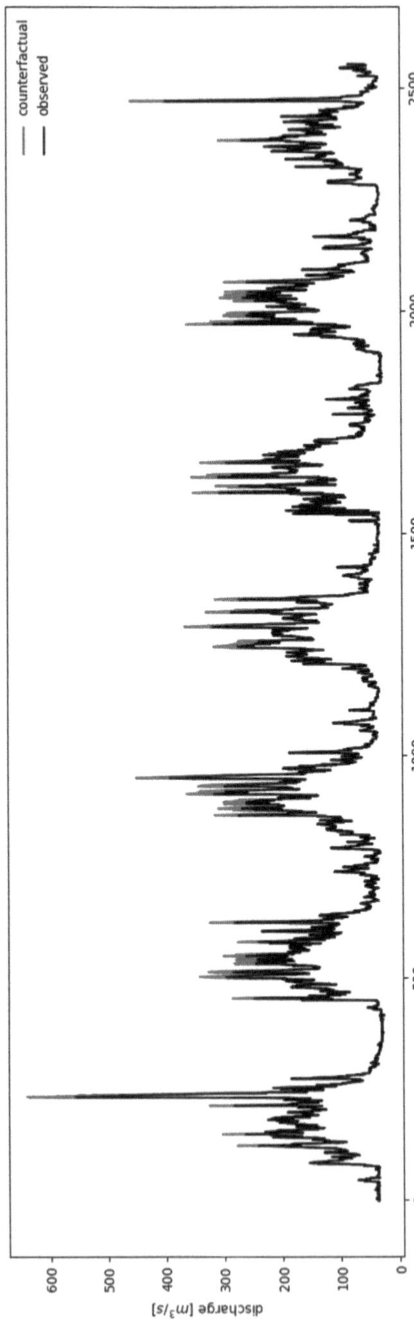

FIGURE 3.54 Hydrograph. Observed (black) and counterfactual (grey) hydrograph for analyzing the sensitivity of riverbed erosion to changes in peak flows.

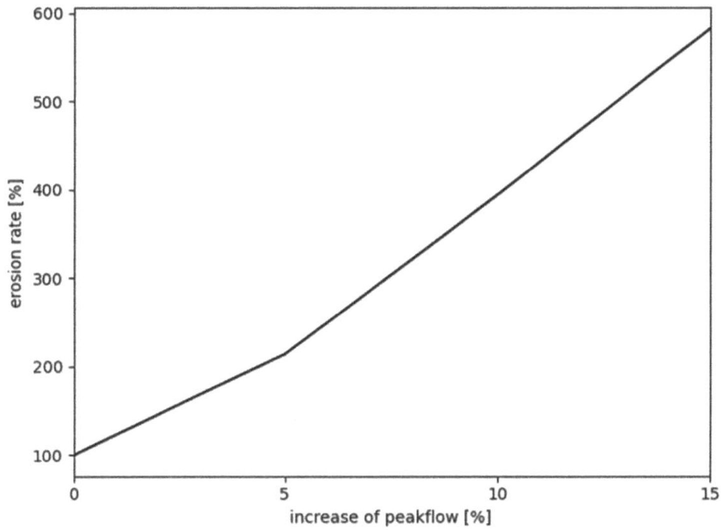

FIGURE 3.55 Sensitivity of erosion rate to peak flow change. Changes in erosion rates depending on an increase in peak flow in the Aare river reach downstream of Thun.

The river discharge capacity here is defined as the maximum peak discharge that does not result in damage to buildings. Therefore, if the discharge capacity is exceeded, damage results from the simulation scenario.

As the dataset for the values at risk, we used the building dataset of Röthlisberger et al. (2018) and the monetary assessment model M5. The vulnerability was calculated with the vulnerability function of Jonkman et al. (2008). We simulated flood maps for return periods of 30, 100, and 300 years for all erosion rates. We overlaid each of the flood maps and calculated the damage to each building depending on the flood depth at the building, its monetary value, and its object-specific vulnerability. For each of the flood maps of the return periods of 30, 100, and 300 years, the damages were summed up. The flood risk curve therefore resulted from these scenarios. The flood risk was calculated by integrating the area under the flood risk curve. For each of the erosion rates, a flood risk value in terms of expected annual damage resulted. The flood risk values of the counterfactual simulations were then compared with the simulation representing the current situation. The expected annual damage from the current situation is ~25′000 CHF/a.

The rate of fluvial erosion or riverbed aggradation markedly alters the discharge capacity. An erosion rate of 0.5 cm/a and 1 cm/a resulted in an increase of the discharge capacity to 122% and 670% of the current discharge capacity, respectively. In contrast, a deposition rate (i.e., a negative erosion rate) of 0.5 and 1 cm/a resulted in a discharge capacity of 56% and 67% of the current discharge capacity, respectively. The change in discharge capacity is highly nonlinear. An erosion rate above 0.5 cm/a suddenly leads to a sharp increase of the discharge capacity. The rate of change is much higher above this threshold than below this threshold (Fig. 3.56, left).

The sensitivity of flood risk to the erosion rate is less than the sensitivity of the discharge capacity. An erosion rate of 0.5 and 1 cm/a resulted in a decrease in flood risk to 79% and 61% of the current flood risk in terms of expected annual damage. In contrast, a riverbed aggradation with a rate of 0.5 and 1 cm/a resulted in an increase of flood risk to 151% and 218% of the current value, respectively. Also, this sensitivity analysis shows a nonlinear behavior.

In summary, the three sensitivity analyses showed that riverbed morphology changes through riverbed incision and aggradation have relevant effects on discharge capacity and therefore on the occurrence probability of flooding in the floodplain. Changes of discharge capacity and the effects of higher peak discharges on riverbed morphology show that the current developments in river morphology can have a nonlinear influence on flood risk over a long time period. The probability of flooding in the floodplains outside of the river channel due to exceeding the discharge capacity will be reduced in the long term if riverbed erosion continues. If only the change in channel morphology is considered, the greater the erosion rate, the greater the decrease in the probability of flooding. Conversely, the effect of sedimentation trends on runoff capacity is somewhat less sensitive. If this sensitivity study is representative, it can be assumed that in the river sections with erosion tendency the flood risk will decrease in the long term due to this negative feedback process. It can be seen that sedimentation in the riverbed would have comparably influence on the flow capacity and flood risk as the erosion process.

Consequently, flood risk is also affected by river morphology changes (Fig. 3.57). The processes behind this morphological change are slow and

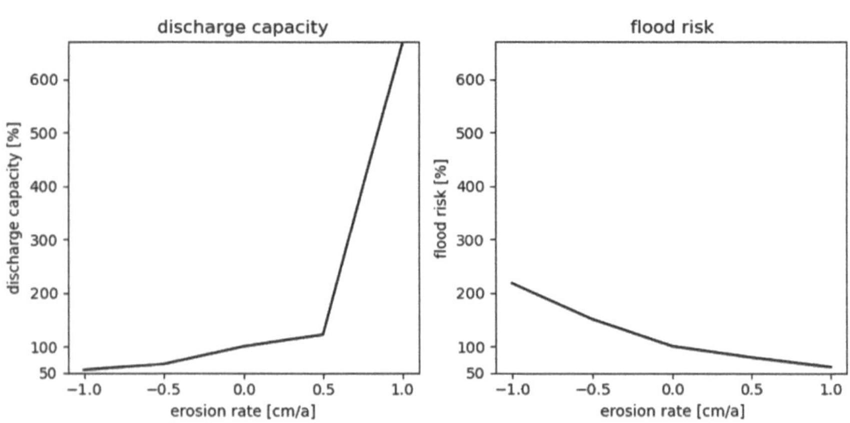

FIGURE 3.56 Discharge capacity and flood risk change. Simulated change in discharge capacity and flood risk after the period 2020–2100 depending on the erosion rate.

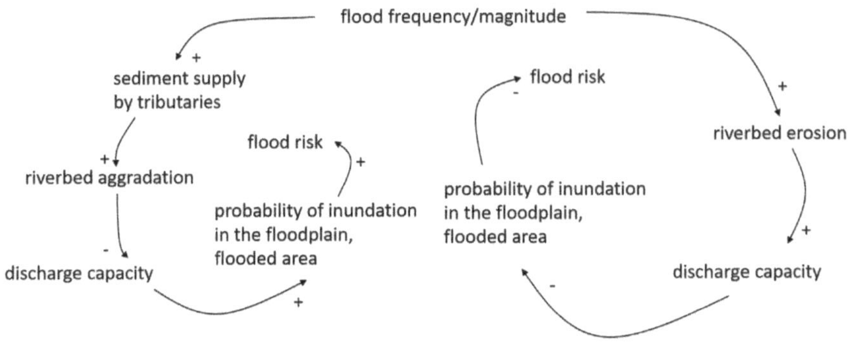

FIGURE 3.57 Feedback effects. Feedback effects between changes in riverbed erosion and aggradation and flood risks.

occur over decades. Thus, flood risk is expected to change slowly over decades. However, extreme flood events can suddenly change the river morphology either by riverbed erosion or sediment deposition. Thus, we must consider two time scales in this process, the long-term and slow change that is superimposed with short-term changes during and after an extreme flood event.

Feedback effects of landscape shaping flood events on hydrological change

In previous sections, we have seen that the riverbeds and the river morphology can be shaped by human and natural processes. Special attention was given to fluvial erosion or riverbed aggradation due to sedimentation. This alteration of the riverbed is relevant to the discharge capacity of the river channel and thus for the occurrence probability of flooding. Riverbed incision and aggradation work both in the long-term and the short-term timescale. The latter happens when an extreme flood event is shaping the river channel morphology either by riverbed erosion, lateral erosion, channel displacement, or by sedimentation. This happens during extreme flood events. Here, I describe one such example where an extreme flood event led to a shift in the system behavior of a river-lake system.

This case study describes the reconstruction of a historic flood event in Switzerland, namely the flood event of October 1868. This flood event was one of the most catastrophic floods in the early days of the new Federal State of Switzerland (Brönnimann et al., 2018). The flood event was a combined event of consequent floods within a short period. After an already rainy September 1868, two heavy precipitation events occurred within 8 days. The first rainfall episode on 27 and 28 September mainly affected the cantons of Ticino, Grisons and St. Gallen; the second phase from 1 to 5 October affected Ticino,

Valais and Uri. The rainfall event had a precipitation sum of 1118 mm on the San Bernardino Pass, accumulated over 8 days. The intense rainfall led to flooding in both the Northern and Southern Alps. The Alpine Rhine Valley and Magadino Plain were heavily flooded. Lake Maggiore reached its highest-ever measured level of 199.98 m above sea level on October 4, 1868 (Brönnimann et al., 2018). The resulting damages were catastrophic and widely spread over Switzerland. A total of 51 people died. Numerous bridges collapsed. The losses exceeded any hitherto known dimension. Thus, the flood event led to adaptation of the practice of governance, legislation, public funding, research, and expert consulting. The adaptation measures triggered by the event range from the faster implementation of previously discussed river engineering measures and a corresponding paradigm shift in the political process of flood management, to the debate on the role of deforestation for floods, which ultimately led to the Swiss Forestry Law. Stucki et al. (2018) reconstructed the weather conditions of this extreme rainfall event, the hydrology of the flood event, as well as the damage. Moreover, they documented the political decisions in the aftermath of the flood event and their long-term meaning for Swiss democracy. In regard to flood risk change, this documentation offers a look at the environmental changes that were triggered by the event itself and at the following human adaptation measures.

The analysis started with detailed weather construction by means of reanalyzing and dynamical downscaling (Stucki et al., 2018). The results of the reanalysis were used as an input to a hydrological model, in this case the PREVAH model (Viviroli, Zappa, Gurtz, & Weingartner, 2009) that was set up for the catchment of the Lago Maggiore lake (6368 km^2). The employed model was originally developed for short-term forecasts on an hourly basis in the Lake Maggiore catchment (Andres et al., 2016). The current complex water routing modules, with many anthropogenic alterations (hydropower), were exchanged with a simpler module that represents the catchment conditions before any relevant human interventions to the river system. With this simplified model, runoff for the time period 1961–2010 was simulated and compared with measured data. Only a small offset from the original model was observed. Finally, a lake-level/lake-outflow relationship was established, which was validated using the runoff measurement station below Lake Maggiore (Sesto Calende). This relationship is important to be able to correctly simulate lake level. It describes how much water discharges at a given lake level.

The reconstructed heavy precipitation events of September and October 1868 led to a sharp rise in the lake levels in two stages. Both precipitation reconstructions had the same amount of rainfall sum (278 and 279 mm) but differed in their timing. However, both reconstructions of the lake levels by hydrological simulation clearly led to an underestimation of the lake level. Therefore, several alternative scenarios and parameter settings were tested, including parameter and input data perturbations, a rather unlikely blockage of the outflow of the lake, or the consideration of the decimated forest area of 1868 (50% decrease) in comparison to today. The most likely reason for the large underestimation was an altered lake-level/lake-outflow relationship, which occurred from erosion of the lake outflow during the flood event (Ambrosetti et al., 1994).

According to the modeling study and the analysis of the historical documents, the mismatch between the modeled and observed lake levels highlighted an interesting feedback mechanism. The extremely high lake level of the Lago Maggiore lake during this heavy precipitation event resulted in a high outflow discharge at the lake outflow, the Ticino river. The extreme discharge led to a massive erosion of the riverbed at the lake outflow and thus to an increased lake outflow capacity. Therefore, the flood event directly led to the adaptation of the lake-level/lake-outflow system. This negative feedback set the new system conditions and consequently the current lake-level/lake-outflow relationship implemented in the current hydrological model did represent the former system state. Only the implementation of the estimated former lake-level/lake-outflow relationship into the model resulted in the correct reconstruction of the observed lake levels during the flood event of 1868 by the hydrological model. The observed maximum lake level was 199.98 m a.s.l., and the maximum simulated lake level with the current lake-level/lake-outflow relationship was 196.9 m a.s.l. Therefore, the negative feedback mechanism resulted in a reduction of the lake level by nearly 3 m. This means that the same amount of rainfall today cannot lead anymore to the same devastating floods as in the flood event of October 1868. This is an example where a flood event reduced the flood risk remarkably.

To quantify this feedback effect on the flood damage in the Magadino floodplain between Bellinzona and the Lago Maggiore lake, we reconstructed the digital terrain model by interpreting historical maps and reconstructed the historical digital terrain model as described in the previous sections of this chapter. The historical digital elevation model was implemented in a 2D hydrodynamic model. Again, we used the 2D version of the BASEMENT model (Vetsch et al., 2017). We simulated two different situations. The first simulation used the current digital terrain model that represents the current river morphology of the Ticino river with its massive structural flood protection measures and the discharge of the Ticino river near Bellinzona that was reconstructed by the hydrological model using weather reanalysis. In this simulation, the lake level of the Lago Maggiore lake was computed with the current lake-level/lake-outflow relationship. The second simulation used the reconstructed digital terrain model with the natural river morphology of the Ticino river (ca. 1864 AD). In this simulation, we used the same discharge of the Ticino river near Bellinzona that was reconstructed by the hydrological model using weather reanalysis. In contrast to the first simulation, we used the observed lake level in the second simulation. Thus, the difference between the flooding of both simulations showed (a) the effect of the altered lake-level/

lake-outflow relationship, and (b) the effect of the structural flood protection measurements on the Ticino river. To quantify the effects, we overlaid the flood simulations with the dataset of the buildings and counted the number of flooded buildings in both simulations. To disentangle the effect of settlement growth, we historized the building age by attributing the year of construction of each building. This data attribution is based on historical maps.

The modeling experiment showed that the high lake level observed in the past cannot be attained with today's lake-level/lake-outflow relationship. This means that the lake cannot be dammed to the same extent today as in the past. The modeling experiment also showed that the river corrections strongly reduced the flooded areas in the Magadino Plain between Bellinzona and Lago Maggiore lake. The example showed how different the effects of the same rainfall event can be on the ground and that changes in river morphology need to be considered in the interpretation of historical sources. Fig. 3.58 shows the flooded areas triggered by the 1868 rainfall event in the reconstructed historical and the current digital elevation model. Without structural flood protection measures, nearly the whole floodplain was flooded during the October 1868 rainfall event.

Since 1868, the settled area has changed, as well as the river course and the lake outflow. In October 1868, 436 buildings downstream of Bellinzona and along the Swiss shoreline of Lago Maggiore lake were located in the reconstructed flood area of the measured lake level and of the simulated runoff. When looking at the existing buildings in 2016, the same flooded area (historical digital terrain model) would affect 3934 buildings. This means that the settled area has increased by a factor of 9. If today's river course is taken into account (current digital terrain model, simulated lake level and simulated runoff in Bellinzona), 944 buildings were exposed to the flood event. This shows the effectiveness of the structural flood protection measures and lateral levees of the Ticino river and the increase of the lake outflow due to riverbed erosion during the event of 1868. Today, 2990 buildings benefit from these intended and unintended changes in the lake-river system.

This model experiment showed another potential use of reconstructing past flood events and of retromodel experiments. Reconstructions of past flood events can thereby represent an important foundation for the assessment of flood frequency and magnitude. Furthermore, they can provide a basis for the analysis of the spatial and temporal dynamics in the development of flood risks. The reconstruction of historical flood events can supplement records from the instrumental measurement period and can therefore produce important findings for the assessment of extreme events. In this case, only the reconstruction of the historical situation and the comparison with the current situation allowed us to analyze and quantify the feedback mechanism of flood events that alter the river morphology and consequently the course of all following flood events. The identified and analyzed feedback mechanism is shown in Fig. 3.59.

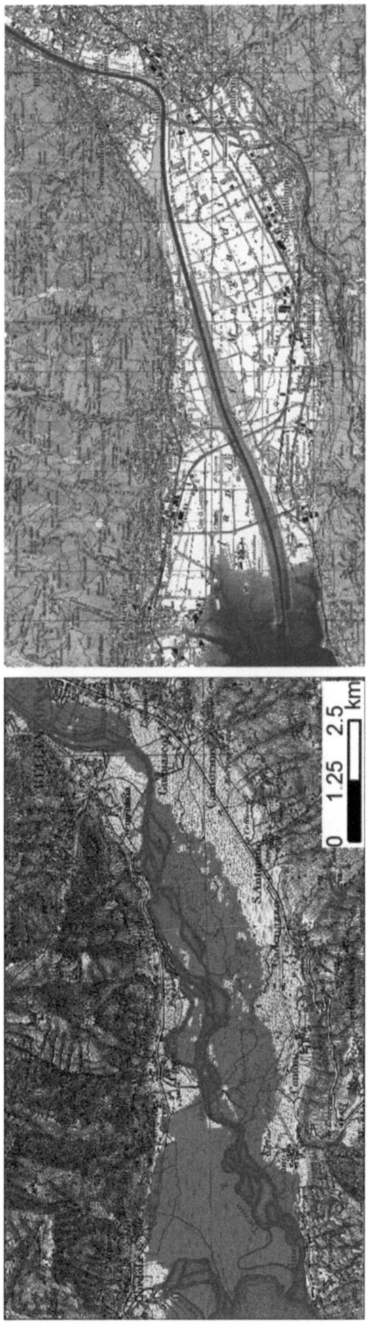

FIGURE 3.58 Flooded areas. Flooded areas triggered by the 1868 rainfall event simulated with the reconstructed historical (left) and the current digital elevation model (right). *Modified from Brönnimann, S., Rohr, C., Stucki, P., Summermatter, S., Bandhauer, M., Barton, Y., Fischer, A., Froidevaux, P., Germann, U., Grosjean, M., Hupfer, F., Ingold, K., Isotta, F., Keiler, M., Martius, O., Messmer, M., Mülchi, R., Panziera, L., Pfister, L., ... Zischg, A. (2018). 1868 – the flood that changed Switzerland: Causes, consequences and lessons for the future (Vol. 94).*

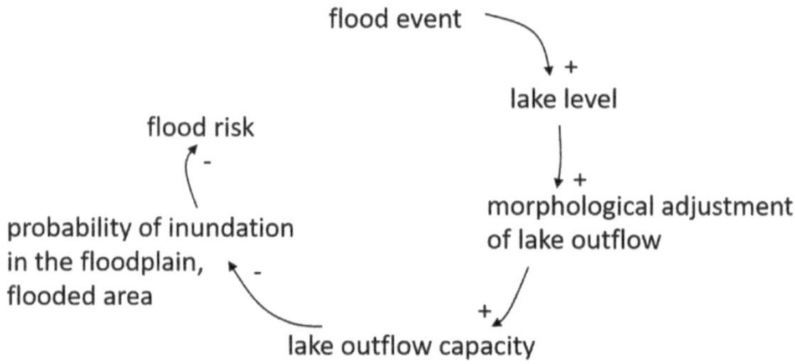

FIGURE 3.59 **Feedback effect.** Feedback mechanism between a landscape-shaping flood event and flood risk.

Feedback effects of flood events on the initiation of flood protection projects

In previous sections, we have seen that flood events can alter the river morphology by erosion or sedimentation of the river channel. Thus, a flood event can have a feedback effect on river morphology. This is a natural feedback mechanism. However, there exists another feedback mechanism between flood events and river morphology. In many cases, a flood event triggers societal adaptation to the flood risk that had been manifested by the event. The societal reaction can range from relocation of settlements, e.g., leaving the flood prone area, to the implementation of flood risk reduction measures. The latter often consist of structural flood protection measures such as heightening the existing lateral levees, or creating new levees, or shaping the geometry of the river channel with hard engineering work, e.g., by deepening the riverbed. These measures substantially modify river morphology with the aim of increasing river discharge capacity and, consequently, decreasing the probability of flooding outside of the river channel. Hence, this is a similar feedback loop to a flood event that shapes the river morphology by riverbed incision (Fig. 3.60). However, in contrast to the previous example, it is a societal feedback mechanism.

Suter et al. (2016) analyzed how often a flood event triggered the initialization of flood protection projects. In this study, we evaluated the role of Swiss communities in the process of planning and implementing flood control projects. We analyzed 71 Swiss flood control projects by evaluating technical reports, online surveys, and interviews with communal project managers. 53% of all recently implemented flood control measures in these communities were planned in response to a flood event. These projects therefore can be considered as a societal adaptation to a flood event. 31% of the analyzed projects were started after finishing the hazard map. These projects can be defined as prevention projects. The existence of the hazard information guided

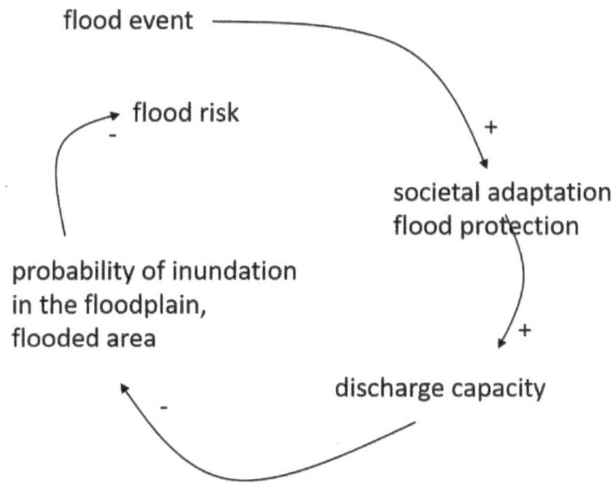

FIGURE 3.60 Feedback effect. Feedback mechanism between the flood event and societal adaptation by implementing structural flood protection measures.

the communal decision-makers to initiate the planning of a flood protection project. The rest of the projects were initiated for other reasons, such as river revitalization, the construction of a relevant industrial facility, or for substituting old flood protection measures that were in decay. These findings highlight that there is still a long way forward to prioritize flood prevention against reaction to flood events.

The levee effect

This feedback effect of flood events on societal adaptation is often also termed as the "levee effect" (Tobin & Montz, 1994; White, 1945). Society is learning from previous flood events and adapting its environment according to new knowledge. Sometimes, this results in heightening already existing levees or on avoiding the spread of settlements in previously flooded areas for a certain time period (Collenteur et al., 2015). As a feedback process from the human intervention to the natural system back to the human sphere, levees might increase societal vulnerability by instilling a sense of safety (D'Angelo et al., 2020). The levee heightening decreases the flood risk as shown above due to the increased discharge capacity. But the lowered flood risk will increase the subjective feeling of safety and, therefore, the now better protected floodplains will be used for settlement, infrastructure, and other economic activities. This in turn will increase the residual risk. This can also be termed as a boomerang effect or rebound effect in flood risk management (Zischg, Hofer, Mosimann, et al., 2018). The heightening of lateral levees in upstream river reaches has another unintended consequence and feedback process: the loss of floodplain retention effects upstream by levee heightening may worsen the peak discharge of downstream river reaches and thus may influence flood risk downstream (Di Baldassarre et al., 2009). Hence, this effect must not only be considered at the scale of a single river reach but also at the scale of the river system.

The societal memory effect on house construction activities in flood-prone areas

The previously mentioned levee effect describes the mechanism where levees are heightened after a flood event that led to inundation in the floodplain to prevent flooding in future. In contrast to the levee effect, not all flood events trigger human adaptation in the form of planning and implementing structural flood risk protection measures. Sometimes, a flood event remains in the collective societal memory of a community and the previously flooded areas are avoided for being used for immobile structures, e.g., the construction of houses or infrastructure. A case study in Bangladesh observed the differing behavior of settlement development along river reaches with or without levees (Ferdous et al., 2020). The higher the level of flood protection, the higher the increase in population density over the past decades, as well as the number of assets exposed to flooding. The study also showed that flood mortality rates associated with the 2017 flooding in Bangladesh were lower in the areas with lower protection levels. The houses located in the river reach with lower safety levels had a higher distance from the river than the houses located in floodplains with higher safety levels. Richert et al. (2019) found that structural flood protection measures reduce the probability of having or taking individual adaptation measures. This observation is also confirmed by an analysis of satellite nighttime light data (Mård et al., 2018) at a global level. Societies with low protection levels tend to resettle further away from the river after damaging flood events and societies with high protection levels show no significant changes in human proximity to rivers. Keeping settlements distant from rivers is mostly governed by experiences made by flood events and is thus an effect or spatial imprint of collective societal memory. However, this memory is declining and the rate of decline depends on how the memory is cultivated and kept alive within society (Song et al., 2021). Fanta et al. (2019) observed in a high number of settlement growth analyses that for a period of one generation after each flood, new settlements appeared preferentially in safer places. The collective memory effect decreased after one generation and the settlement growth was more directed toward the rivers.

From a complexity perspective, these observations are signs of the effects of time delays on flood risk change and of path dependency. This means that a flood event can lead to a period of about 30 years where flood risk is not increasing substantially because the formerly flooded zones are kept free from new exposure. This is a delay or time lag in the flood risk evolution pathway; the rate of buildings being constructed in hazard zones is declined for a while and construction sites are pushed more toward the outer borders of the previously flooded area. After this time period, the effect of societal learning on land use regulation diminishes, and the resulting situation of the past period will set the stage for the following period. If flood memory is lacking, the community perspective may focus to valorize the path-dependent situation where highly valuable arable land or potential construction sites are free from previous constructions and "available" for being used. Thus, the new situation now builds on the development path of the old situation. Viglione et al. (2014) conducted a stylized model experiment to test this hypothesis. They found that a low collective memory in combination with a high trust in flood safety measures can lead a high risk-taking attitude. Following flood events can lead to disasters and limit community development because of high damages caused by flooding.

To analyze the effect of the collective flood memory on the construction of buildings in flood prone areas, we overlaid the buildings dataset with attributed construction dates with a geodata set of past flood events. The buildings in the main floodplains of the Aare river basin upstream of Bern were systematically classified by adding the minimum year on which each building appeared for the first time in topographic maps. The basic dataset was taken from the Swiss digital landscape model (Swisstopo, 2021). The historic maps came also from the Federal Office for Topography (Swisstopo, 2020). This step resulted in a building footprint polygons dataset with the approximated year of construction. This dataset was overlaid with the event documentation dataset of the Canton of Bern (Kanton Bern, 2021). This dataset identifies and delimits the flooded areas of all known past flood events in the region. Each flood event is represented by a number of polygons that show the extension of the flooded areas and are attributed with the date of the event. The event documentation data contains flood events starting from the year 1000 AD. However, we focused on flood events after 1940 to avoid the information availability bias in the dataset.

Fig. 3.61 shows how long it takes for a building to be constructed at a specific location after being flooded during the most recent flood event. The analysis showed that many buildings constructed in flood prone areas were constructed within one or two decades after a flood event, but the median period between the most recent flood event and the construction year of the house is about 28 years and the average is about 3 years. This confirms that the collective memory holds on for about one generation in the study area. The average memory period decreases with time (Fig. 3.62). The average memory is about 100–200 years until the 1990s and has been decreasing since the end of the 20th century.

However, this observation might show a combined effect of both: the sharp increase in construction of buildings since the 1960s and the bias of the event documentation dataset toward the mapping of more recent flood events than of older events due to information availability. This information bias is also resulting from the fact that flood events in floodplains previously not used for

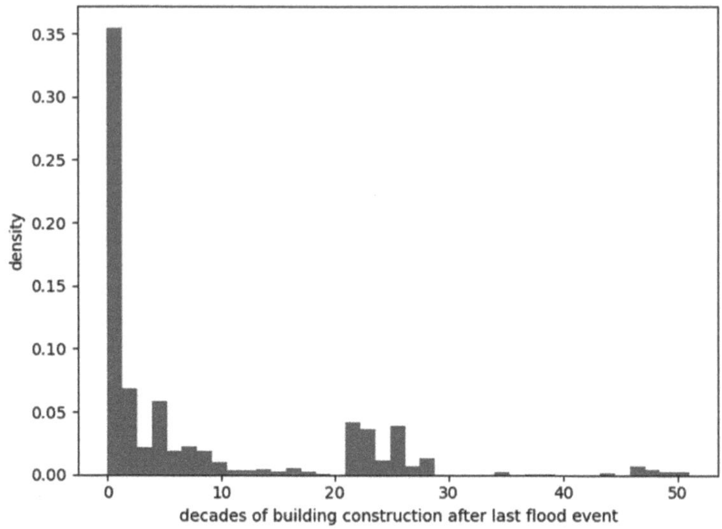

FIGURE 3.61 Societal memory. Frequency of the duration in decades between a flood event and the construction of a building at the same place.

FIGURE 3.62 Societal memory. Average period between the construction of a house and the most recent flood event at the location.

settlement had not been noticed and therefore not mapped. Thus, the more the settlements grow toward the floodplains, the more attention is given to map the flood events. While during the period 1940–2009, 100–230 buildings were flooded per decade, more than 500 buildings were flooded per decade after 2010.

Time lags in the implementation of flood risk management measures

Flood hazard maps and their implementation in spatial planning are key factors in flood risk management. To avoid new susceptible constructions in flood prone areas, the consideration of flood hazard maps in land use planning is widely seen as an effective instrument for managing flood risks. We investigated the impact of a federal framework law on building construction in Switzerland by spatiotemporal analyses of new residential buildings within and outside of flood prone areas. We overlaid object-specific building data on year of construction with large-scale flood hazard maps, which are both nationwide available. The elaboration of these datasets is described in the previous chapters. We quantitatively assessed and compared the regional differences in years of completion of the flood hazard maps and in years of their implementation into the legally binding land use plan. We quantified the effects of land use regulation by the share and absolute numbers of newly constructed buildings within flood-prone areas (i.e., the mapped hazard zones) and by the spatial patterns of exposure.

In terms of annual newly constructed buildings over whole Switzerland, the period 1971–80 shows the highest amounts in all analyzed buildings as well as in all buildings assessed in a flood hazard map Fig. 3.63. The exposure ratios (i.e., the number of newly constructed buildings within hazardous areas compared to the total number of newly constructed buildings within the perimeter of a flood hazard map) during this period of construction are lower than in any other period. It is noticed that the amount of annual newly constructed buildings and the share of exposed buildings show a negative correlation. When interpreting the data of consecutive time periods as time series, one detects a decrease in exposure ratios for low and medium hazard levels (and consequently summarized over all hazard levels) from 1919 to 1970, whereas from 1981 to 2012, the same exposure ratios increase. Interestingly, the share of the newly constructed buildings within hazardous areas compared to the total number of newly constructed buildings is increasing since the introduction of the hazard maps in the communal land use regulation. However, it cannot be assessed how this share would have been if the hazard zone mapping had not been introduced.

From the first national law on flood hazard mapping in 1991, it took almost 25 years to a mapping coverage ratio of at least 50% throughout Switzerland. In 2016, 12 Cantons had a coverage of more than 80% of the building stock,

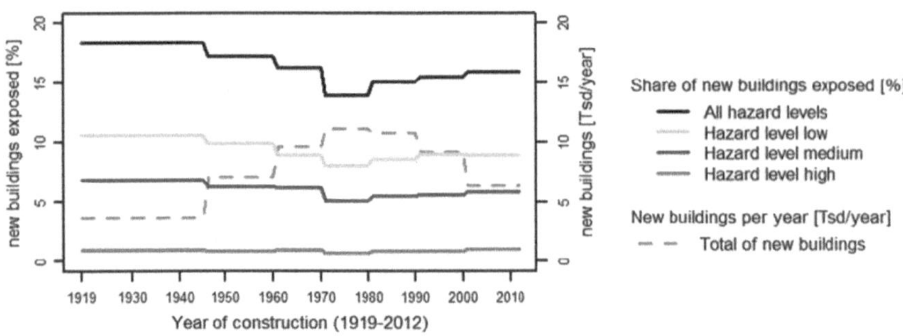

FIGURE 3.63 Flood exposure. Newly constructed buildings per construction period. The share of new buildings exposed is the ratio of the number of newly constructed building that are potentially exposed to floods (at respective hazard level) to the number of total newly constructed buildings for which a flood hazard map exists. The new buildings per year are calculated by dividing the total number of buildings newly constructed within a particular time period by the length of the respective time period, figures are presented in thousands of buildings per year. *From Röthlisberger, V., Zischg, A., & Keiler, M. (2016). Spatiotemporal aspects of flood exposure in Switzerland. In* E3S web of conferences *(Vol. 7). EDP Sciences. https://doi.org/10.1051/e3sconf/20160708008.*

five below 50%, and in the remaining nine Cantons, the coverage was between 54% and 78% (Bruchez, 2017). Fig. 3.64 shows the number of municipalities that are implementing the hazard zone maps in their land use regulation plans. This underlines the long time delay between the introduction of a new legislative framework or planning instrument and its implementation into practice. Similar time delays must probably be considered in future for climate change adaptation measures.

However, in most Cantons, the mapping itself (i.e., the elaboration of the hazard map) does not have an impact on the location of new residential buildings with respect to flood exposure. The construction rate of new buildings within and outside after the elaboration of hazard maps does not differ significantly. The impact on the share and absolute figures of newly constructed buildings with residential purpose is only evident after the legal implementation of the hazard zone maps in local buildings codes, yet only with respect to the highest hazard level and not in all Cantons. Only the implementation of the communal land use plan makes the hazard maps legally binding for the authorities and homeowners. The pure existence of the hazard information is not enough to guide land development. These results highlight the importance of temporal and institutional factors on the effect of spatial planning measures on flood risk management.

Furthermore, we identified differences in spatial patterns of exposure. While the share of newly constructed buildings within flood-prone areas is significantly clustered in mountainous regions, the absolute number of exposed new buildings with residential purpose shows significant clusters in the

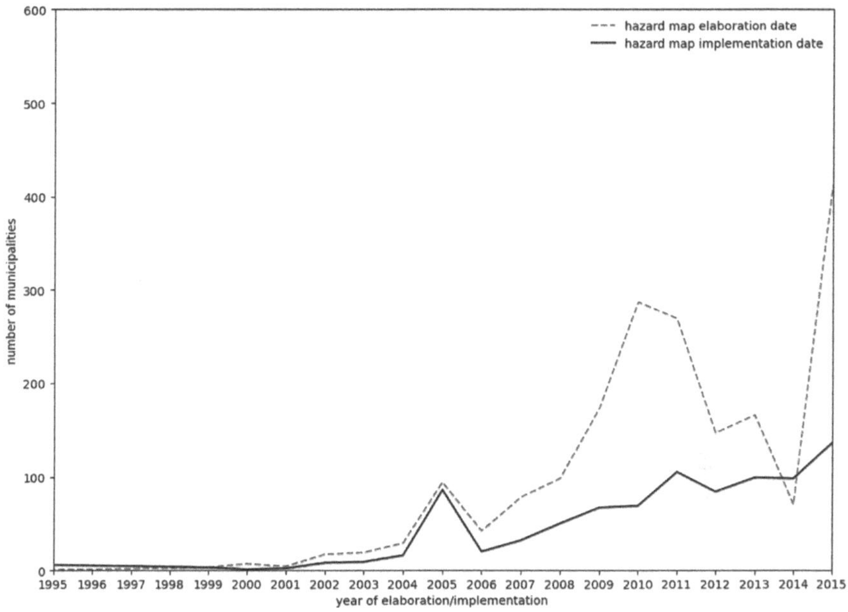

FIGURE 3.64 Implementation of hazard maps. Number of municipalities elaborating hazard maps per year and their implementation in communal land use regulation plans. *Data from Bruchez, L. (2017). Effet des cartes de danger de crues sur la construction de nouveaux bâtiments d'habitation dans les communes de Suisse.*

metropolitan areas of the Swiss plateau. The absolute number and the share respectively of newly exposed buildings are two different types of exposure units and therefore support two different risk management strategies. While high absolute numbers indicate priority areas for cost-efficient strategies, high shares can suggest areas of interest for strategies focused on the most vulnerable regions.

Other aspects that are contributing to the complexity of flood risk management

In addition to the above quantified aspects that are making flood risk management a complex issue, there are several others that are waiting to be analyzed in detail. Thus, I am mentioning a few examples here. Merely, I present some hypotheses that have not been tested yet. An example of another feedback process is an unintended consequence of the clear water legislation on flood risk. Until the late 1960s, many rivers of Switzerland were heavily polluted with wastewater. Swimming in the rivers was partially forbidden because of health risks. After the implementation of water purification plants and sewage treatment plants, the water quality improved. Moreover, since the

1990s, a new paradigm has been manifested in river engineering. Many river revitalization projects have been implemented in the past 30 years. Some urban governments opened the newly emerging and attractive waterfronts of rivers and lakes and guided urban development to consider the water landscapes. These developments led to a shift in the subjective awareness of planners and citizens to the role of water courses for well-being. The first examples raised proved the high quality of life of recreational places along the rivers and further increased the requests for such recreational places in other cities. Especially in urban areas, where the coevolving deindustrialization left large spaces of formerly industrial areas along the rivers that became new opportunities for urban transformation and development. This newly available space was often used for residential construction and the creation of leisure areas. All these trends have led to a highly demanded space for living near water courses. Consequently, the housing prices and the quality standards (i.e., costs per square meter) increased in the direct neighborhood of water fronts. This bias of relatively higher insurance values in the direct neighborhood of rivers is now slowly becoming visible in the data. This is an example of a possible unintended consequence of the water quality legislation for flood risk management and thus a positive feedback between two governmental tasks. However, it is still open if and how much this influences flood risk.

A clearly negative feedback is the improvement of early warning systems. Since the first years of the 21st century, the weather forecasts have been markedly improved. The improved numerical weather forecasts also enabled hydrological forecasts. These in turn enable a dedicated and location-specific warning against flood events. On the basis of the warnings, the local intervention forces can prepare their actions (e.g., evacuation) and house-owners can install property-level flood protection measures. This has led to a reduction in fatalities during flood events. The effectiveness of these new possibilities in terms of saved lives is still under investigation but is becoming more and more evident.

In mountainous areas, the altitudinal vegetation belts will shift upward (Zischg et al., 2021a). Thus, there is a change in the tree composition to be expected. During this transformation, some tree species will die out and make place for others. During the transformation period, the dying of trees that are not adapted to the new climate will contribute to the recruitment of large wood in rivers. Thus, the temporary increased tree mortality will lead to a temporarily increased supply of dead wood and large wood that can lead to an increase of flood risk due to the occlusion of bridges and related backwater effects or surge waves. More feedback of vegetation change to flood risks is expected with the progress of climatic changes. For example, the spread of invasive species may alter erosion and thus sediment balance in rivers (Gilbert & Wilcox, 2021).

An emergent phenomenon that often arises after huge flood events is the increase of awareness of flood risk by the population and the government. This raised attention is often guiding and triggering the planning and implementation of flood protection measures, of property-level adaptation measures, the optimization of early warning systems, the optimization of the organization of catastrophe management, or sometimes even of changes in legislation and financing. This shows that flood risk will be dynamic in space and time also in the future, and new feedback effects and phenomena will emerge that will contribute to the complexity of flood risk management.

Acknowledgments

I thank Guido Felder for the hydrological modeling with PREVAH.

References

Adams, R., Western, A. W., & Seed, A. W. (2012). An analysis of the impact of spatial variability in rainfall on runoff and sediment predictions from a distributed model. *Hydrological Processes, 26*(21), 3263–3280. https://doi.org/10.1002/hyp.8435

Ambrosetti, W., Barbanti, L., de Bernardi, R., Libera, V., & Rolla, A. (1994). *La piena del Lago Maggiore dell'autunno 1993: un evento di portata secolare, Documenta dell'Istituto Italiano di Idrobiologia, 45, Istituto Italiano di Idrobiologia.*

Andres, N., Lieberherr, G., Sideris, I. V., Jordan, F., & Zappa, M. (2016). From calibration to real-time operations: An assessment of three precipitation benchmarks for a Swiss river system. *Meteorological Applications, 23*(3), 448–461. https://doi.org/10.1002/met.1569

Aspinall, W., & Woo, G. (2019). Counterfactual analysis of runaway volcanic explosions. *Frontiers in Earth Science, 7*. https://doi.org/10.3389/feart.2019.00222

Attems, M., Thaler, T., Genovese, E., & Fuchs, S. (2019). Implementation of property-level flood risk adaptation (PLFRA) measures: Choices and decisions. *WIREs Water, 7*(1). https://doi.org/10.1002/wat2.1404

Bates, P. D., & De Roo, A. P. J. (2000). A simple raster based model for flood inundation simulation. *Journal of Hydrology, 236*(1–2), 54–77. https://doi.org/10.1016/S0022-1694(00)00278-X

Bermúdez, M., Neal, J. C., Bates, P. D., Coxon, G., Freer, J. E., Cea, L., & Puertas, J. (2017). Quantifying local rainfall dynamics and uncertain boundary conditions into a nested regional-local flood modeling system. *Water Resources Research, 53*(4), 2770–2785. https://doi.org/10.1002/2016WR019903

BFS. (2021a). *Bilanz der ständigen Wohnbevölkerung, 1861–2019*. Bundesamt für Statistik. https://www.bfs.admin.ch/bfs/de/home/statistiken/bevoelkerung/stand-entwicklung/bevoelkerung.assetdetail.13707405.html.

BFS. (2021b). *Bruttoinlandprodukt, lange serie*. Bundesamt für Statistik. https://www.bfs.admin.ch/bfs/de/home/statistiken/volkswirtschaft/volkswirtschaftliche-gesamtrechnung/bruttoinlandprodukt.assetdetail.14347493.html.

BFS. (2021c). *Eidg. Gebäude- und Wohnungsregister GWR*. Bundesamt für Statistik.

Bhuyian, M., Kalyanapu, A., & Hossain, F. (2017). Evaluating conveyance-based DEM correction technique on NED and SRTM DEMs for flood impact assessment of the 2010 Cumberland river flood. *Geosciences, 7*(4), 132. https://doi.org/10.3390/geosciences7040132

Biggs, R., Vos, A., de, Preiser, R., Clements, H., Maciejewski, K., & Schlüter, M. (2021). *The Routledge handbook of research methods for social-ecological systems*. Routledge International Handbooks.

Blair, P., & Buytaert, W. (2016). Socio-hydrological modelling: A review asking "why, what and how?". *Hydrology and Earth System Sciences, 20*(1), 443–478. https://doi.org/10.5194/hess-20-443-2016

Bodmer, S. (n.d.). Plan und Grundriss von dem so genannten Cander Wasser.

Bögli, T. (2018). *Der prägende Einfluss der Kander auf die Region Thun.*

Bouwer, L. M. (2013). Projections of future extreme weather losses under changes in climate and exposure. *Risk Analysis, 33*(5), 915–930. https://doi.org/10.1111/j.1539-6924.2012.01880.x

Bressan, F., Mantilla, R., Schilling, K. E., Palmer, J. A., & Weber, L. (2020). Hydrologic-hydraulic modeling of sediment transport along the main stem of a watershed: Role of tributaries and channel geometry. *Hydrological Sciences Journal, 65*(2), 183–199. https://doi.org/10.1080/02626667.2019.1687897

Brestolani, F., Solari, L., Rinaldi, M., & Lollino, G. (2015). On the morphological impacts of gravel mining: The case of the Orco river. In *Engineering geology for society and territory - volume 3: River basins, reservoir sedimentation and water resources* (pp. 319–322). Springer International Publishing. https://doi.org/10.1007/978-3-319-09054-2_66

Broderick, C., Murphy, C., Wilby, R. L., Matthews, T., Prudhomme, C., & Adamson, M. (2019). *Using a scenario-neutral framework to avoid potential maladaptation to future flood risk.* Water Resources Research. https://doi.org/10.1029/2018WR023623

Brönnimann, S., Rohr, C., Stucki, P., Summermatter, S., Bandhauer, M., Barton, Y., Fischer, A., Froidevaux, P., Germann, U., Grosjean, M., Hupfer, F., Ingold, K., Isotta, F., Keiler, M., Martius, O., Messmer, M., Mülchi, R., Panziera, L., Pfister, L., … Zischg, A. (2018). 1868 – the flood that changed Switzerland: Causes, consequences and lessons for the future (Vol. 94).

Bruchez, L. (2017). Effet des cartes de danger de crues sur la construction de nouveaux bâtiments d'habitation dans les communes de Suisse.

BWW, B., & BUWAL. (1997). *Berücksichtigung der Hochwassergefahren bei raumwirksamen Tätigkeiten.*

Calvin, K., & Bond-Lamberty, B. (2018). Integrated human-earth system modeling—state of the science and future directions. *Environmental Research Letters, 13*(6). https://doi.org/10.1088/1748-9326/aac642

Castellarin, A., Di Baldassarre, G., Bates, P. D., & Brath, A. (2009). Optimal cross-sectional spacing in Preissmann scheme 1D hydrodynamic models. *Journal of Hydraulic Engineering, 135*(2), 96–105. https://doi.org/10.1061/(ASCE)0733-9429(2009)135:2(96)

Castelletti, A., Galelli, S., Ratto, M., Soncini-Sessa, R., & Young, P. C. (2012). A general framework for dynamic emulation modelling in environmental problems. *Environmental Modelling and Software, 34*, 5–18. https://doi.org/10.1016/j.envsoft.2012.01.002

Castelletti, A., Galelli, S., Restelli, M., & Soncini-Sessa, R. (2012). Data-driven dynamic emulation modelling for the optimal management of environmental systems. *Environmental Modelling and Software, 34*, 30–43. https://doi.org/10.1016/j.envsoft.2011.09.003

CH2018 Project team. (2018). *CH2018—climate scenarios for Switzerland.* MeteoSwiss.

Ciullo, A., De Bruijn, K. M., Kwakkel, J. H., & Klijn, F. (2019). Systemic flood risk management: The challenge of accounting for hydraulic interactions. *Water, 11*(12). https://doi.org/10.3390/w11122530, 2530.

Clark, M. P., Fan, Y., Lawrence, D. M., Adam, J. C., Bolster, D., Gochis, D. J., Hooper, R. P., Kumar, M., Leung, L. R., Mackay, D. S., Maxwell, R. M., Shen, C., Swenson, S. C., & Zeng, X. (2015). Improving the representation of hydrologic processes in earth system models. *Water Resources Research, 51*(8), 5929–5956. https://doi.org/10.1002/2015WR017096

Collenteur, R. A., de Moel, H., Jongman, B., & Di Baldassarre, G. (2015). The failed-levee effect: Do societies learn from flood disasters? *Natural Hazards, 76*(1), 373–388. https://doi.org/10.1007/s11069-014-1496-6

Culley, S., Maier, H. R., Westra, S., & Bennett, B. (2021). Identifying critical climate conditions for use in scenario-neutral climate impact assessments. *Environmental Modelling & Software, 136*, 104948. https://doi.org/10.1016/j.envsoft.2020.104948

D'Angelo, C., Fiori, A., & Volpi, E. (2020). Structural, dynamic and anthropic conditions that trigger the emergence of the levee effect: Insight from a simplified risk-based framework. *Hydrological Sciences Journal, 65*(6), 914–927. https://doi.org/10.1080/02626667.2020.1729985

deBruijn, K. M., Lips, N., Gersonius, B., & Middelkoop, H. (2016). The storyline approach: A new way to analyse and improve flood event management. *Natural Hazards, 81*(1), 99–121. https://doi.org/10.1007/s11069-015-2074-2

Deltares. (2012). *SOBEK suite.*

Di Baldassarre, G., Castellarin, A., & Brath, A. (2009). Analyse des effets de la surélévation des levées sur la propagation des crues: Exemple du Fleuve Pô, Italie. *Hydrological Sciences Journal, 54*(6), 1007–1017. https://doi.org/10.1623/hysj.54.6.1007

Di Baldassarre, G., Kooy, M., Kemerink, J. S., & Brandimarte, L. (2013). Towards understanding the dynamic behaviour of floodplains as human-water systems. *Hydrology and Earth System Sciences, 17*(8), 3235–3244. https://doi.org/10.5194/hess-17-3235-2013

Donges, J. F., Heitzig, J., Barfuss, W., Wiedermann, M., Kassel, J. A., Kittel, T., Kolb, J. J., Kolster, T., Müller-Hansen, F., Otto, I. M., Zimmerer, K. B., & Lucht, W. (2020). Earth system modeling with endogenous and dynamic human societies: The copan:CORE open World-Earth modeling framework. *Earth System Dynamics, 11*(2), 395–413. https://doi.org/10.5194/esd-11-395-2020

Donges, J. F., Winkelmann, R., Lucht, W., Cornell, S. E., Dyke, J. G., Rockström, J., Heitzig, J., & Schellnhuber, H. J. (2017). Closing the loop: Reconnecting human dynamics to Earth System science. *Anthropocene Review, 4*(2), 151–157. https://doi.org/10.1177/2053019617725537

Dottori, F., Szewczyk, W., Ciscar, J. C., Zhao, F., Alfieri, L., Hirabayashi, Y., Bianchi, A., Mongelli, I., Frieler, K., Betts, R. A., & Feyen, L. (2018). Increased human and economic losses from river flooding with anthropogenic warming. *Nature Climate Change, 8*(9), 781–786. https://doi.org/10.1038/s41558-018-0257-z

Elmer, F., Hoymann, J., Düthmann, D., Vorogushyn, S., & Kreibich, H. (2012). Drivers of flood risk change in residential areas. *Natural Hazards and Earth System Science, 12*(5), 1641–1657. https://doi.org/10.5194/nhess-12-1641-2012

Emmanuel, I., Andrieu, H., Leblois, E., Janey, N., & Payrastre, O. (2015). Influence of rainfall spatial variability on rainfall-runoff modelling: Benefit of a simulation approach? *Journal of Hydrology, 531*, 337–348. https://doi.org/10.1016/j.jhydrol.2015.04.058

Emmanuel, I., Payrastre, O., Andrieu, H., Zuber, F., Lang, M., Klijn, F., & Samuels, P. (2016). Influence of the spatial variability of rainfall on hydrograph modelling at catchment outlet: A case study in the Cevennes region. *France. E3S Web of Conferences, 7*. https://doi.org/10.1051/e3sconf/20160718004, 18004.

Fanta, V., Šálek, M., & Sklenicka, P. (2019). How long do floods throughout the millennium remain in the collective memory? *Nature Communications, 10*(1). https://doi.org/10.1038/s41467-019-09102-3

Felder, G., Gómez-Navarro, J. J., Zischg, A. P., Raible, C. C., Röthlisberger, V., Bozhinova, D., Martius, O., & Weingartner, R. (2018). From global circulation to local flood loss: Coupling models across the scales. *Science of the Total Environment, 635*, 1225–1239. https://doi.org/10.1016/j.scitotenv.2018.04.170

Felder, G., & Weingartner, R. (2016). An approach for the determination of precipitation input for worst-case flood modelling. *Hydrological Sciences Journal, 61*(14), 2600–2609. https://doi.org/10.1080/02626667.2016.1151980

Felder, G., & Weingartner, R. (2017). Assessment of deterministic PMF modelling approaches. *Hydrological Sciences Journal, 62*(10), 1591–1602. https://doi.org/10.1080/02626667.2017.1319065

Felder, G., Zischg, A., & Weingartner, R. (2017). The effect of coupling hydrologic and hydrodynamic models on probable maximum flood estimation. *Journal of Hydrology, 550*, 157–165. https://doi.org/10.1016/j.jhydrol.2017.04.052

Ferdous, M. R., Di Baldassarre, G., Brandimarte, L., & Wesselink, A. (2020). The interplay between structural flood protection, population density, and flood mortality along the Jamuna River, Bangladesh. *Regional Environmental Change, 20*(1). https://doi.org/10.1007/s10113-020-01600-1

Flato, G. M. (2011). Earth system models: An overview. *Wiley Interdisciplinary Reviews: Climate Change, 2*(6), 783–800. https://doi.org/10.1002/wcc.148

Fleischmann, A., Paiva, R., & Collischonn, W. (2019). Can regional to continental river hydrodynamic models be locally relevant? A cross-scale comparison. *Journal of Hydrology X, 3*. https://doi.org/10.1016/j.hydroa.2019.100027, 100027.

FOEN. (2018a). *Cross-sections and river bathymetrie for the main river network in Switzerland, Federal Office for the Environment*. Bundesamt für Umwelt.

FOEN. (2018b). *River discharge measurements in Switzerland*. Bundesamt für Umwelt.

FOEN. (2021). *Hydrological data and extreme value statistics of the Swiss river network*. Bundesamt für Umwelt.

Forrester, J. W. (1969). *Urban dynamics*. Massachusetts Institute of Technology Press.

Fuchs, S., Birkmann, J., & Glade, T. (2012). Vulnerability assessment in natural hazard and risk analysis: Current approaches and future challenges. *Natural Hazards, 64*(3), 1969–1975. https://doi.org/10.1007/s11069-012-0352-9

Fuchs, S., Keiler, M., & Zischg, A. (2015). A spatiotemporal multi-hazard exposure assessment based on property data. *Natural Hazards and Earth System Sciences, 15*(9), 2127–2142. https://doi.org/10.5194/nhess-15-2127-2015

Gilbert, J. T., & Wilcox, A. C. (2021). An ecogeomorphic framework coupling sediment modeling with invasive riparian vegetation dynamics. *Journal of Geophysical Research: Earth Surface, 126*(6). https://doi.org/10.1029/2021JF006071

Giorgi, F., & Avissar, R. (1997). Representation of heterogeneity effects in earth system modeling: Experience from land surface modeling. *Reviews of Geophysics, 35*(4), 413–437. https://doi.org/10.1029/97rg01754

Grames, J., Prskawetz, A., Grass, D., & Blöschl, G. (2015). Modelling the interaction between flooding events and economic growth. In *IAHS-AISH proceedings and reports* (Vol. 369, pp. 3–6). Copernicus GmbH. https://doi.org/10.5194/piahs-369-3-2015

Grimaldi, S., Li, Y., Walker, J. P., & Pauwels, V. R. N. (2018). Effective representation of river geometry in hydraulic flood forecast models. *Water Resources Research, 54*(2), 1031–1057. https://doi.org/10.1002/2017WR021765

Grimaldi, S., Schumann, G. J. P., Shokri, A., Walker, J. P., & Pauwels, V. R. N. (2019). Challenges, opportunities, and pitfalls for global coupled hydrologic-hydraulic modeling of floods. *Water Resources Research, 55*(7), 5277–5300. https://doi.org/10.1029/2018WR024289

Grosjean, G. (1962). *Die Ableitung der Kander in den Thunersee vor 250 Jahren*.

Guerrin, J., Bouleau, G., & Grelot, F. (2014). "Functional fit" versus "politics of scale" in the governance of floodplain retention capacity. *Journal of Hydrology, 519*, 2405–2414. https://doi.org/10.1016/j.jhydrol.2014.08.024

Guo, D., Westra, S., & Maier, H. R. (2017). Use of a scenario-neutral approach to identify the key hydro-meteorological attributes that impact runoff from a natural catchment. *Journal of Hydrology, 554*, 317–330. https://doi.org/10.1016/j.jhydrol.2017.09.021

Guo, D., Westra, S., & Maier, H. R. (2018). An inverse approach to perturb historical rainfall data for scenario-neutral climate impact studies. *Journal of Hydrology*, 877–890. https://doi.org/10.1016/j.jhydrol.2016.03.025

Gusyev, M., Gädeke, A., Cullmann, J., Magome, J., Sugiura, A., Sawano, H., & Takeuchi, K. (2016). Connecting global- and local-scale flood risk assessment: A case study of the Rhine river basin flood hazard. *Journal of Flood Risk Management, 9*(4), 343–354. https://doi.org/10.1111/jfr3.12243. Blackwell Publishing Inc.

Hallegatte, S., Green, C., Nicholls, R. J., & Corfee-Morlot, J. (2013). Future flood losses in major coastal cities. *Nature Climate Change, 3*(9), 802–806. https://doi.org/10.1038/nclimate1979

Harris, C., Davies, M. C. R., & Etzelmüller, B. (2001). The assessment of potential geotechnical hazards associated with mountain permafrost in a warming global climate. *Permafrost and Periglacial Processes, 12*(1), 145–156. https://doi.org/10.1002/ppp.376

Hazeleger, W., Van Den Hurk, B. J. J. M., Min, E., Van Oldenborgh, G. J., Petersen, A. C., Stainforth, D. A., Vasileiadou, E., & Smith, L. A. (2015). Tales of future weather. *Nature Climate Change, 5*(2), 107–113. https://doi.org/10.1038/nclimate2450

Hirabayashi, Y., Mahendran, R., Koirala, S., Konoshima, L., Yamazaki, D., Watanabe, S., Kim, H., & Kanae, S. (2013). Global flood risk under climate change. *Nature Climate Change, 3*(9), 816–821. https://doi.org/10.1038/nclimate1911

Hoch, J. M., Eilander, D., Ikeuchi, H., Baart, F., & Winsemius, H. C. (2019). Evaluating the impact of model complexity on flood wave propagation and inundation extent with a hydrologic-hydrodynamic model coupling framework. *Natural Hazards and Earth System Sciences, 19*(8), 1723–1735. https://doi.org/10.5194/nhess-19-1723-2019

Hoch, J. M., Neal, J. C., Baart, F., Van Beek, R., Winsemius, H. C., Bates, P. D., & Bierkens, M. F. P. (2017). GLOFRIM v1.0-A globally applicable computational framework for integrated hydrological-hydrodynamic modelling. *Geoscientific Model Development, 10*(10), 3913–3929. https://doi.org/10.5194/gmd-10-3913-2017

Huber, B., Gubelmann, P., Zischg, A., Augustin, S., & Frehner, M. (2019). Modellierung der vegetationshöhenstufen und der areale von buche und tanne für die Schweiz. *Schweizerische Zeitschrift fur Forstwesen, 170*(6), 326–337. https://doi.org/10.3188/szf.2019.0326

Hutton, N. S., Tobin, G. A., & Montz, B. E. (2019). The levee effect revisited: Processes and policies enabling development in Yuba County, California. *Journal of Flood Risk Management, 12*(3). https://doi.org/10.1111/jfr3.12469. e12469.

IPCC. (2012). Managing the risks of extreme events and disasters to advance climate change adaptation. In *Special report of the intergovernmental panel on climate change*.

Jongman, B., Koks, E. E., Husby, T. G., & Ward, P. J. (2014). Increasing flood exposure in The Netherlands: Implications for risk financing. *Natural Hazards and Earth System Sciences, 14*(5), 1245–1255. https://doi.org/10.5194/nhess-14-1245-2014

Jonkman, S. N., Bočkarjova, M., Kok, M., & Bernardini, P. (2008). Integrated hydrodynamic and economic modelling of flood damage in The Netherlands. *Ecological Economics, 66*(1), 77–90. https://doi.org/10.1016/j.ecolecon.2007.12.022

Jordan, C., Visscher, J., Viet Dung, N., Apel, H., & Schlurmann, T. (2020). Impacts of human activity and global changes on future morphodynamics within the tien river, vietnamese mekong delta. *Water, 12*(8). https://doi.org/10.3390/w12082204, 2204.

Kanton Bern. (2021). *NGKAT: Ereigniskataster der Naturgefahrenprozesse Lawine, Rutschung, Sturz, Wasser/Murgang, Doline des Kantons Bern. Kanton Bern.*

KAWA. (2011). *LiDAR Bern—airborne Laser Scanning. Gesamtbericht Befliegung—Befliegung Kanton. Amt für Wald des Kantons Bern.*

Keller, L., Rössler, O., Martius, O., & Weingartner, R. (2019). Comparison of scenario-neutral approaches for estimation of climate change impacts on flood characteristics. *Hydrological Processes, 33*(4), 535–550. https://doi.org/10.1002/hyp.13341

Keller, L., Zischg, A. P., Mosimann, M., Rössler, O., Weingartner, R., & Martius, O. (2019). Large ensemble flood loss modelling and uncertainty assessment for future climate conditions for a Swiss pre-alpine catchment. *Science of the Total Environment, 693*. https://doi.org/10.1016/j.scitotenv.2019.07.206

Kelly, R. A., Jakeman, A. J., Barreteau, O., Borsuk, M. E., ElSawah, S., Hamilton, S. H., Henriksen, H. J., Kuikka, S., Maier, H. R., Rizzoli, A. E., van Delden, H., & Voinov, A. A. (2013). Selecting among five common modelling approaches for integrated environmental assessment and management. *Environmental Modelling and Software, 47*, 159–181. https://doi.org/10.1016/j.envsoft.2013.05.005

Koch, C. I. I. (1826). *Bericht der Schwellen-Commission über die Correktion der"nAar von Thun bis Bern*.

KOHS. (2007). *Auswirkungen der Klimaänderung auf den Hochwasserschutz in der Schweiz, Wasser, Energie, Luft. Wasser, Energie, Luft* (Vol. 99, pp. 55–60).

Kreibich, H., Blauhut, V., Aerts, J. C. J. H., Bouwer, L. M., Van Lanen, H. A. J., Mejia, A., Mens, M., & Van Loon, A. F. (2019). How to improve attribution of changes in drought and flood impacts. *Hydrological Sciences Journal, 64*(1), 1–18. https://doi.org/10.1080/02626667.2018.1558367

Lin, Y. C., Jenkins, S. F., Chow, J. R., Biass, S., Woo, G., & Lallemant, D. (2020). Modeling downward counterfactual events: Unrealized disasters and why they matter. *Frontiers in Earth Science, 8*. https://doi.org/10.3389/feart.2020.575048

Liu, J., Dietz, T., Carpenter, S. R., Alberti, M., Folke, C., Moran, E., Pell, A. N., Deadman, P., Kratz, T., Lubchenco, J., Ostrom, E., Ouyang, Z., Provencher, W., Redman, C. L., Schneider, S. H., & Taylor, W. W. (2007). Complexity of coupled human and natural systems. *Science, 317*, 1513–1516. https://doi.org/10.1126/science.1144004, 5844.

Liu, J., Dietz, T., Carpenter, S. R., Folke, C., Alberti, M., Redman, C. L., Schneider, S. H., Ostrom, E., Pell, A. N., Lubchenco, J., Taylor, W. W., Ouyang, Z., Deadman, P., Kratz, T., & Provencher, W. (2007). Coupled human and natural systems. *Ambio, 36*(8), 639–649. https://doi.org/10.1579/0044-7447(2007)36[639:CHANS]2.0.CO;2

Llena, M., Vericat, D., Martínez-Casasnovas, J. A., & Smith, M. W. (2020). Geomorphic adjustments to multi-scale disturbances in a mountain river: A century of observations. *Catena, 192*(104584), 1–19. https://doi.org/10.1016/j.catena.2020.104584

Luhmann, N. (1987). *Soziale Systeme, Suhrkamp Taschenbuch Wissenschaft* (Vol. 666).

Mård, J., Di Baldassarre, G., & Mazzoleni, M. (2018). Nighttime light data reveal how flood protection shapes human proximity to rivers. *Science Advances, 4*(8). https://doi.org/10.1126/sciadv.aar5779. eaar5779.

Martín-Vide, J. P., Ferrer-Boix, C., & Ollero, A. (2010). Incision due to gravel mining: Modeling a case study from the Gállego River, Spain. *Geomorphology, 117*(3–4), 261–271. https://doi.org/10.1016/j.geomorph.2009.01.019

Mateo, R. C., Yamazaki, D., Kim, H., Champathong, A., Vaze, J., & Oki, T. (2017). Impacts of spatial resolution and representation of flow connectivity on large-scale simulation of floods. *Hydrology and Earth System Sciences, 21*(10), 5143–5163. https://doi.org/10.5194/hess-21-5143-2017

Montz, B. E., & Tobin, G. A. (2008). Livin'large with levees: Lessons learned and lost. *Natural Hazards Review, 9*(3), 150–157. https://doi.org/10.1061/(ASCE)1527-6988(2008)9:3(150)

Müller-Hansen, F., Schlüter, M., Mäs, M., Donges, J. F., Kolb, J. J., Thonicke, K., & Heitzig, J. (2017). Towards representing human behavior and decision making in Earth system models—an overview of techniques and approaches. *Earth System Dynamics, 8*(4), 977–1007. https://doi.org/10.5194/esd-8-977-2017

Munoz, S. E., Giosan, L., Therrell, M. D., Remo, J. W. F., Shen, Z., Sullivan, R. M., Wiman, C., O'Donnell, M., & Donnelly, J. P. (2018). Climatic control of Mississippi River flood hazard amplified by river engineering. *Nature, 556*, 95–98. https://doi.org/10.1038/nature26145, 7699.

Nazemi, A., & Wheater, H. S. (2015). On inclusion of water resource management in Earth system models—Part 1: Problem definition and representation of water demand. *Hydrology and Earth System Sciences, 19*(1), 33–61. https://doi.org/10.5194/hess-19-33-2015

Neal, J., Keef, C., Bates, P., Beven, K., & Leedal, D. (2013). Probabilistic flood risk mapping including spatial dependence. *Hydrological Processes, 27*(9), 1349–1363. https://doi.org/10.1002/hyp.9572

Neuwirth, C. (2017). System dynamics simulations for data-intensive applications. *Environmental Modelling and Software, 96*, 140–145. https://doi.org/10.1016/j.envsoft.2017.06.017

Neuwirth, C., Peck, A., & Simonović, S. P. (2015). Modeling structural change in spatial system dynamics: A daisyworld example. *Environmental Modelling and Software, 65*, 30–40. https://doi.org/10.1016/j.envsoft.2014.11.026

Nicótina, L., Alessi Celegon, E., Rinaldo, A., & Marani, M. (2008). On the impact of rainfall patterns on the hydrologic response. *Water Resources Research, 44*(12). https://doi.org/10.1029/2007WR006654

Nikolopoulos, E. I., Borga, M., Zoccatelli, D., & Anagnostou, E. N. (2014). Vitesse d'averse à l'échelle du bassin: Quantification, dépendance d'échelle et effets sur la crue correspondante. *Hydrological Sciences Journal, 59*(7), 1363–1376. https://doi.org/10.1080/02626667.2014.923889

Ochoa-Rodriguez, S., Wang, L. P., Gires, A., Pina, R. D., Reinoso-Rondinel, R., Bruni, G., Ichiba, A., Gaitan, S., Cristiano, E., Van Assel, J., Kroll, S., Murlà-Tuyls, D., Tisserand, B., Schertzer, D., Tchiguirinskaia, I., Onof, C., Willems, P., & Ten Veldhuis, M. C. (2015). Impact of spatial and temporal resolution of rainfall inputs on urban hydrodynamic modelling outputs: A multi-catchment investigation. *Journal of Hydrology, 531*, 389–407. https://doi.org/10.1016/j.jhydrol.2015.05.035

Palmer, P. I., & Smith, M. J. (2014). Earth systems: Model human adaptation to climate change. *Nature, 512*, 365–366. https://doi.org/10.1038/512365a, 7515.

Pappenberger, F., Dutra, E., Wetterhall, F., & Cloke, H. L. (2012). Deriving global flood hazard maps of fluvial floods through a physical model cascade. *Hydrology and Earth System Sciences, 16*(11), 4143–4156. https://doi.org/10.5194/hess-16-4143-2012

Paschalis, A., Fatichi, S., Molnar, P., Rimkus, S., & Burlando, P. (2014). On the effects of small scale space-time variability of rainfall on basin flood response. *Journal of Hydrology, 514*, 313–327. https://doi.org/10.1016/j.jhydrol.2014.04.014

Pattison, I., Lane, S. N., Hardy, R. J., & Reaney, S. M. (2014). The role of tributary relative timing and sequencing in controlling large floods. *Water Resources Research, 50*(7), 5444–5458. https://doi.org/10.1002/2013WR014067

Prudhomme, C., Reynard, N., & Crooks, S. (2002). Downscaling of global climate models for flood frequency analysis: Where are we now? *Hydrological Processes, 16*(6), 1137–1150. https://doi.org/10.1002/hyp.1054

Prudhomme, C., Wilby, R. L., Crooks, S., Kay, A. L., & Reynard, N. S. (2010). Scenario-neutral approach to climate change impact studies: Application to flood risk. *Journal of Hydrology, 390*(3–4), 198–209. https://doi.org/10.1016/j.jhydrol.2010.06.043

Ramirez, J. A., Zischg, A. P., Schürmann, S., Zimmermann, M., Weingartner, R., Coulthard, T., & Keiler, M. (2020). Modeling the geomorphic response to early river engineering works using CAESAR-Lisflood. *Anthropocene, 32*. https://doi.org/10.1016/j.ancene.2020.100266

Ratto, M., Castelletti, A., & Pagano, A. (2012). Emulation techniques for the reduction and sensitivity analysis of complex environmental models. *Environmental Modelling and Software, 34*, 1–4. https://doi.org/10.1016/j.envsoft.2011.11.003

Razavi, S., Tolson, B. A., & Burn, D. H. (2012). Numerical assessment of metamodelling strategies in computationally intensive optimization. *Environmental Modelling and Software, 34*, 67–86. https://doi.org/10.1016/j.envsoft.2011.09.010

Remo, J. W. F., & Pinter, N. (2007). Retro-modeling the middle Mississippi river. *Journal of Hydrology, 337*(3–4), 421–435. https://doi.org/10.1016/j.jhydrol.2007.02.008

Remo, J. W. F., Pinter, N., & Heine, R. (2009). The use of retro- and scenario-modeling to assess effects of 100+ years river of engineering and land-cover change on Middle and Lower Mississippi River flood stages. *Journal of Hydrology, 376*(3–4), 403–416. https://doi.org/10.1016/j.jhydrol.2009.07.049

Richert, C., Erdlenbruch, K., & Grelot, F. (2019). The impact of flood management policies on individual adaptation actions: Insights from a French case study. *Ecological Economics, 165*, 106387. https://doi.org/10.1016/j.ecolecon.2019.106387

Roese, N. J. (1997). Counterfactual thinking. *Psychological Bulletin, 121*(1), 133–148. https://doi.org/10.1037/0033-2909.121.1.133

Roese, N. J., & Olson, J. M. (1995). *What might have been: The social psychology of counterfactual thinking. Lawrence Erlbaum Ass.*

Roos, M. M. D., Hartmann, T. T., Spit, T. T. J. M., & Johann, G. G. (2017). Constructing risks–internalisation of flood risks in the flood risk management plan. *Environmental Science and Policy, 74*, 23–29. https://doi.org/10.1016/j.envsci.2017.04.007

Röthlisberger, V., Zischg, A., & Keiler, M. (2016). Spatiotemporal aspects of flood exposure in Switzerland. In *E3S Web of Conferences* (Vol. 7)EDP Sciences. https://doi.org/10.1051/e3sconf/20160708008

Röthlisberger, V., Zischg, A. P., & Keiler, M. (2018). A comparison of building value models for flood risk analysis. *Natural Hazards and Earth System Sciences, 18*(9), 2431–2453. https://doi.org/10.5194/nhess-18-2431-2018

Rougé, C., Mathias, J. D., & Deffuant, G. (2015). Vulnerability: From the conceptual to the operational using a dynamical system perspective. *Environmental Modelling and Software, 73*, 218–230. https://doi.org/10.1016/j.envsoft.2015.07.018

Serinaldi, F., & Grimaldi, S. (2011). Synthetic design hydrographs based on distribution functions with finite support. *Journal of Hydrologic Engineering, 16*(5), 434–446. https://doi.org/10.1061/(ASCE)HE.1943-5584.0000339

Shepherd, T. G., Boyd, E., Calel, R. A., Chapman, S. C., Dessai, S., Dima-West, I. M., Fowler, H. J., James, R., Maraun, D., Martius, O., Senior, C. A., Sobel, A. H., Stainforth, D. A., Tett, S. F. B., Trenberth, K. E., van den Hurk, B. J. J. M., Watkins, N. W., Wilby, R. L., & Zenghelis, D. A. (2018). Storylines: An alternative approach to representing uncertainty in physical aspects of climate change. *Climatic Change, 151*(3–4), 555–571. https://doi.org/10.1007/s10584-018-2317-9

Simonovic, S. P. (2009). Managing flood risk, reliability and vulnerability. *Journal of Flood Risk Management, 2*(4), 230–231. https://doi.org/10.1111/j.1753-318X.2009.01040.x

Simonovic, S. P. (2011). *Systems approach to management of disasters: Methods and applications*, 1. Aufl.

Singh, V. P. (1997). Effect of spatial and temporal variability in rainfall and watershed characteristics on stream flow hydrograph. *Hydrological Processes, 11*(12), 1649–1669.

Slater, L. J., Singer, M. B., & Kirchner, J. W. (2015). Hydrologic versus geomorphic drivers of trends in flood hazard. *Geophysical Research Letters, 42*(2), 370–376. https://doi.org/10.1002/2014GL062482

Smith, A., Freer, J., Bates, P., & Sampson, C. (2014). Comparing ensemble projections of flooding against flood estimation by continuous simulation. *Journal of Hydrology, 511*, 205–219. https://doi.org/10.1016/j.jhydrol.2014.01.045

Song, S., Wang, S., Fu, B., Dong, Y., Liu, Y., Chen, H., & Wang, Y. (2021). Improving representation of collective memory in socio-hydrological models and new insights into flood risk management. *Journal of Flood Risk Management, 14*(1). https://doi.org/10.1111/jfr3.12679

Staffler, H., Pollinger, R., Zischg, A., & Mani, P. (2008). Spatial variability and potential impacts of climate change on flood and debris flow hazard zone mapping and implications for risk management. *Natural Hazards and Earth System Science, 8*(3), 539–558. https://doi.org/10.5194/nhess-8-539-2008

Steinschneider, S., McCrary, R., Mearns, L. O., & Brown, C. (2015). The effects of climate model similarity on probabilistic climate projections and the implications for local, risk-based adaptation planning. *Geophysical Research Letters, 42*(12), 5014–5022. https://doi.org/10.1002/2015GL064529

Stucki, P., Bandhauer, M., Heikkilä, U., Rössler, O., Zappa, M., Pfister, L., Salvisberg, M., Froidevaux, P., Martius, O., Panziera, L., & Brönnimann, S. (2018). Reconstruction and simulation of an extreme flood event in the Lago Maggiore catchment in 1868. *Natural Hazards and Earth System Sciences, 18*(10), 2717–2739. https://doi.org/10.5194/nhess-18-2717-2018

Suter, H., Thomi, L., Weingartner, R., & Zischg, A. (2016). Was macht Hochwasserschutzprojekte erfolgreich?. In *Wasser Energie Luft* (Vol. 108, pp. 115–120).

SVKG-SEK/SVIT. (2012). *Das Schweizerische Schätzerhandbuch. Schweizerische Vereinigung kantonaler Grundstückbetwertungsexperten.*

Swisstopo. (2017). *swissBUILDINGS3D, Swisstopo, Bern.* Bundesamt für Topographie.

Swisstopo. (2020). *Siegfriedkarte 1870–1926.* Bundesamt für Topographie.

Swisstopo. (2021). *The topographic landscape model TLM.* Bundesamt für Topographie.

Syvitski, J. P. M., Vörösmarty, C. J., Kettner, A. J., & Green, P. (2005). Impact of humans on the flux of terrestrial sediment to the global coastal ocean. *Science, 308*, 376–380. https://doi.org/10.1126/science.1109454, 5720.

Thomi, L. (2010). *Rôle des paramètres sociopolitiques et des connaissances dans la gestion des risques hydrologiques.*

Thomi, L., Zischg, A., & Suter, H. (2015). *Was macht Hochwasserschutzprojekte erfolgreich?: Eine Evaluation der Risikoentwicklung, des Nutzens und der Rolle privater Geldgeber.* Geographisches Institut.

Tobin, G. A. (1995). The levee love affair: A stormy relationship? *JAWRA Journal of the American Water Resources Association, 31*(3), 359–367. https://doi.org/10.1111/j.1752-1688.1995.tb04025.x

Tobin, G. A., & Montz, B. E. (1994). The flood hazard and dynamics OF the urban residential land market. *JAWRA Journal of the American Water Resources Association, 30*(4), 673–685. https://doi.org/10.1111/j.1752-1688.1994.tb03322.x

Trigg, M. A., Birch, C. E., Neal, J. C., Bates, P. D., Smith, A., Sampson, C. C., Yamazaki, D., Hirabayashi, Y., Pappenberger, F., Dutra, E., Ward, P. J., Winsemius, H. C., Salamon, P., Dottori, F., Rudari, R., Kappes, M. S., Simpson, A. L., Hadzilacos, G., & Fewtrell, T. J. (2016). The credibility challenge for global fluvial flood risk analysis. *Environmental Research Letters, 11*(9). https://doi.org/10.1088/1748-9326/11/9/094014

US Army Corps of Engineers, Hydrologic Engineering Center. (2012). *Hydrologic modeling system (HEC-HMS).*

Valentová, J., Valenta, P., & Weyskrabová, L. (2010). Assessing the retention capacity of a floodplain using a 2D numerical model. *Journal of Hydrology and Hydromechanics, 58*(4), 221–232. https://doi.org/10.2478/v10098-010-0021-1

Vetsch, D., Siviglia, A., Ehrbar, D., Facchini, M., Gerber, M., Kammerer, S., Peter, S., Vonwiler, L., Volz, C., Farshi, D., Mueller, R., Rousselot, P., Veprek, R., & Faeh, R. (2017). *BASEMENT—basic simulation environment for computation of environmental flow and natural hazard simulation.*

Viglione, A., Di Baldassarre, G., Brandimarte, L., Kuil, L., Carr, G., Salinas, J. L., Scolobig, A., & Blöschl, G. (2014). Insights from socio-hydrology modelling on dealing with flood risk—roles of collective memory, risk-taking attitude and trust. *Journal of Hydrology, 518*, 71—82. https://doi.org/10.1016/j.jhydrol.2014.01.018

Vischer, D. (2003). Die Geschichte des Hochwasserschutzes in der Schweiz. *Berichte Des BWG, Serie Wasser, 5.*

Vischer, D., & Fankhauser, U. (1990). Jahre Kanderumleitung. *Wasser Energie Luft, 82*, 17—25.

Viviroli, D., Zappa, M., Gurtz, J., & Weingartner, R. (2009). An introduction to the hydrological modelling system PREVAH and its pre- and post-processing-tools. *Environmental Modelling and Software, 24*(10), 1209—1222. https://doi.org/10.1016/j.envsoft.2009.04.001

Viviroli, D., Zappa, M., Schwanbeck, J., Gurtz, J., & Weingartner, R. (2009). Continuous simulation for flood estimation in ungauged mesoscale catchments of Switzerland—Part I: Modelling framework and calibration results. *Journal of Hydrology, 377*(1—2), 191—207. https://doi.org/10.1016/j.jhydrol.2009.08.023

Voinov, A., & Shugart, H. H. (2013). "Integronsters", integral and integrated modeling. *Environmental Modelling and Software, 39*, 149—158. https://doi.org/10.1016/j.envsoft.2012.05.014

Ward, P. J., Jongman, B., Salamon, P., Simpson, A., Bates, P., De Groeve, T., Muis, S., De Perez, E. C., Rudari, R., Trigg, M. A., & Winsemius, H. C. (2015). Usefulness and limitations of global flood risk models. *Nature Climate Change, 5*(8), 712—715. https://doi.org/10.1038/nclimate2742

Ward, P. J., Jongman, B., Weiland, F. S., Bouwman, A., Van Beek, R., Bierkens, M. F. P., Ligtvoet, W., & Winsemius, H. C. (2013). Assessing flood risk at the global scale: Model setup, results, and sensitivity. *Environmental Research Letters, 8*(4). https://doi.org/10.1088/1748-9326/8/4/044019

Werner, B. T., & McNamara, D. E. (2007). Dynamics of coupled human-landscape systems. *Geomorphology, 91*(3—4), 393—407. https://doi.org/10.1016/j.geomorph.2007.04.020

White, G. (1945). *Human adjustment to floods*. Department of Geography, University of Chicago.

Wingo, P., Brookes, A., & Bolte, J. (2017). Modular and spatially explicit: A novel approach to system dynamics. *Environmental Modelling and Software, 94*, 48—62. https://doi.org/10.1016/j.envsoft.2017.03.012

Winsemius, H. C., Van Beek, L. P. H., Jongman, B., Ward, P. J., & Bouwman, A. (2013). A framework for global river flood risk assessments. *Hydrology and Earth System Sciences, 17*(5), 1871—1892. https://doi.org/10.5194/hess-17-1871-2013

Wirth, S. B., Girardclos, S., Rellstab, C., & Anselmetti, F. S. (2011). The sedimentary response to a pioneer geo-engineering project: Tracking the Kander River deviation in the sediments of Lake Thun (CHE). *Sedimentology, 58*(7), 1737—1761. https://doi.org/10.1111/j.1365-3091.2011.01237.x

WMO. (2009). Manual on estimation of probable maximum precipitation (PMP). In *WMO reports* (Vol. 1045)World Meteorological Organization.

Woo, G. (2019). Downward counterfactual search for extreme events. *Frontiers in Earth Science, 7*. https://doi.org/10.3389/feart.2019.00340

Zhou, Y., Li, D., Lu, J., Yao, S., Yan, X., Jin, Z., Liu, L., & Lu, X. X. (2020). Distinguishing the multiple controls on the decreased sediment flux in the Jialing River basin of the Yangtze River, Southwestern China. *Catena, 193*. https://doi.org/10.1016/j.catena.2020.104593

Zischg, A. (2016). River corrections and long-term changes in flood risk in the Aare valley, Switzerland. *E3S Web of Conferences, 7*. https://doi.org/10.1051/e3sconf/20160711010, 11010.

Zischg, A. (2018). Floodplains and complex adaptive systems—perspectives on connecting the dots in flood risk assessment with coupled component models. *Systems, 6*(2), 9. https://doi.org/10.3390/systems6020009

Zischg, A. P., & Bermúdez, M. (2020). Mapping the sensitivity of population exposure to changes in flood magnitude: Prospective application from local to global scale. *Frontiers in Earth Science, 8*. https://doi.org/10.3389/feart.2020.534735

Zischg, A. P., Felder, G., Mosimann, M., Röthlisberger, V., & Weingartner, R. (2018). Extending coupled hydrological-hydraulic model chains with a surrogate model for the estimation of flood losses. *Environmental Modelling and Software, 108*, 174–185. https://doi.org/10.1016/j.envsoft.2018.08.009

Zischg, A. P., Felder, G., Weingartner, R., Quinn, N., Coxon, G., Neal, J., Freer, J., & Bates, P. (2018). Effects of variability in probable maximum precipitation patterns on flood losses. *Hydrology and Earth System Sciences, 22*(5), 2759–2773. https://doi.org/10.5194/hess-22-2759-2018

Zischg, A. P., Hofer, P., Mosimann, M., Röthlisberger, V., Ramirez, J. A., Keiler, M., & Weingartner, R. (2018). Flood risk (d)evolution: Disentangling key drivers of flood risk change with a retro-model experiment. *Science of the Total Environment, 639*, 195–207. https://doi.org/10.1016/j.scitotenv.2018.05.056

Zischg, A., Mair, V., & Lang, K. (2012). PROALP-KARTIERUNG und monitoring von Permafrost in der Autonomen Provinz Bozen Südtirol, Italien. In *Conference proceedings INTERPRAEVENT 2012* (Vol. 12, pp. 1–12).

Zischg, A. P., Mosimann, M., Bernet, D. B., & Röthlisberger, V. (2018). Validation of 2D flood models with insurance claims. *Journal of Hydrology, 557*, 350–361. https://doi.org/10.1016/j.jhydrol.2017.12.042

Zischg, A., Schober, S., Sereinig, N., Rauter, M., Seymann, C., Goldschmidt, F., Bäk, R., & Schleicher, E. (2013). Monitoring the temporal development of natural hazard risks as a basis indicator for climate change adaptation. *Natural Hazards, 67*(3), 1045–1058. https://doi.org/10.1007/s11069-011-9927-0

Zischg, A., Frehner, M., Gubelmann, P., Augustin, S., Brang, P., & Huber, B. (2021a). Participatory modelling of upward shifts of altitudinal vegetation belts for assessing site type transformation in Swiss forests due to climate change. *Applied Vegetation Science, 24*(4), Article e12621. https://doi.org/10.1111/avsc.12621

Zischg, A. P., Röthlisberger, V., Mosimann, M., Profico-Kaltenrieder, R. N., Bresch, D. N., Fuchs, S., … Keiler, M. (2021b). Evaluating targeted heuristics for vulnerability assessment in flood impact model chains. *Journal of Flood Risk Management, 14*(4), e12736. https://doi.org/10.1111/jfr3.12736

Zoccatelli, D., Borga, M., Viglione, A., Chirico, G. B., & Blöschl, G. (2011). Spatial moments of catchment rainfall: Rainfall spatial organisation, basin morphology, and flood response. *Hydrology and Earth System Sciences, 15*(12), 3767–3783. https://doi.org/10.5194/hess-15-3767-2011

Chapter 4

Rivers and floodplains as complex adaptive systems?

In the previous chapter, we have seen how variable flood risk is over time and how flood risk is changing with the underlying changes of the components of risk, namely hazard, exposure, and vulnerability. The components of flood risk are coevolving and their current and future development paths are dependent on the historical situation. Moreover, we have learned from a few examples how the natural and social systems are intertwined and how human adaptation to the changing environment or to extreme flood events amplifies or dampens the evolution of flood risk. The development of flood risk in the past lets us imagine how flood risk will unfold in the future and how complex the control of flood risk might be. To confront this complexity when analyzing and developing climate change adaptation measures, we need a theoretical framework that goes beyond the frameworks of the different disciplines that are studying flood risk (change). A theoretical framework that understands the world as complex and interconnected and thus enables us to approach the wicked problems our society is facing today is needed. Boulton et al. (2015) envision a complexity world view that is required to study the complexity and uncertainty of our times. A complexity world view or a complexity perspective focuses on the "becoming" of patterns, emergences, trends, collapse, or events rather than on "being" (Boulton et al., 2015). The complexity perspective supports overcoming the reductionist point of view, which understands the functioning of the world as a machine that is built from parts. Complexity science and complexity theory offer a valuable perspective for looking at flood risk change and for developing the flood risk management practice of the 21st century. It offers a methodological framework for interdisciplinary and transdisciplinary collaborations. The core of complexity science is the study of complex systems. Before entering in detail on the complex systems perspective of flood risk change, I am describing the characteristics of complex systems.

Characteristics of complex systems

Complex systems are not strictly defined and every discipline uses its own definition. Physical and natural sciences define complex systems mostly from

Flood Risk Change. https://doi.org/10.1016/B978-0-12-822011-5.00002-8

the viewpoint of nonliving systems. In contrast, social sciences look at complex systems from a human-centered perspective. In the past decades, a new science has evolved that focuses on complex systems, namely complexity science. Complexity science is a collection of models and theories that enable the study of different features of complex systems and across different kinds of systems in a common way (Ladyman & Wiesner, 2020). This allows us to go beyond domain-specific theories and models and enables a framework for interdisciplinary or transdisciplinary research.

Complex systems are described by a set of characteristics. As the study of flood risks involves both scientific disciplines, the natural sciences and social sciences, here I am defining and summarizing some characteristics of complex systems from both disciplinary perspectives. Common to most of the definitions of complex systems is the mentioning of multiple interactions between many different system components. Some definitions go beyond and state that complex systems are open systems that adapt or react to their boundary conditions. The adaptation or reaction to changing environmental conditions leads to the emergence or self-organization of new patterns, structures, or behavior. Emergence may occur at another scale than the scale of the interactions of the system components. After Boulton et al. (2015), a "complex system comprises many unique elements that interact in multiple ways. The elements themselves can change, learn, and adapt. The connection can change, loosen, reform, and the boundaries of the system can also shift over time."

In the following, a list of characteristics of complex systems is summarized from the books of Ladyman and Wiesner (2020) and Boulton et al. (2015). It is nonexhaustive and focuses on those features of complexity that are relevant to the topic of flood risk change. I am breaking down the universal characteristics of complex systems for the topic of the book. As flood risk is a measure of the relationship between human and the natural systems, features from both living and nonliving complex systems are listed.

Numerosity and diversity

As stated above, complex systems consist of a large number of parts and subsystems and many interactions between these components. Moreover, complex systems have a high diversity of system components with different behavior. This means that systems with only a very few components and few interactions between themselves cannot be complex. Looking at the examples and case studies shown in the previous chapter, we can observe that flood risk is composed by the interactions of many parts. For example, the hazard is composed of numerous physical processes that act together and form local predisposition and susceptibility, ranging from atmospheric processes to earth surface processes and their interaction. Exposure is composed of thousands of buildings that all have their own vulnerability against flood hazards. Moreover, the exposure of buildings is connected with the exposure of mobility along the traffic networks and with the exposure of other infrastructure such as the network of lifelines. All elements of risk (hazard, exposure, and vulnerability) interact together on the local scale. The state of the exposed elements (vulnerability) depends on the magnitude of the hazard. The example of buildings also shows the high diversity of the floodplains. Depending on cultural, geological, environmental, and economic settings, buildings can be constructed from wood, concrete, stone, or clay and thus can exhibit a high diversity of physical vulnerability. The physical variability is also highly variable within a specific region. A high diversity of subcomponents enhances the adaptation capacity of the whole system to external shocks or changes to the system's boundary conditions. Hence, diversity is a variation of system components.

Order and disorder

In regard to flood risk, we can observe patterns of order or disorder in space and time. The spatial footprint of settlements and the topological relationship between settlement and rivers is a result of past urban development. For a long time in human history, with the exception of specific epochs, settlement growth was disordered in terms of spontaneous occupation of places for constructing buildings and infrastructure. This development is lacking centralized control. In contrast, some epochs and regions exhibited an ordered (i.e., planned) settlement growth. Here, a centralized instance controls the development of some subsystems. The results of these examples are cities with a rectangular and regular street network and an "artificial" topology between the city and the water bodies. A certain degree of spontaneous order results from the interactions of the components on an aggregated scale. In addition to spatial expressions, order or disorder can also arise on a temporal scale. In particular, societal trends lead to spontaneous developments. For example, if technological development changes the way and manner how buildings are being constructed, the temporal development (temporal order) of construction techniques manifests in space (spatial order) by the contemporaneous pop-up of homogeneous settlement types and of spatial clusters with the same type of buildings. Another example of an order that results from the interaction of system components in floodplains is the historical pattern of the road network. In Switzerland, the spatial footprint of the road network at the beginning of industrialization was shaped by the flood hazard in the main valleys. The main roads were mainly located at the lateral borders of the valleys at a safe distance from the regularly flooded areas.

Nested structure, modularity, and multiple scales

The components of a complex system are interrelated. However, system components can operate at different scales. Some components operate slowly on a very large timescale, while others fluctuate or vary on short timescales. This is called structural modularity. Each component has another function for the whole system. Thus, the components interact either at the same scale or across scales. Components of higher order can be connected with components of lower order. This means that some lower-order components are nested in a higher-order structure. In flood risk management, an example of a nested structure is the hierarchical organization of local, cantonal, and national governments in a federal state that are responsible for the planning and implementation of flood protection measures. The cantonal level (supra-communal level or regional level) elaborates on risk analysis and sets the priorities for investments in flood protection measures. This is the hierarchical level with the most specific expertise. It approaches the communal administration in case of safety deficiencies and is consulting the communities in the planning process. The communal level (local level) is responsible for planning and implementation of flood protection measures. The communities are responsible for the maintenance of the structural protection measures. The role of the federal government (national level) is in financing and in defining the overall strategies for flood risk management. In Switzerland, the projects for flood risk protection measures are financed by all three hierarchical levels, in many cases by equally dividing the costs. The nested structure is evident, the highest-order level is the strategical level, the second-order level is following the regional implementation of the strategy, and the lowest-order level is the community that is implementing it. In terms of geography, the hierarchical levels manifest at different spatial scales: the national scale, the regional scale, and the local scale. The involvement of three hierarchical levels is the simplest case. In many projects there are other hierarchical levels involved, such as nation-wide operators of network infrastructures who have a special interest in a local site that affects the functionality of the whole network. The owners of land and buildings are an additional entity, the most molecular level. Another example of a nested structure is the road network in mountainous areas. The remote villages in the headwater catchments and upstream parts of the Alpine valleys are in many cases accessible only via one main road. If this road is interrupted by a natural hazard event, the road network breaks into distinct parts. The villages behind the interruptions are disconnected from the main network but remain connected. The mobility between these villages does not stop, but the exchange between this subsystem and the main network is limited. This can happen in many locations in the Alps, and thus the network can have many hierarchical clusters during extreme weather situations.

Feedback

In the previous chapter, we have learned about how feedback plays a role in evolving flood risks. Feedback arises when the interactions between the components of complex systems are iterated. This means feedback arises from a previous interaction. Parts of the system interact as a result of the previous interactions of other system components. Feedback can also have a time lag and thus can occur at different timescales. Feedbacks are relevant as they may lead to emergence and order or disorder. Moreover, it is an important mechanism that enables the manifestation of adaptation, e.g., by social learning. In flood risk research and management, the safety paradox or the so-called levee effect is an example of a feedback mechanism. The spatial footprint of urban spread is influenced by the implementation of flood protection measures. In this case, the settlements develop toward the rivers. In contrast, the periodical experience of flooding feeds collective societal memory and influences the spatial footprint of urban areas spreading in the opposite direction. Periodically flooded areas are kept free from development. Feedbacks thus can have two directions. Negative feedback works in the opposite of other forces or drivers of change and thus can have a corrective or stabilizing effect. Land use regulation is introducing a controlling effect and is thus to be considered as negative feedback. In contrast, positive feedback can amplify the forces of other drivers of change.

Nonlinearity

Feedback may not only be positive or negative, but the interactions between system components can also be linear or nonlinear. Nonlinearity is often called a main characteristic of complex systems. In contrast to linear correlations between two variables that make the behavior of interacting elements of a system relatively comprehensible, nonlinear correlations make predictions more difficult and require computational experiments to assess the behavior or the state space of a system. The mechanical dose–response thinking is failing when nonlinear behavior of elements and a feedback between those elements are present. In nonlinear correlations between variables of a system, the grade of change of one element does not correspond to the grade of change of the other. Besides the nonlinearity of the correlations between system elements, also the behavior of the whole system can be nonlinear as a result from the many interactions of the system elements. An example of nonlinearity in flood risk management is shown in the risk curve. The risk curve describes the relationship between the frequency/magnitude of a flood event and its impact. This curve depends on the floodplain topography and the topological relationship between the exposed values and infrastructure and the hazard. Thus, the risk curve is site-specific, and can be linear, convex, or concave. Hence, an increase of peak discharge, for example, by 10% due to climatic changes can lead to an increase by 100% of the flood impacts in the case of high-frequency floods, but to an increase by 8% of the flood impacts in the case of very extreme and low-frequency floods. In another floodplain, this relationship between flood magnitude and flood impacts can be very different. Moreover, the increase of rainfall intensity can lead to a nonlinear increase of river discharge depending on the catchment characteristics. Both sensitivities, the exposure sensitivity and the catchment sensitivity, must be combined. Thus, the sensitivity of floodplains to changing boundary conditions due to climatic changes is the result of site-specific factors and nonlinear correlations between the sensitivities to change of subsystems. This can make the assessment of the effects of climatic changes on flood risk complex.

Nonequilibrium, fluctuations, episodical changes

An open system has influxes and outfluxes of energy or matter. Thus, open systems can be externally modified or react to external influences. If net influx and net outflux are not in balance, the system is in nonequilibrium and is either gaining or losing matter or energy. The reaction of the system, or of system components, to changing environmental conditions can be immediate or can cause a time lag. This leads to fluctuations or to episodical changes. An example of fluctuation from the flood risk management perspective can be the so called "catastrophe gap" (Pfister, 1999). In Switzerland, the 20th century was a period with only a very few extreme flood events. This relatively "quiet" period after a century with a high number of flood events led to a relative stabilization of flood risk management practice. Since the end of the 20th century, the frequency of flood events is rising and with it the effort in the implementation of flood protection measurements is rising too.

Emergence, self-organization, self-regulation

The interactions of system components occur at specific scales. Out of these interactions, sometimes new patterns or a change in the system behavior may result at a higher-order level or at another scale. Emergence can result in terms of changing system behavior, system structure, or system dynamics. The aggregated behavior and the order of the whole system can result from a very large number of disordered and uncoordinated interactions between elements. This is also called self-organization. Herein, it has to be distinguished by the order within the system and the order of the system. An example of an emergent phenomenon is the change of the hydrological regime of a watershed after radical changes of the water balance, e.g., mountain torrents running dry in summer months after the disappearance of glaciers and thus changing from perennial rivers to nonperennial rivers (Messager et al., 2021). This change leads to emergent patterns in the habitat quality and thus to new ecosystems. An example of self-organization or self-regulation is the issue of new legislation after catastrophic flood events. Funding for flood protection projects is eased after flood events. The so-called windows of opportunities after flood events make previously not-accepted flood protection projects acceptable to the public and politics. Moreover, flood events may sometimes foster a paradigm shift in flood risk management practice. After such an emergence of new practices, the flood risk will change at another pace or in another direction than before. The emergent pattern results from the individual parts of the system. However, from the emergent phenomenon at the higher-order scale, often it cannot be concluded backward to the contribution of individual system components to the overall change. Emergent patterns can lead to a decrease in the state space of the system and thus to a reduction of adaptability.

History and memory, path dependency

Open systems can react to perturbations or changes to their environmental conditions. However, the reaction to the same amount of change can differ dependent on the system state. The latter is a result of historical developments and previous adaptations. The previous development of a system is expressed in the current order or structure of the system. The current order can also be defined as a memory of the system. If an external shock or perturbation is influencing an early system state that has had no chance to adapt to similar perturbations before, the reaction of this system may be very different from the reaction of a system that is already adapted to shocks of similar magnitude. Thus, the effect of memory or adaptation also leads to different future pathways of development. A poorly adapted system will develop differently under future perturbations and changing boundary conditions than an already well-adapted system. In flood risk management, we find this path dependency in many situations. We have seen in the previous chapter that potentially flooded areas are kept free from urban spread for a long period if a flood event happens and can be observed. On the other hand, we can observe that settlements that did not experience extreme flood events in the past may be much more exposed to flooding than others because of the lack of experience. Also, the vulnerability of houses can decline after the experience of flood events (Elmer et al., 2010; Kreibich et al., 2005; Thieken et al., 2007). This past adaptation sets the current status of the system and will be a benefit if flood events increase in the future. The path dependency is an important concept for assessing the options to confront the effects of climatic changes on flood risks. The current situation of the topological relationship between the hazard (river morphology and flood protection structures) and the exposure/vulnerability may limit the degree of freedom for climate adaptation, e.g., a settlement having a high building density along the river that makes a river widening as a flood protection measure impossible.

Robustness

One of the emergent properties of a complex system can be robustness. This can result from many negative feedback effects between the subsystems. These regulate the system toward stability and may lead to the ability to absorb external shocks. Robustness also may result from decentralized or distributed regulation of the behavior of subsystems. The lack of a centralized controlling instance that is reacting to an external perturbation in one or the other way and thus is influencing the whole system leads to avoiding overregulation or maladaptation. If a system is robust, it is able to absorb external shocks and to recover from perturbations. In this sense, a robust system can be a resilient system, although the latter term is defined more widely. A resilient system is able to correct errors in its structure. Thus, it adapts to changing boundary conditions by adding or removing elements of the system, or even by reorganizing the connections between the elements. This results in a changed order. In contrast to robustness, a system that gains from shocks can be defined as antifragile (Taleb, 2012). An example of a robust system from the risk management perspective is a flood protection structure that is designed to protect settlements or infrastructure from floods up to a certain flood magnitude. If the flood exceeds the design event, the structure does not collapse and is guiding the overflowing water toward those areas that have the least damage potential. This leads to smart behavior in the overload case.

Adaptive behavior

A question about the robustness of complex systems is whether the system can be robust without changing the fundamental structure or not. The reorganization of its own structure and order of a system, or even the interactions between the system elements after an external shock or for recovery, is defined as an adaptive behavior. Mechanical systems with strongly predefined ranges of functions cannot adapt organically. In contrast, complex systems are able to reorganize structure and order, as well as functions and interconnections of system elements. The term "adaptive behavior" is mostly used for living systems and rarely for physical systems. From a strategical perspective in flood risk management, an adaptive behavior could be a paradigm shift from a strategy that focuses purely on structural flood protection to a strategy that focuses on vulnerability or resilience. The first strategy influences the hazard, while the second influences the vulnerability of elements exposed to floods but lets the hazard happen. When adaptive behavior is at the heart of studying complex systems, often the term "complex adaptive systems" (CASs) is used. CASs are "systems that involve many components that adapt or learn as they interact" (Holland, 2006). They have many components, often called agents, that interact and adapt or learn. The adaptation of the system can manifest as an abrupt change in behavior, often termed as a regime shift or as a slow change. The change of a system is sometimes accompanied by a critical slowing down or by fluctuating behavior and abrupt shift when a tipping point is approached. A robust system is restating the system behavior very fast after an external shock. However, the time needed for recovery can increase if the number of shocks increases. The effect of each external perturbation or shock is added to the previous ones and is thus cumulative. An increased time required for recovery might be an early warning signal that the system may be approaching a tipping point with a subsequent system adaptation or regime shift (Gunderson et al., 2017). The functional dependency of the recovery time on the magnitude of the external shock or perturbation is a signature of a tipping point. In contrast to the critical slowing down, another signal of a system that approaches a tipping point is the increase of fluctuations or oscillations. An example of a system that approaches a tipping point from the flood risk management perspective is nuisance flooding. Nuisance flooding is a high frequency of small flood events. After Karegar et al. (2017), nuisance flooding is an early indicator of areas that will eventually experience increased catastrophic flooding and land loss. The cumulative effects of a high number and increasing number of small-magnitude floods may finally lead to human adaptation, e.g., by implementing flood protection measures (Moftakhari et al., 2017).

Uncertainty

The characteristics of complex systems, especially the features of nonlinearity, emergence, and the presence of tipping points that may lead to changes in the overall system behavior, make predictions highly uncertain. The lack of determinism in complex systems leads to the fact that disturbances or external shocks of similar magnitudes but at different times and or at different system states may lead to very different outcomes. Hence, often there can be multiple pathways for future development. Uncertainty is limited knowledge about future, past, or current events. Regarding flood risk management practice, uncertainty refers to the gap between the available knowledge and the knowledge decision-makers need for the best decision. If the external context of a complex system is uncertain, and if there is limited knowledge of how the system works and reacts to changing boundary conditions, we encounter deep uncertainty (Marchau et al., 2019).

Long-term evolution of floodplains seen from a complex systems perspective

Flood risk arises in floodplains. In these areas, the flood hazards overlap with the human sphere. Floodplains are areas of low-lying ground adjacent to a river. A floodplain stretches from the banks of the river to the outer edges of the valley and is periodically flooded by the water exiting the river channel. Floodplain features are landforms produced by stream erosion, sediment transport, and deposition, such as point bars, oxbow lakes, and terraces. Thus, floodplains are formed mainly of river sediments and thus are relatively flat. Due to this characteristic, floodplains provide preferred land and space for human development. From the ecological perspective, floodplains link terrestrial and aquatic environments and provide several important functions, such as providing habitats and regulating water and sediment fluxes.

In the previous chapter, we have seen how an early river engineering project altered the long-term evolution of a floodplain. The Kander river deviation of 1714 AD triggered several adaptation processes of the lake-river system and its floodplains of the Aare river basin upstream of Bern, Switzerland. The deviation of the Kander river into Lake Thun led to an adaptation of the lake/lake-outflow system. The lake outflow adapted to the new hydrological regime of the lake by erosion and was adapted by human interventions, followed in the three centuries after the Kander correction. The first human intervention triggered a cascade of unintended consequences. A new river delta emerged in Lake Thun and the downstream sediment balance was heavily altered. This in turn led to erosion and riverbed incision. Consequently, the connection between the river channel of the Aare river downstream of Thun and its floodplains was disturbed, as well as the groundwater regime. The ecological habitats along the lake shores and the river changed and new ecosystems raised. Overall, a new behavior of the whole system of lakes, rivers, and floodplains emerged in terms of the hydrological regime of the river basin. Moreover, the initial intervention into the river system triggered a cascade of social adaptation measures that vice versa influenced the natural system, and this fed back again to the human system. Since 1714, society has gained from huge technological innovations, and these shaped society and its relationship to nature. For example, after the Kander river deviation, society entered into the phase of industrialization and modernization. The most recent human intervention into the river system that aimed at correcting the unintended consequences of the earliest intervention took place in a very different societal organization. The ability to alter the human environment is of orders of magnitudes higher nowadays than during the times of the first river corrections. From the perspective of a contemporary decision-maker in the early 18th century, all these innovations and societal changes could not have been foreseen. The knowledge of river erosion was lacking and thus the decision-makers confronted many unknown unknowns.

Nowadays, we are confronting other uncertainties, especially regarding future societal and technological changes, but also regarding climatic changes. We will face deep changes in the energy systems, meaning, for example, an increased exploitation of water resources for power generation. We do not know what the future flood risk safety standards will be, in which direction the economy and society will develop and many other uncertainties. Therefore, we have similar short sight toward future developments of the floodplains and must confront the complexity of future development paths.

All in all, the case study of the Kander river deviation of the early 18th century and the story of human adaptation and managing the unintended consequences of the first human intervention to the natural system over three centuries teach us that we can find many features and characteristics of complex systems in floodplains. We find numerous elements and processes that exhibit a nested structure. The system elements are interconnected and react to each other at different spatial and temporal scales and with time lags. We find hierarchical organizations in human adaptation. We have seen that the river-lake-floodplain system is in nonequilibrium and that perturbations of the system boundaries meet a specific system status. The case study shows memory effects and path dependency. Overall, we can state that the studied system, i.e., the river system of the Aare river upstream of Bern including the lakes and floodplains, adapted over centuries to a new system behavior (here in terms of hydrological behavior and flood risk change). Societal change is influencing environmental change and the latter feeds back to the first (Fig. 4.1). As a consequence, society is adapting to environmental changes and some adaptation measures in turn change the environment again. Therefore, the case study showed us that river systems and floodplains can be defined as complex adaptive systems. This leads to the conclusion that complexity science and the complexity perspective will provide a useful framework to understand the evolution of floodplains and of flood risk.

Complexity of flood risk change

The quantitative analysis of flood risks is a challenging task because it requires combining different methodological approaches for analyzing and quantifying all three main factors of flood risk, i.e., hazard, exposure, and vulnerability. The hazard analysis requires statistical analyses, hydrological and geomorphological analyses, terrain surveying, and modeling. The analysis of exposure and vulnerability requires engineering approaches, data science and data mining approaches, and approaches from social sciences. Once the basic data is preprocessed and available as geodata, the quantification of flood risk is conducted by overlaying and intersecting the data by means of geographical information systems. This gives a spatiotemporal snapshot of the current flood risk. Nevertheless, because of many uncertainties and complicated issues, it is a relatively straightforward and simple process.

However, when we are looking at the flood risk change, i.e., when the focus is laid on the trajectory or pathways of flood risk change, or when we are aiming to predict future flood risk changes, the analysis process becomes

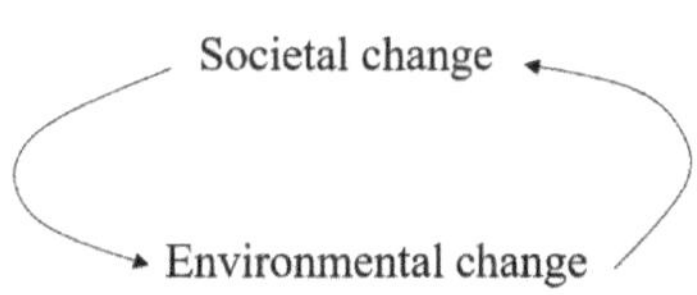

FIGURE 4.1 Floodplains as complex adaptive systems. Floodplains can be interpreted as complex adaptive systems.

complex (Haimes, 2009). Only in this case of focusing on flood risk change, the complexities of all drivers of change and their interconnections are becoming relevant. To be able to predict the future evolution of flood risks, we must predict all relevant drivers of change and feedback between some of these drivers. We must consider time lags, nonlinear sensitivities to change, emergence, and the adaptation of the human systems to changes in the environmental systems. Hence, the complex systems perspective considers the evolution of floodplains as complex adaptive systems. The benefits of this perspective are that it provides a theoretical foundation to study dynamic systems, and it centers human action and human adaptation in contrast to a pure hazard perspective. It enables systemic and synergistic thinking, and thus it fosters inter- and transdisciplinary collaboration in flood risk management, and, finally, by anticipating emergence, the complexity perspective avoids being surprised by unintended outcomes of human interventions in coupled human and natural systems.

Trajectories and pathways of flood risk evolution

In the previous chapter and in the sections above, we have learned how the drivers of flood risk changed over decades and centuries. In the following, I am summarizing the overall flood risk evolution in Switzerland. It is a very generalized and schematized picture that can locally be very different from the overall mean. The generalization is drawn from the case studies shown in the previous chapter and from a case study described in the following chapter, namely the case study of flood risk change in the Aare river basin (Zischg, 2016), in the Lago Maggore and Magadino floodplain near Bellinzona (Brönnimann et al., 2018, 2019), and the case study of flood risk change in the floodplain of the Emme river near Burgdorf (Zischg et al., 2018).

In the early 18th century, the population and economy in Switzerland were relatively stable. The growth rates were low. Thus, flood risk these days has not grown significantly. Exposure to flood risks, at least to the floods of the main rivers, was avoided. Settlements were preferable located on elevated terrain or on the alluvial fans of the lateral tributaries. The remaining risk was due to flooding of small torrents and tributaries. With the beginning of industrialization, the need for energy for production facilities increased. The attractivity of using hydropower for these facilities led to locating new facilities along the water courses. Moreover, the upcoming need for mobility, e.g., the construction of railways and roads, has led to an increased exposure of the traffic infrastructure to floods. Consequently, flood risks increased during this era. However, the industrialization period has coevolved with the rise of engineering sciences, natural sciences, and technological advancement. These trends have led to huge possibilities to shape the human environment. In the case of flood risks, technological advancement and rising knowledge allowed us to correct the main rivers and to install kilometers of river levees and lateral dikes. Hence, flood risk decreased in the mid-19th century despite the population growth in that era. The high occurrence of extreme flood events in the 19th century triggered massive intervention measures, in addition to river corrections. The forest protection and water management laws were issued; the national authorities were enabled to financially fund and plan large flood protection projects, and the first public insurance institutions were founded. Nevertheless, the population growth in late industrialization led to an increase in flood risks. After 1900, flood protection structures were massively implemented also in the secondary Swiss river network and thus flood risks decreased again. Since the mid-20th century, several large hydropower reservoirs have been constructed. These reduced the flood risks downstream due to the retention effect of floodwaters on the downstream peak discharges. Since the early 20th century, a decade-long "catastrophe gap" has occurred in Switzerland. This climate variability led to a remarkable decrease in the number and magnitude of flood events. Thus, society has adapted to a stable situation without external shocks. In addition, after World War II, strong economic growth and population growth led to a significant increase in flood exposure. Only since the end of the 20th century, flood events are occurring again with a higher probability of occurrence and magnitude. The widespread flooding events of 1987, 1999, and 2005 triggered a number of adaptation measures, e.g., the implementation of land use regulation by means of hazard mapping, the increased effort in planning and implementing flood protection works, the paradigm change from a hazard-centered management perspective to a risk-based perspective, and the optimization of early warning systems. Thus, the growth of flood risk evolution has slowed since the beginning of the 21st century. However, the evolution of flood risks in the near and far future is very uncertain. We know that the probability and intensity of heavy precipitation events will rise due to climatic changes, and we might assume that population growth and urbanization will continue. These are relevant drivers of change that will lead to increasing flood risks in the future. However, we don't know if the effort in flood risk management will be able to control the rising risks and keep them constant or even if future risk management will lower flood risk in future (Fig. 4.2). We have seen that there are many positive and negative feedbacks between the drivers of change and thus, a prediction of future flood risks is a complex task.

Sensitivity to climatic changes—short-term future pathways

The prediction of future flood risks is complex due to the many interacting elements of hazard, exposure, and vulnerability, as well as their interactions. However, the evolution of floodplains and flood risks depends on historical development and the current flood risk in terms of the current situation of hazard, exposure, and vulnerability. This dependence of floodplains allows us to estimate the short-term development of future flood risks despite the

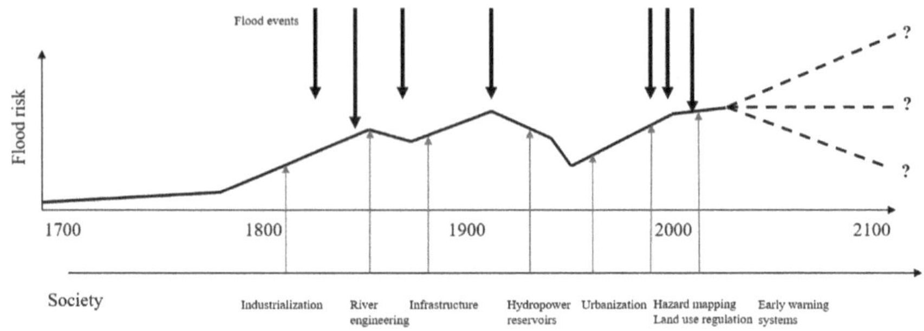

FIGURE 4.2 Flood risk evolution. Schematized and generalized pathway of flood risk evolution in Switzerland in the past three centuries.

inherent complexity. Thus, the analysis of the current snapshot of flood risks allows us to predict flood risk at least for the next years or decade. However, this requires analyzing the sensitivity of a floodplain to changes in the boundary conditions. In terms of flood risk change, the sensitivity of a floodplain can be how exposure changes if the hazard changes, for example, because of increasing flood magnitudes due to climatic changes. In the previous chapter, we have seen how the "system behavior" of a floodplain in terms of the topological relationship between hazard and exposure can be summarized by surrogate models. These meta-models or surrogate models generalize and synthesize a high number of flood simulations with the related overlay between the flood maps and the values at risk, including their vulnerability to flooding. The most simplified version of such surrogate models is a table or a relationship that lists the number of exposed buildings, kilometers of affected roads, exposed persons, the monetary value of flood damage, etc., in relation to each flood scenario of a certain magnitude. The relationship between the increasing flood magnitude and the increasing flood exposure or flood losses is often termed as the "flood risk curve" of a floodplain. This relationship can be used to identify the sensitivity of the floodplain against an increase in flood magnitudes (e.g., increased peak discharges).

In the following, I explain the concept of the sensitivity of floodplains to an increase in flood magnitudes. The concept was developed and applied by (Zischg & Bermúdez, 2020). The main principle of this sensitivity analysis is the geographical overlay of the spatial extent of a basic flood event and of a flood event with a higher magnitude with spatial data of socioeconomic exposure. The latter can include people, infrastructure, cultural sites, or any other socioeconomic asset. Here, I am showing only the sensitivity of population exposure to increasing flood magnitudes. In this study, we mapped this sensitivity for the global river network. We used the flood maps from the Joint Research Centre of the European Commission JRC (Alfieri et al., 2017; Dottori et al., 2016, 2018). These consist of flood scenarios with return periods of 10, 20, 50, 100, 200, and 500 years. These maps did not consider the flood protection measures and thus are overestimating the flooded areas for scenarios below the regionally different safety standards. Thus, we corrected the flood maps with data from the FLOPROS dataset (Scussolini et al., 2016). This database globally maps the flood protection standards of river flood protection by indicating a return period of the design event. Thus, inundated areas from the JRC flood maps are deleted if the return period of the flood scenario is lower than the return period of the protection standard. As for the exposure, we used the Gridded Population of the World (GPW) v4 dataset from the Center for International Earth Science Information Network at Columbia University (CIESIN- Facebook Connectivity Lab and Center for International Earth Science Information Network, Columbia University, 2016). Each of the corrected flood maps was overlaid with the population dataset and the number of exposed people was extracted for each floodplain. We calculated the exposure footprint of each floodplain.

The population exposure footprint of a floodplain can be visualized by plotting a two-dimensional graph with the exceedance probability of a flood hazard scenario on the x-axis and the related number of exposed people on the y-axis. The slope of the interpolated curve between the base points of the hazard scenarios shows how an increase in the probability of a flood relates to an increase in the exposed population, and the range of flood magnitudes over which the sensitivity of the floodplain is highest can be identified. For example, a river reach or a floodplain without lateral levees or flood protection might exhibit a population exposure with a return period as low as 50 years, but the rise in exposure with increasing flood magnitude will be relatively linear (Fig. 4.3A). In contrast, a river with a high standard of flood protection, e.g., with lateral dikes protecting the floodplain from floods with return periods up to 100 years, will not be sensitive to an increase in the frequency of floods of lower magnitudes. In contrast, a significant increase in the affected population will occur when the flood magnitude of the river system exceeds the standard of protection (Fig. 4.3B). These floodplains are sensitive to changes in flooding over the middle range of magnitudes. The third example shows a restored river with a large riverbed (Fig. 4.3C). In this example, an increase in the flood magnitude will only lead to a slight increase in the area inundated, and a steep increase in flood exposure will only occur in the case of an extreme flood. This means that this floodplain is only sensitive to changing flood magnitudes in the upper extremes. These characteristics of the exposure functions can be simplified by stating whether a curve is convex, concave, or linear. A concave curve (e.g., Fig. 4.3B) indicates that the sensitivity of the exposure of a floodplain to changes in flood magnitude is higher in the lower range, while a convex curve (e.g., Fig. 4.3C) indicates a higher sensitivity in the upper range of magnitudes. A floodplain with a nearly linear form (e.g., Fig. 4.3A) of the exposure footprint function shows the same sensitivity across all flood magnitudes. The curve also allows the identification of the range of

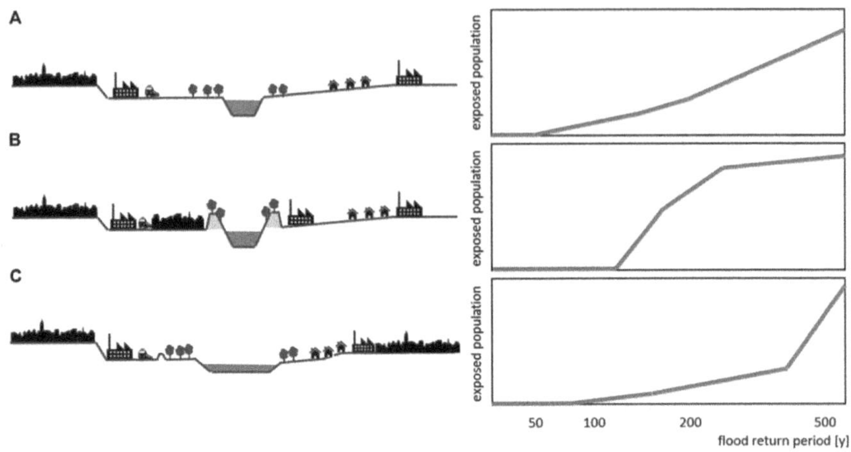

FIGURE 4.3 Sensitivity of flood exposure to increasing flood magnitudes. Schema of different characteristics of the exposure footprint of the floodplain. (A) A nearly linear increase in the exposed population with increasing flood magnitude. (B) A concave shape, meaning that exposure increases sharply following levee overtopping. This floodplain is sensitive to changes in floods with return periods around the standard of protection. (C) A relatively robust behavior to changes in lower flood magnitudes. This floodplain is mostly sensitive to flood magnitudes in the upper extremes. *From Zischg, A. P., & Bermúdez, M. (2020). Mapping the sensitivity of population exposure to changes in flood magnitude: Prospective application from local to global scale.* Frontiers in Earth Science, 8. *https://doi.org/10.3389/feart.2020.534735.*

flood magnitudes at which the gradient is steepest, as well as the magnitude at which abrupt changes in the gradient occur.

The gradient between increasing flood magnitude and increasing exposure can be described as the sensitivity of the exposure footprint of a floodplain to increasing flood magnitude. The steeper the slope, the higher the sensitivity of the floodplain to floods of increasing magnitude. Differences in this gradient indicate whether a floodplain exhibits a higher sensitivity to changes in the extremes or to changes in the lower magnitudes. Gradients that are calculated using normalized values of magnitude and exposure can be used to compare the sensitivity of one floodplain with another. Consequently, the normalized gradients can be used to map and compare the sensitivities of multiple floodplains. The gradient is normalized as follows:

$$g_{norm} = \frac{(exp_{dc} - exp_{bs})/exp_{bs}}{(RP_{dc} - RP_{bs})/RP_{bs}}$$

where g_{norm} is the normalized gradient, exp_{bs} the number of exposed population of the basic scenario, exp_{dc} the exposure of the downward counterfactual scenario (i.e., the next higher flood magnitude), RP_{bs} the return period of the basic scenario, and RP_{dc} the return period of the downward counterfactual scenario.

In the application on a global scale, the basic scenario was a design event with a return period of 10 years (RP10). We calculated the normalized gradients along the whole flood risk curve from the basic scenario to the most extreme scenario. The following pairs of scenarios were considered: RP10/20, RP20/50, RP50/100, RP100/200, and RP200/500. The gradients vary remarkably across the globe (Fig. 4.4). In the lower range of flood magnitudes, the highest gradients are found in the rivers in South America, Africa, Asia, and Eastern Europe. The Mississippi River system, most European rivers, Indian, and Australian rivers do not show any increase in exposed population when the flood magnitude increases within this range, because these rivers have a high flood protection standard. In contrast, the Indian and European rivers show a higher sensitivity in the upper extremes of the flood magnitudes. These flood magnitudes are generally above the standard of protection according to the FLOPROS database. Exceptions are river systems in the Western and Eastern United States that have a protection standard for floods of up to a 1-in-200 years design event. Fig. 4.4 synthesizes the analyses by highlighting the flood magnitude range with the highest sensitivity gradient. Rivers in South America, Equatorial and Sub-Saharan Africa, Eastern Europe, Middle East, and Southeast Asia exhibit the highest sensitivity to changes in flood magnitudes in the lower range (RP10/20, RP20/50). These river reaches have relatively low protection standards and their flood exposure function is concave in shape. In contrast, the rivers of India, North America, Europe, Australia, Asia, and South Africa are more sensitive to changes in the upper range of flood magnitudes (RP100/200, RP200/500) and thus have convex exposure functions. The rivers of Central and East Asia are most sensitive to changes in the upper range of flood magnitudes since they have protection standards in the middle of the flood magnitude range.

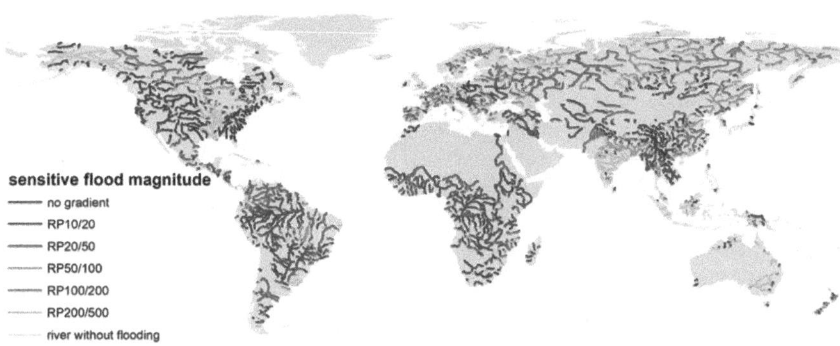

FIGURE 4.4 The sensitivity of rivers to changes in flood magnitude. Range of flood magnitudes showing the highest relative gradient for sensitivity to change. *From Zischg, A. P., & Bermúdez, M. (2020). Mapping the sensitivity of population exposure to changes in flood magnitude: Prospective application from local to global scale.* Frontiers in Earth Science, 8. *https://doi.org/10.3389/feart.2020.534735.*

The knowledge of the regionally varying sensitivities of the floodplains to climatic changes allows the estimation of how flood risk will change in the near future. Of course, this sensitivity per se will not remain stable and will change itself when the circumstances change. Interestingly, the form of the flood risk curve also allows getting a first impression on how the rate of change of flood exposure will change in the future. If we remember Fig. 4.3, the normalized gradient changes along the flood magnitudes indicate that flood risks in some floodplains may change suddenly with a minor change of flood magnitudes, but with a higher increase in flood magnitudes, the gradient may be lower and vice versa. The example furthermore shows the spatial complexity of these sensitivities. A high regional variability is therefore expected for the impacts of climatic changes on flood risk. The example also shows how bottom-up analyses of the sensitivity of systems to change in environmental conditions can be analyzed. If this bottom-up analysis approach is once combined with a top-down analysis approach, e.g., modeling the process chain from climate, weather, rainfall, discharge, to flood impacts, it may enhance the understanding of the process behind changing climate impacts (Zischg, 2018). The maps of the most sensitive floodplains furthermore could support identifying those areas where flood adaptation is needed in a framework of vulnerability-centered flood risk management strategies.

References

Alfieri, L., Bisselink, B., Dottori, F., Naumann, G., de Roo, A., Salamon, P., Wyser, K., & Feyen, L. (2017). Global projections of river flood risk in a warmer world. *Earth's Future, 5*(2), 171–182. https://doi.org/10.1002/2016EF000485

Boulton, J. G., Allen, P. M., & Bowman, C. (2015). *Embracing complexity: Strategic perspectives for an age of turbulence*. Oxford University Press.

Brönnimann, S., Stucki, P., & Zischg, A. (2019). *Simulation des Hochwassers von 1868 und Lehren für die Zukunft* (Vol. 78, pp. 13–20).

Brönnimann, S., Rohr, C., Stucki, P., Summermatter, S., Bandhauer, M., Barton, Y., Fischer, A., Froidevaux, P., Germann, U., Grosjean, M., Hupfer, F., Ingold, K., Isotta, F., Keiler, M., Martius, O., Messmer, M., Mülchi, R., Panziera, L., Pfister, L., ... Zischg, A. (2018). *1868 – the flood that changed Switzerland: Causes, consequences and lessons for the future* (Vol. 94).

CIESIN- Facebook Connectivity Lab and Center for International Earth Science Information Network, Columbia University. (2016). *High resolution settlement layer (HRSL)*. Columbia University.

Dottori, F., Salamon, P., Bianchi, A., Alfieri, L., Hirpa, F. A., & Feyen, L. (2016). Development and evaluation of a framework for global flood hazard mapping. *Advances in Water Resources, 94*, 87–102. https://doi.org/10.1016/j.advwatres.2016.05.002

Dottori, F., Szewczyk, W., Ciscar, J. C., Zhao, F., Alfieri, L., Hirabayashi, Y., Bianchi, A., Mongelli, I., Frieler, K., Betts, R. A., & Feyen, L. (2018). Increased human and economic losses from river flooding with anthropogenic warming. *Nature Climate Change, 8*(9), 781–786. https://doi.org/10.1038/s41558-018-0257-z

Elmer, F., Thieken, A. H., Pech, I., & Kreibich, H. (2010). Influence of flood frequency on residential building losses. *Natural Hazards and Earth System Sciences, 10*(10), 2145–2159. https://doi.org/10.5194/nhess-10-2145-2010

Gunderson, L., Cosens, B. A., Chaffin, B. C., Arnold, C. A. T., Fremier, A. K., Garmestani, A. S., Craig, R. K., Gosnell, H., Birge, H. E., Allen, C. R., Benson, M. H., Morrison, R. R., Stone, M. C., Hamm, J. A., Nemec, K., Schlager, E., & Llewellyn, D. (2017). Regime shifts and panarchies in regional scale social-ecological water systems. *Ecology and Society, 22*(1). https://doi.org/10.5751/ES-08879-220131

Haimes, Y. Y. (2009). On the complex definition of risk: A systems-based approach. *Risk Analysis, 29*(12), 1647–1654. https://doi.org/10.1111/j.1539-6924.2009.01310.x

Holland, J. H. (2006). Studying complex adaptive systems. *Journal of Systems Science and Complexity, 19*(1), 1–8. https://doi.org/10.1007/s11424-006-0001-z

Karegar, M. A., Dixon, T. H., Malservisi, R., Kusche, J., & Engelhart, S. E. (2017). Nuisance flooding and relative sea-level rise: The importance of present-day land motion. *Scientific Reports, 7*(1). https://doi.org/10.1038/s41598-017-11544-y

Kreibich, H., Thieken, A. H., Petrow, T., Müller, M., & Merz, B. (2005). Flood loss reduction of private households due to building precautionary measures—lessons learned from the Elbe flood in August 2002. *Natural Hazards and Earth System Sciences, 5*(1), 117–126. https://doi.org/10.5194/nhess-5-117-2005

Ladyman, J., & Wiesner, K. (2020). *What is a complex system*. Yale University Press.

Marchau, V. A. W. J., Walker, W. E., Bloemen, P. J. T. M., & Popper, S. W. (2019). *Decision making under deep uncertainty: From theory to practice*.

Messager, M. L., Lehner, B., Cockburn, C., Lamouroux, N., Pella, H., Snelder, T., Tockner, K., Trautmann, T., Watt, C., & Datry, T. (2021). Global prevalence of non-perennial rivers and streams. *Nature, 594*(7863), 391–397. https://doi.org/10.1038/s41586-021-03565-5

Moftakhari, H. R., AghaKouchak, A., Sanders, B. F., & Matthew, R. A. (2017). Cumulative hazard: The case of nuisance flooding. *Earth's Future, 5*(2), 214–223. https://doi.org/10.1002/2016EF000494

Pfister, C. (1999). *Wetternachhersage: 500 Jahre Klimavariationen und Naturkatastrophen*. Bern: Haupt.

Scussolini, P., Aerts, J. C. J. H., Jongman, B., Bouwer, L. M., Winsemius, H. C., De Moel, H., & Ward, P. J. (2016). FLOPROS: An evolving global database of flood protection standards. *Natural Hazards and Earth System Sciences, 16*(5), 1049–1061. https://doi.org/10.5194/nhess-16-1049-2016

Taleb, N. N. (2012). *Antifragile: Things that gain from disorder/Nassim Nicholas Taleb*. Random House.

Thieken, A. H., Kreibich, H., Müller, M., & Merz, B. (2007). Coping with floods: Preparedness, response and recovery of flood-affected residents in Germany in 2002. *Hydrological Sciences Journal, 52*(5), 1016–1037. https://doi.org/10.1623/hysj.52.5.1016

Zischg, A. (2016). River corrections and long-term changes in flood risk in the Aare valley, Switzerland. *E3S Web of Conferences, 7*, 7. https://doi.org/10.1051/e3sconf/20160711010

Zischg, A. (2018). Floodplains and complex adaptive systems—perspectives on connecting the dots in flood risk assessment with coupled component models. *Systems, 6*(2), 9. https://doi.org/10.3390/systems6020009

Zischg, A. P., & Bermúdez, M. (2020). Mapping the sensitivity of population exposure to changes in flood magnitude: Prospective application from local to global scale. *Frontiers in Earth Science, 8*. https://doi.org/10.3389/feart.2020.534735

Zischg, A. P., Hofer, P., Mosimann, M., Röthlisberger, V., Ramirez, J. A., Keiler, M., & Weingartner, R. (2018). Flood risk (d) evolution: Disentangling key drivers of flood risk change with a retro-model experiment. *Science of the Total Environment, 639*, 195–207. https://doi.org/10.1016/j.scitotenv.2018.05.056

Chapter 5

Modeling spatiotemporal dynamics of flood risk change

Four out of five Swiss communities have been affected by floods in the past 40 years. They not only caused a lot of suffering for the affected population, but also resulted in high costs. Floods are responsible for more than two-thirds of the damage caused by natural events. Around 300′000 buildings are located in flood-prone areas, which is 13% of all buildings. The replacement value of all buildings in hazardous areas is around 500 billion Swiss Francs. Because building land is a very rare commodity in Switzerland, more and more buildings have been built in flood-prone areas in recent decades. This has an impact on the damage. Every seventh person lives in a building at risk of flooding. In addition to residential buildings, numerous infrastructure structures such as roads, railroads, and hospitals are also affected by floods.

In Switzerland, the principle of subsidiarity applies when dealing with natural hazards. It is therefore primarily the municipalities that are responsible for the defense against natural hazards. Together with the canton, they ensure that the hazard is recognized and that it causes as little damage as possible. To this end, they take appropriate structural, organizational, and spatial planning measures. To ensure that damage does not increase in the future, the municipalities must take natural hazards into account in activities with a spatial impact, such as building densification. They do this by integrating hazard maps into land use planning. Hazard maps are the central instrument of risk-based spatial planning. However, dividing the municipal territory into hazard zones is not enough—the hazard maps must be interpreted from a risk perspective. In this task of widening the perspective from a hazard-centered view to a risk-centered management approach, the governments of municipalities need support. As seen in the previous chapters, governing flood risk change is a complex task. To support the understanding of the complexity of flood risk change and to support the paradigm change from a pure hazard defense and emergency response approach to a risk-based management approach, the Mobiliar Lab for Natural Risks of the University of Bern developed interactive online tools. Moreover, these tools aim at supporting risk communications and rising risk awareness activities. The Mobiliar Lab for Natural Risks (www.mobiliarlab.unibe.ch) is a joint research initiative at the

Flood Risk Change. https://doi.org/10.1016/B978-0-12-822011-5.00003-X

University of Bern's Oeschger Centre for Climate Change Research and Swiss Mobiliar. The Mobiliar Lab's research work focuses mainly on floods, storms, and hail as well as the associated loss and damage potential, with strong emphasis placed on the practical application of findings. The Mobiliar Lab was established in 2013. The partners jointly defined the scope of the research's subject matter. Since then, researchers have further expanded their focus for practical use.

In the following, I will show a few examples of model experiments, analyses, and research outcomes that have been implemented in publicly accessible and interactive web-based tools. The tools are a main outcome of the project "Research Initiative Flood Risk—from Understanding to Action." With this research initiative, the Mobiliar Lab for Natural Risks, the Oeschger Centre for Climate Change Research and the Unit Geomorphology of the Institute of Geography at the University of Bern complemented traditional process research with the aspects of damage, vulnerability, and flood risk change. In this initiative, several interactive online tools and maps have been developed that can be used as decision support tools or for risk communication. The tools are published on the web platform www.floodrisk.ch. Special emphasis was placed on innovative cartographic visualizations that can be used by authorities in their risk communication.

Examples of flood risk change analysis

I am presenting five interactive online tools for visualizing natural risks and their changes in space and time. They all have different purposes, show different aspects of flood risk, and have different levels of complexity. First, I will show the image database "Collective Flood Memory" that makes geolocated images of past flood events accessible to the interested public. Second, and third, I will describe the tools "Damage Potential" and "Damage Simulator." These are decision support tools for authorities of municipalities and cantons to identify where major property damage could occur. The first of these two tools focuses on flood exposure. It answers the question of what types of assets, values, and infrastructure are exposed to floods in Switzerland. The second focuses on flood damage that may occur in a flood event. This tool is projecting future changes in flood risk due to changes in hazard, exposure, and vulnerability. The fourth tool is focusing on the dynamics of flood risk change in the past two centuries and thus is named "Flood risk dynamics." This tool valorizes a model experiment on disentangling the effects of the main drivers of risk on flood risk change. The fifth tool shows the spatiotemporal evolution of selected extreme flood events. It is based on the coupled component model atmosphere-hydrology-hydraulics-impacts and is named "Flood dynamics."

The tool "Flood memory"

As demonstrated in the previous chapter, society is learning from previous flood events and adapting its environment according to new knowledge (Collenteur et al., 2015; Tobin, 1995; Tobin & Montz, 1994; White et al., 2001). As long as society is aware of the potential hazard represented by a past flood event, the previously flooded areas are prevented from being used for immobile structures. The societal memory of the flood event is declining after time, unless it is kept alive by memorials or periodical cultural ceremonies. This period of awareness lasts about one generation (Fanta et al., 2019).

Societal learning, i.e., the societal memory of a flood event, is an important driver to keep flood risk constant. The previously flooded areas are kept free from settlements for a certain time period. This is an important time delay effect that is slowing or stopping the urban spreading toward flood-prone areas for a certain period. Moreover, the societal memory effect influences the path dependency of future flood risk change. The areas that have been kept free for a certain period are determining the future flood risk development pathway. In contrast to already occupied areas, these areas will offer more degrees of freedom for future generations. The number of opportunities for flood risk management and thus the number of options to deal with flood risks will be much higher than in other cases. For example, the option for future river widening to enhance discharge capacity and reduce flooding will only be possible under such circumstances. Thus, sustainable flood risk management aims at keeping the societal memory of flood events alive as long as possible.

This aim of keeping awareness and societal memory alive was the motivation for the development of the tool "Collective flood memory" (www.ueberschwemmungsgedaechtnis.ch). The tool is a database for images and photos of flood events. The photos are georeferenced, meaning that each photo has coordinates that allow for geolocating the photo. Moreover, each photo is attributed with the date and principal information about the content. The graphical user interface of the tool consists of a map that shows the location of all photos in the database throughout Switzerland. The dataset can be constrained by selecting a certain time period with a time slider. The map is located in the center of the graphical user interface, and the time slider and the header are located at the top. On the right hand, all photos within the map extent are listed and previewed. They are chronologically ordered. Searching for photos can be done in two main ways. The first approach is to zoom in to a specific location on the map. Consequently, all photos are selected and listed in the preview that is located within the map extent. The second option for searching is to enter the name of the municipality, the name of the location, or the name of the river. A filter allows for selecting images with specific copyrights, e.g., images uploaded under the creative commons license. Fig. 5.1 shows the graphical user interface of the tool.

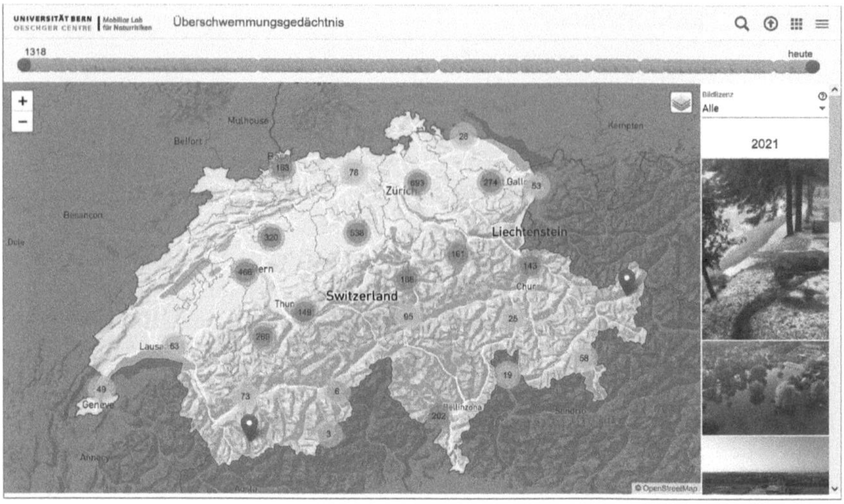

FIGURE 5.1 **Flood memory.** Graphical user interface of the tool "Collective Flood Memory" for collecting and visualizing images and photos of flood events to keep societal memory updated (www.ueberschwemmungsgedaechtnis.ch).

The main innovation of the tool is that each citizen can upload photos of flood events. Therefore, the tool is growing with the community effort. With this characteristic, the tool is a contribution to citizen science approaches. An upload service asks the user for information about the image location and content, it enables an easy georeferentiation if the photo itself has no coordinates and about the contact address and license. All uploaded photos are reviewed by an editor. Only after a quality check by the editor, will the image be published. After a few years of experience, the main advantage of the tool is the possibility to search for photos by geographical location. Moreover, the upload functionality is very well used. During rainy periods and after flood events, the tool is highly accessed by Internet users. In June and July 2021, the tool had more than 6000 views, and more than 70 photos have been uploaded by citizens. According to feedback from users, the tool is also used in risk communication, e.g., during information events of municipal governments that are presenting flood risk prevention projects. Overall, we see a high potential for the tool for keeping alive the societal memory of flood events during the next decades. From a complexity perspective, it is a small but important contribution to keep flood risk change low by promoting social awareness and therefore by extending the time lag between a flood event and a potential urbanization of the formerly flooded areas.

The tool "Flood damage potential"

Recently, nearly all municipalities in Switzerland finished the elaboration of hazard maps. Now, the hazard maps are being implemented on the land use zoning and land use regulation maps and therefore are becoming legally binding for the authorities and for landowners in the form of hazard zone maps. These maps will guide urban development for the next decades. The hazard zone maps will be periodically updated, e.g., every 10 years. However, the hazard zone maps give information only about the hazard. The other factors that determine flood risk are missing, i.e., information on exposure or vulnerability. Thus, after the completion of all hazard maps in the Swiss municipalities, a spatially high-resolution overview of flood-prone assets was missing. The extension of flood hazard maps toward flood exposure maps aims at increasing the information content of maps on flood risks.

The federal government, cantons, and municipalities spend hundreds of millions of Swiss Francs each year to protect the population and property from flooding. Flood hazard maps show with high spatial resolution where, how often, and how intense, flood events can occur. If the hazard maps are spatially analyzed in combination with socioeconomic data such as population data or building values, the locations of hotspots with high damage can be estimated. This information is necessary for assessing where a hazard plays an important role for society and where it does not. The information on flood exposure allows a simplified weighting or prioritization of the hazard zones. This might contribute to a better priority setting in taking actions for flood risk reduction.

Flood risk assessments at the local level are now standard practice in the planning of flood protection measures. Some cantons have also prepared risk overviews based on hazard maps (e.g., Canton of Zurich, Canton of Schaffhausen). Switzerland-wide evaluations of flood hazards to different protected assets (e.g., residential population, workplaces) have been conducted by the Federal Office for the Environment (Bezzola & Loat, 2016). These analyses rely on coarse-resolution data of modeled floodplains ("Aquaprotect," FOEN and SwissRe, 2008), which, unlike hazard maps, have not been verified in the field, do not cover small catchments, and do not consider protective structures. Flood risk assessments are also relevant for other institutions such as the Swiss Federal Railways (SBB) and the Federal Roads Office (FEDRO). Both produce such analyses for their infrastructure networks. The efforts of all these actors illustrate the relevance of flood risk assessments. Nevertheless, a nationwide overview of flood-prone assets with a high spatial resolution is still missing. The Mobiliar Lab for Natural Risks aimed to make a significant contribution to close this gap with the tool "Flood damage potential." The term "damage potential" here describes the absolute monetary value or other quantitative characteristics of flood-exposed assets, persons, and infrastructure.

The interactive web-tool "Flood damage potential" (www.schadenpotenzial.ch) is publishing the results of an applied research project on flood exposure in Switzerland. The research project aimed at evaluating different methods and parameters for flood exposure analysis. This should contribute to a better understanding of one of the flood risk drivers, namely exposure. It is essential for the effective and efficient implementation of risk-reduction strategies. In this analysis Röthlisberger et al. (2017) considered two types of exposure indicators and two different spatial aggregation schemes: the density of exposed assets (exposed numbers per km^2) and the ratio of exposed assets (share of exposed assets compared to total amount of assets in a specific region). The spatial units were administrative geographical units such as municipalities or Cantons and grid cells of similar size as the municipalities. While identifying high densities of exposed assets highlights priority areas for cost-efficient strategies, high exposure ratios can suggest areas of interest for strategies focused on the most vulnerable regions, i.e., regions with a low capacity to cope with a disaster. The spatial distribution of high exposure densities and high exposure ratios identified different regions of exposure hotspots. The study found that spatial cluster analysis provides more information for the prioritization of flood protection measures than "simple" maps of spatially aggregated data represented in quantiles. In addition, the study showed that the data aggregation scheme influenced the results of the flood exposure hot spot analysis. The aggregation based on grid cells supports the comparability of different regions better than aggregation based on municipalities and is, thus, more appropriate for nation-wide analyses.

The implementation of this research into an interactive online tool followed primarily the conclusions of this paper. Therefore, the outreach tool aims at visualizing flood exposure in different geographical aggregation units. As most of the stakeholders are working with administrative geographical units, we aggregated the number and monetary values of exposed people and assets in these administrative spatial units, namely cantons, municipalities, and neighborhoods. In addition, the tool is visualizing the flood exposure also in regular spatial units, i.e., in a hexagonal grid. As suggested by the research paper mentioned above, the tool shows the absolute numbers of exposed assets and people but also the share of the exposed assets to the total number of assets within the geographical unit.

The following assets can be selected:

- buildings (total, differentiated by use),
- building values, persons (resident persons, workplaces),
- hospitals,
- cultural assets,
- schools and universities,
- as well as retirement and nursing homes.

In summary, the user of the tool "Flood damage potential" can select four filters: (1) the absolute or relative (share) numbers of exposed assets; (2) the category of exposed people or assets; (3) the hazard class according to the three categories: low, medium, and high hazard level; and (4) the spatial aggregation unit.

The exposure analysis was based on the cantonal hazard maps in Switzerland. The overlay between the flood hazard maps and the exposed assets was made at the original spatial resolution, namely the coordinates of the single buildings. The output of the analysis had to be aggregated to the targeted spatial aggregation unit because of the privacy legislation. Due to this regulation, individual objects cannot be visualized on the maps but have to be aggregated. However, the analysis is based on the overlay of individual objects. In the following, the procedure of exposure analysis is described.

Method

The first step of the preprocessing was the harmonization of the hazard maps from the cantons. This step was done by the Mobiliar insurance company. The hazard maps were harmonized, geometrically adjusted, and merged for the whole of Switzerland. The result was a seamless hazard map that assigns one of the following five hazard levels to each area in Switzerland: "high hazard" (red), "medium hazard" (blue), "low hazard" (yellow), "no hazard" (white) or "residual hazard" (yellow/white), and "no analysis." The category "no analysis" stands for areas that lie outside the investigation perimeter of the hazard map.

The second step was the preparation of a dataset of the buildings. The geometries of the building's footprint come from the digital terrain and landscape model swissTLM3D (Swisstopo, 2021). The buildings were attributed with information about the number of residents, the number of working places, the functionality of the building, the period of construction, and the monetary value of the building (the reconstruction value). The procedure for attributing the building values comes from Röthlisberger et al. (2018). The model parameters were estimated on the basis of more than half a million building insurance contracts in Switzerland. The data come from 11 (out of a total of 19) cantonal building insurance companies, which hold a monopoly within the respective canton. The reconstruction values were calculated using two different methods. Both methods assume average cubic meter prices of above-ground building volumes differentiated by various criteria. The volumes used are identical in both methods and are determined from the swissALTI3D (Swisstopo, 2014) digital elevation and digital terrain surface models. In the first method ("Basis Canton"), a uniform cubic meter price was used per canton. This is based on the determined building volumes and the total insurance sums from the 2017 annual report of the respective cantonal building insurance. For the seven cantons without cantonal building insurance, the value of a building insurance canton with a similar building structure was used. With the second method ("Basis building zone"), the cubic meter price is determined uniformly throughout Switzerland, but differentiated according to the use of the building and the land use planning zone type where the building was located. The flood hazard map was overlaid with the building footprints from the topographic landscape model (swissTLM3D). The highest hazard level of all hazard levels that intersected with the building footprint was determined for each building. Each building was attributed to the number of employees from the statistics of business structure (BFS, 2021a) and the number of inhabitants from the statistics of buildings and dwellings (BFS, 2021b). All buildings with attached flood hazard information were overlaid with the number of persons with main residence per building from the Building and Housing Statistics of the Federal Office for Statistics (BFS, 2021b). In addition to the building dataset, this extended dataset shows the geolocated numbers of residents, the numbers of work places, and the monetary values. The flood hazard map was overlaid with the school and university areas and the hospital areas from the swissTLM3D dataset. For each school and university area or hospital area, the highest hazard level at which at least one building is located was determined. The flood hazard map was overlaid with other cultural assets in the inventory of cultural assets of the Federal Office for Civil Protection. The cultural property protection inventory distinguishes between four types of cultural property: individual buildings, ensembles of buildings, archaeology, and special cases. The number of other cultural assets per hazard level was added for each spatial unit. All buildings with attached flood hazard information were overlaid with the key figures of Swiss nursing homes from the Federal Office of Public Health. For each place/quarter, municipality, canton, and hexagon, the number of old-age and nursing homes per hazard level was summed up.

The number of buildings, persons, cultural assets, and areas per hazard level was summed up for each geographical unit, namely the cantons, the municipalities, the neighborhoods, and the hexagonal grids. The administrative boundaries were taken from the swissBOUNDARIES3D dataset of the Federal Office of Topography. The lake areas of the larger lakes were cut off. The neighborhood and quarter boundaries are taken from the data set swissNAMES3D of the Federal Office of Topography. Quarter data from a total of 27 cities were slightly adjusted. For a neighborhood or a quarter to be taken into account, it must have at least 100 inhabitants or an area of at least 10 ha. In addition, more than 10 buildings must be included in each of the spatial units. In many cases, a political municipality consists of only one neighborhood/quarter. In most cases, however, the perimeters of the two geographical regions did not coincide. While the perimeters of the municipalities cover the whole area, the perimeters of neighborhoods/quarters are limited to the effectively populated area. A neighborhood/quarter was assigned to the municipality in which the center of gravity is located. A grid of hexagons was laid over Switzerland. Each hexagon has an area of 10 km^2. To prevent potential inferences to individuals, the minimum possible value of resident population and working place per geographic region was set to 3. To prevent conclusions about individual companies, the minimum possible value of the number of working places per geographic region was set to 3. The share of elements at risk of flooding to the overall number of elements in each spatial unit was calculated by dividing the sum of exposed elements by the sum of all elements within the same spatial unit.

Cartography

To allow for assessing the numbers of flood exposure visually, we focused on the cartographic design. The cartography should show the absolute and the relative numbers of exposed elements, as well as the combination of both. The size of the circles indicates the number of elements of the selected category, which are located in a flood hazard area. The number of elements adapts to the current configuration (selection of protected elements, hazard, and geographic region). In contrast, colors indicate the proportions of endangered elements of the selected category for the selected spatial units. The proportions adapt to the current configuration (selection of protected elements, hazard category, and spatial unit). The size of the circles and the colors of the relative values represent the combination of both.

Results

The interactive web-tool "Flood damage potential" is available at www.schadenpotenzial.ch. It shows that in Switzerland, about 300,000 buildings are located in flood-prone areas. This corresponds to 13.6% of all buildings. Buildings with workplaces, on the other hand, are located more frequently (17.5%) in flood hazard areas. School and university premises (19.1%), hospital premises (17.8%) as well as retirement and nursing homes (16.6%) are also at above-average risk of flooding. The fact that buildings with residential use are proportionately less at risk of flooding than those with workplaces is also reflected in the number of people affected. While 15.8% (1.3 of a total of 8.4 million) of the population live in a flood-prone building, almost 21.8% (1.1 of a total of 5.1 million) of workplaces are located in a flood hazard area. If the monetary values of the flood-exposed buildings are considered, then for both value determination methods (basis canton or construction zone), the share at risk is 18.5%; this corresponds to exposed new construction values of around CHF 500 billion. The share of cultural assets at risk of flooding is comparable to the share of buildings at risk of flooding. It is 14% (cultural assets of individual buildings) compared to 11.6% (other cultural assets). In total, 1.75 million of 2.2 million buildings in Switzerland are currently covered by a flood hazard map. This corresponds to a coverage rate of 79.1%. The majority of the buildings outside the hazard map perimeter are buildings outside the building zone, e.g., agricultural buildings. The degree of coverage of the protected assets under consideration ranges from 67.6% (buildings without residential use and workplaces) to 93.8% (school and university premises).

The map view Fig. 5.2 shows that in the Alpine arc (e.g., in St. Gallen Rhine Valley, Bernese Oberland, and Valais), a high share of buildings, as well as of the residential and working population are exposed to floods. In 200 of these municipalities, more than 30% of the buildings are located in flood-prone areas. In absolute numbers, the large urban centers of the Central Plateau stand out; in particular the city of Zurich, where 3800 buildings, 86,000 workplaces,

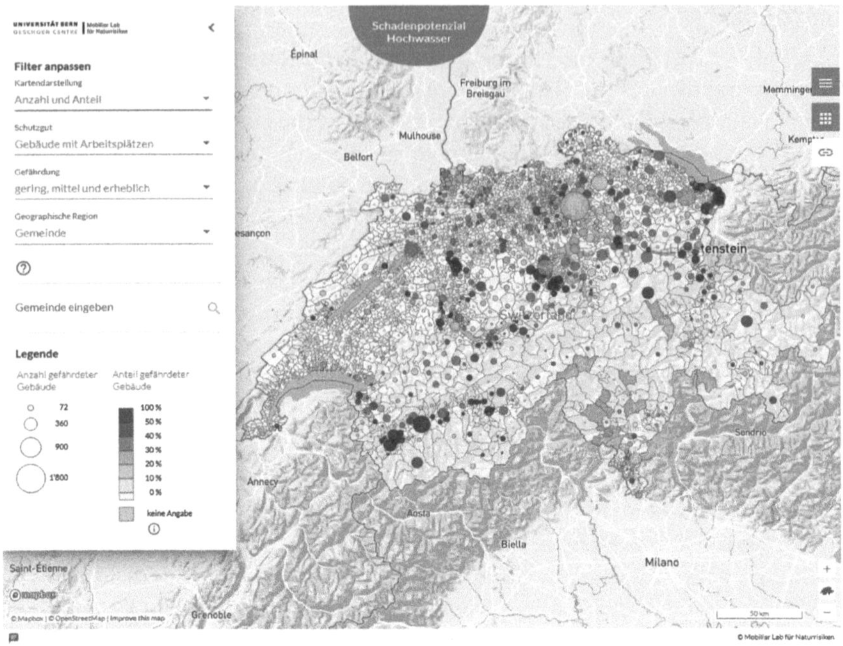

FIGURE 5.2 **Flood damage potential.** Screenshot of the interactive web-tool "Flood damage potential."

and 80,000 inhabitants are at risk from flooding. Particularly affected, in terms of both proportion and number, are the St. Gallen Rhine Valley, the Linth Plain, the regions around Lucerne and Interlaken, and the Rhone Valley between Brig and Lake Geneva.

The tool allows to query detailed information for selected geographical units. The data aggregated at municipality level can be compared with the data aggregated at cantonal and national level. With this ability, the user can assess if the selected municipality has a higher or lower flood exposure than the municipalities in the same canton or than all other municipalities in Switzerland. This ranking can be made on the basis of the flood hazard level and for each parameter of exposure, e.g., for residents, workplaces, numbers and values of buildings, etc. Figs. 5.3 and 5.4 exemplarily compare the municipality of Kandersteg in the Bernese Alps with the Canton of Bern and of the whole Switzerland.

Conclusions from the complexity perspective

Effective and efficient flood risk reduction strategies require sound knowledge of key risk components (hazard, exposure, and vulnerability). In the case of a risk reduction strategy focusing on cost efficiency, areas with a high absolute number and a high density of exposed assets imply a high priority for

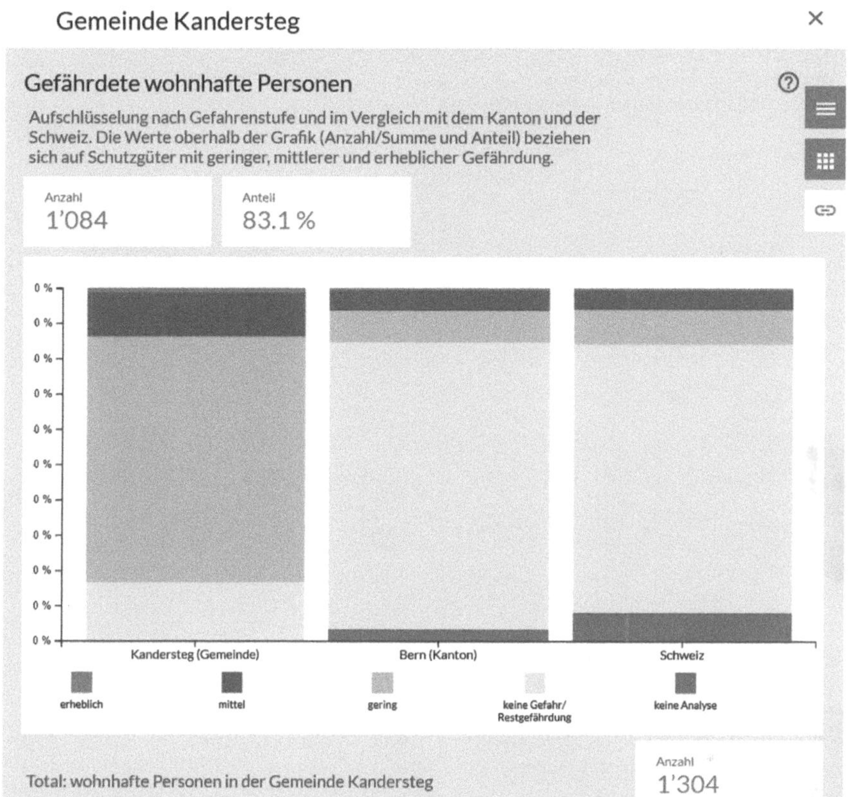

FIGURE 5.3 **Comparison of damage potential.** Screenshot of the query for the number and share of residents that are exposed to flood hazards in the municipality of Kandersteg. In total, 1084 of 1304 residents are exposed to floods. This corresponds to a share of 83.1%. In comparison to the Canton of Bern and Switzerland, the municipality of Kandersteg has a much higher share of exposed residents.

protective measures. Areas with a high proportion of exposed elements, on the other hand, are at the center of strategies that prioritize the protection of areas with high vulnerability, i.e., with low coping capacity in case of an event because a high share of the population will be affected by an event. In Switzerland, areas with a high density of exposed elements hardly overlap with areas with a high proportion of exposed elements. A high density of exposed elements is found mainly in the eastern Central Plateau and in the river plains of the Rhone and Rhine close to the Alps. A high proportion of exposure, on the other hand, is found in the northern Alpine regions. The tool makes it possible for the first time to identify the damage potential of floods, i.e., the flood exposure, down to the level of individual neighborhoods throughout Switzerland. This makes it possible to show how many people or important assets are exposed to flood hazards. With different analysis tools, the

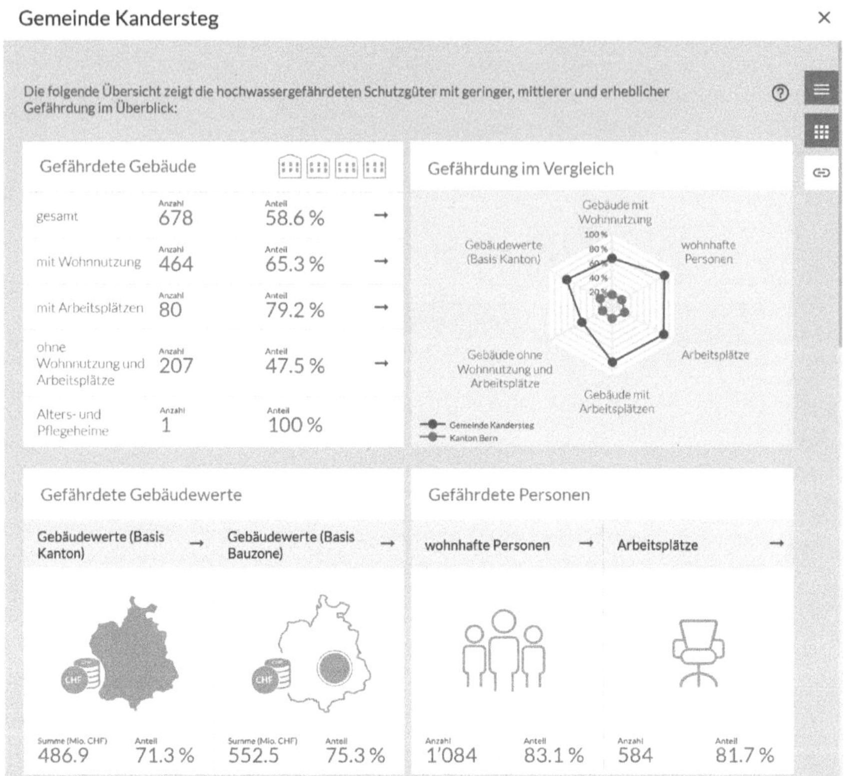

FIGURE 5.4 Summary of exposure. Screenshot of the query for the number and share of residents, buildings, workplaces, and other facilities that are exposed to flood hazards in the municipality of Kandersteg. In total, 678 buildings are exposed to floods. This corresponds to a share of 58.6%. The reconstruction values of these buildings correspond to 486.9—552.5 mio Swiss Francs. Eighty buildings with workplaces and one retirement and nursing home are exposed. The spider graph at the upper right shows the exposure profile of the municipality (red (dark gray in print)) in comparison to the canton (turquoise (light gray in print)). In comparison to the Canton of Bern, the municipality of Kandersteg has a much higher share of exposure for all categories of assets at risk.

absolute and relative values of the flood endangered assets can be compared easily. From the complexity perspective, the tool shows that the spatial patterns of hotspots of exposure are complex and heavily depend on the metric used for looking at exposure. It furthermore allows us to assess where different prioritization strategies and philosophies for the planning of flood prevention projects will shape the future risk landscape.

The tool "Flood damage simulator"

The next step from highlighting flood exposure toward flood risk mapping is to calculate potential impacts of floods by considering the vulnerability of the exposed assets. This step was implemented in the interactive web-tool "Flood damage simulator" (damagesimulator.ch). It builds upon the previously described tool "Flood damage potential," but extends the latter with information about the expected damage. The damages are expressed in monetary units.

The main reasoning behind this tool is to make a distinction between the damage in areas with a low risk of flooding (yellow hazard zones, i.e., a low hazard level) and damage in the other hazard zones (blue and red hazard zones, respectively, the medium and high hazard levels). The previous tool showed that the number of exposed buildings is very high in yellow hazard zones. These land use zones identify and point out potential hazards but do not have binding regulations for homeowners. This means that the construction of buildings in yellow hazard zones is not regulated in terms of flood-prone construction techniques. Therefore, these zones have only little regulatory impact. In contrast, new buildings in blue hazard zones can only be constructed if they are structurally adapted to the hazard level, meaning that they must have permanent property-level flood protection measures to reduce flood vulnerability. The question therefore arises if the relatively higher number of exposed buildings in the yellow hazard zone results in a higher flood risk than the relatively lower number of exposed buildings in blue hazard zones, nevertheless of the lower flood magnitudes.

In addition to the goal of quantifying and visualizing potential flood damage, a second aim of the "Flood damage simulator" tool was to show the influence of property-level object protection on flood risk. The user can select the ratio between buildings with property-level flood protection measures and all buildings within the geographical unit. The higher the ratio, i.e., the higher the implementation level of object protection measures, the lower the flood damage will be in case of a flood event.

Moreover, this tool quantifies the future development of flood damage depending on the future densification of the settlements. The areas where the communal land use zoning and land use regulation plans dedicate land reserves for future construction of buildings were virtually filled with buildings according to the local building codes. This "filling" of land within the current settlements that is currently free from buildings but is reserved for construction and urban development is called "densification." The aim of urban densification is to avoid urban sprawl and thus to foster a sustainable use of land resources. The question was if this sustainability goal of urban densification could have negative consequences for flood risk management, as for example, by increasing flood exposure and flood risks.

The main metric of the "Flood damage simulator" tool is the "possible extent of damage." The possible extent of damage corresponds to the expected damage to buildings per municipality or hexagon for the selected scenario. There are three main scenarios that can be preselected by the user: (a) the current situation of hazard, exposure, and vulnerability; (b) the future situation of hazard, exposure, and vulnerability (around the year 2040); and (c) a user-defined scenario with different combinations of hazard, exposure, and vulnerability. The possible extent of damage is determined by the selected scenario settings as well as an assumed flood event that inundates the yellow, blue, and red zones of the hazard map. The possible extent of damage is indicated as a monetary value in Swiss Francs for the selected scenario. This value therefore does not correspond to the risk, which, in addition to the extent of damage, also considers the probability of a flood event occurring and thus indicates the expected damage for a certain period of time (e.g., per year). The possible extent of damage to a potentially affected building is calculated as the product of the value of the building and its vulnerability, with the value corresponding to the reconstruction value of the building. The vulnerability of the elements at risk depends on the hazard category.

Method

The "Flood damage simulator" shows the possible extent of damage to buildings from flood inundation in Switzerland. The assessments are summarized per municipality or hexagon. The possible extent of damage was modeled for the whole of Switzerland and is therefore based on data available nationwide, such as flood hazard maps, the values of buildings exposed to flooding, and data on documented flood damage from past damaging events. The hazard maps were described in the previous section of this chapter. The flood hazard maps were overlaid with the building footprints from the topographic landscape model of Switzerland (swissTLM3D, Swisstopo, 2021). The value of buildings was assessed based on the replacement costs per cubic meter and the above-ground building volume. The volumes were calculated using the digital elevation and surface model of Switzerland. A uniform replacement cost per cubic meter was used per canton. This is based on the building volumes of all buildings and the total insurance sums given in the 2017 annual report of the respective cantonal buildings' insurance companies (Röthlisberger et al., 2018).

The main advantage of the tool "Flood damage simulator" against the previously described tool is the consideration of the vulnerability of buildings. Here, vulnerability expresses how strongly an object is impacted by a hazardous process. It assumes a value between 0 (no damage) and 1 (complete destruction). Applied to a single building, this means that a flood will cause either no damage (vulnerability equals 0), a certain degree of damage (vulnerability between 0 and 1), or complete destruction (vulnerability equal to 1). To calculate the possible extent of damage to all buildings in Switzerland, we first derived the degree of damage per hazard level (yellow, blue, red) for each of the documented flood insurance claims. This was done by overlaying the geolocated insurance claim data with the hazard maps. In 19 out of 26 cantons, the insurance companies cover building insurance as monopoly institutions, while in the remaining cantons, private insurance companies insure buildings. In 13 of the 19 cantons with public insurance companies, complete information on flood loss claims was available, including the amount of loss, the corresponding insured building value, the date and location of the flood event (for further details, see Bernet et al., 2017, 2019).

For each insurance claim, we calculated the ratio between the monetary damage and the building value, which corresponds to the degree of damage or the vulnerability value per claim. A total of 40′000 claims were evaluated. Of these, 38% related to damage that occurred in a yellow (low hazard level), blue (medium hazard level), or red hazard zone (high hazard level). The flood damage taken into account was reported by 13 cantonal buildings insurance companies from 1979 to 2013. Since the extent to which a building is damaged depends strongly on the intensity of the flood, the degree of damage is primarily related to the intensity of the flood. Thus, we intersected the documented flood insurance claims with the hazard maps in order to only select those claims that occurred in one of the three hazard zones, resulting in a dataset with 14′176 claims (Fig. 5.5). Once the degree of damage from all observations was calculated, we derived a specific degree of damage distribution per hazard level and tested for further influence variables. Besides the hazard level, the building value had the highest importance in modeling the degree of damage. Thus, in addition to the hazard levels, we specified the degree of damage for three different building value categories. The thresholds for classifying the building value categories were identified by a cluster analysis. To verify the model parameter selection, we examined the relationship between the degree of damage, the hazard maps, and the building value. First, we tested whether the degree of damage differed significantly between the individual hazard levels, where the significance level for this study is set at $P = .05$. Secondly, we performed a Spearman correlation test between degree of damage and building values.

To select the best parameter combinations for the model that calculates the degree of loss, we conducted a validation test by calculating the mean absolute error (MAE) between the observed and modeled extent of damage. To perform

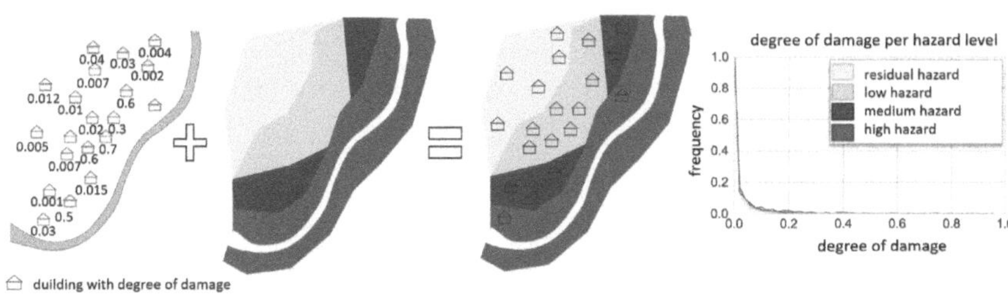

FIGURE 5.5 Degree of damage. The degree of damage was derived from the damage data and the building value. The overlay between the georeferenced data of the degree of damage and the hazard map resulted in specific distributions of degrees of damage per hazard level.

the model validation, we randomly divided the dataset into training and test subsets, containing 90% and 10% of the data, respectively. While we used the test dataset ultimately to test the selected model, we further split the train dataset into a train (70%) and validation (20%) subset. To perform cross-validation, we repeated the selection of train and validation subsets 100 times. We calculated the modeled damage extent by multiplying the building values with the modeled degree of damage. To this end, we drew the degree of damage values from the distributions. In order to cover the entire range of the distributions, we selected five statistical characteristics for each model, namely the 0.25 quantile (referred to as Q25), the 0.5 quantile (referred to as Q50 or median), 0.75 quantile (referred to as Q75), and the 0.95 quantile (referred to as Q95) as well as the mean. Subsequently, we calculated an averaged MAE for each statistical characteristic value per model by averaging the values over the 100-fold repetition. Based on the validation test, we selected the degree of damage values of the model and statistical characteristics for the lowest MAE score. We proceeded to calculate the average of the 100 different degrees of damage values, resulting in one specific degree of damage value per hazard level and building value cluster. Finally, we retested the selected model by calculating the extent of damage based on the modeled averaged degree of damage values and comparing it to the test subset. If we consider all the assessed cases, we notice that the distribution in all three hazard levels is right-skewed. This means that there are many cases of low vulnerability, but fewer cases of high vulnerability. Measured against the building values, the greater part of the claims considered involved minor damage, whereas fewer claims involved major damage. Table 5.1 and Fig. 5.6 show the resulting values of the degree of damage. Table 5.2 shows the degrees of damage depending on the building's value.

TABLE 5.1 Quantiles of the degrees of damage depending on the location in the hazard zones.

Hazard zone	Degree of damage (25%-quantile)	Degree of damage (50%-quantile)	Degree of damage (75%-quantile)	Degree of damage (95%-quantile)	Degree of damage mean
Low (yellow)	0.002	0.006	0.018	0.084	0.019
Medium (blue)	0.002	0.008	0.028	0.115	0.026
High (red)	0.003	0.009	0.032	0.162	0.034

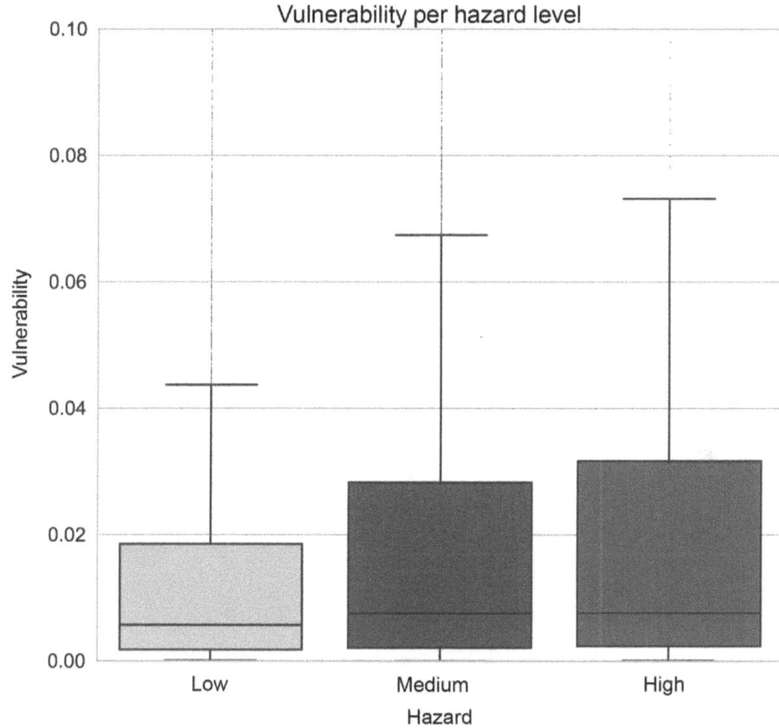

FIGURE 5.6 Vulnerability of buildings. Degree of damage, i.e., vulnerability of buildings in the three hazard zone levels.

We applied the final modeled degrees of damage values to calculate the expected extent of damage to all exposed buildings. Based on the building value and at which hazard level the building was located, we assessed the damage extent by multiplying the building value with the selected degree of damage. The three possible extents of damage are described in the tool as "possible extent of damage for a moderately damaging event" (median), "possible extent of damage for a highly damaging event" (75% quantile), and "possible extent of damage for a very highly damaging event" (95% quantile). Statistically, the calculations using the median—i.e., the 50% quantile—best reproduce the flood damage documented by the cantonal building insurance companies. However, as mentioned above, these contain very many instances of minor damage, resulting from numerous minimally damaging events. It can therefore be assumed that the calculated median underestimates vulnerability in the case of a highly damaging event, as in this instance higher vulnerability occurs more frequently at all hazard levels. For this reason, the damage simulator uses the 75% quantile of the calculated vulnerabilities to visualize the maps. To indicate nevertheless that the possible extent of damage covers a

TABLE 5.2 Quantiles of the degrees of damage depending on the location in the hazard zones in relation to the building value.

Building category	Low building value			Moderate building value		High building value		
Hazard level	**Low**	**Medium**	**High**	**Low**	**Medium**	**Low**	**Medium**	**High**
50% quantile	0.0094	0.0132	0.0145	0.0032	0.0041	0.0011	0.0018	0.0044
75% quantile	0.0259	0.0408	0.0624	0.0113	0.016	0.0046	0.0089	0.0166
95% quantile	0.0997	0.1389	0.2213	0.0695	0.0882	0.0416	0.0615	0.1035

wide range of low to very high vulnerability, the overview analysis shows not only the possible extent of damage based on the 75% quantile, but also those based on the median (50% quantile) and the 95% quantile.

Factors that determine flood risk change

The tool has three functions that allow for assessing the effects of three influencing factors of flood risk change. These are (i) urban densification (scenario "urban densification"), (ii) property-level flood protection measures (scenario "property level flood protection"), and (iii) the effect of flood protection measures (scenario "flood protection") on hazard zoning. We elaborated these scenarios by assuming specific rates of change of these drivers of flood risk change. The effect of urban densification was analyzed by altering the buildings dataset while the effect of property level flood protection on flood risk change was analyzed by assuming different implementation levels of this kind of flood prevention. The effect of flood protection measures was analyzed by changing the hazard zones.

Urban densification

Urban densification is defined as the inward development of the settlement by optimizing the available land reserves within the building zones. The land reserves considered in this study are the undeveloped reserves within and at the outer border of the settlements, such as gaps between existing buildings or undeveloped properties of building land located at the periphery of settlements (Di Carlo et al., 2017). Fig. 5.7 visualizes the concept of urban densification.

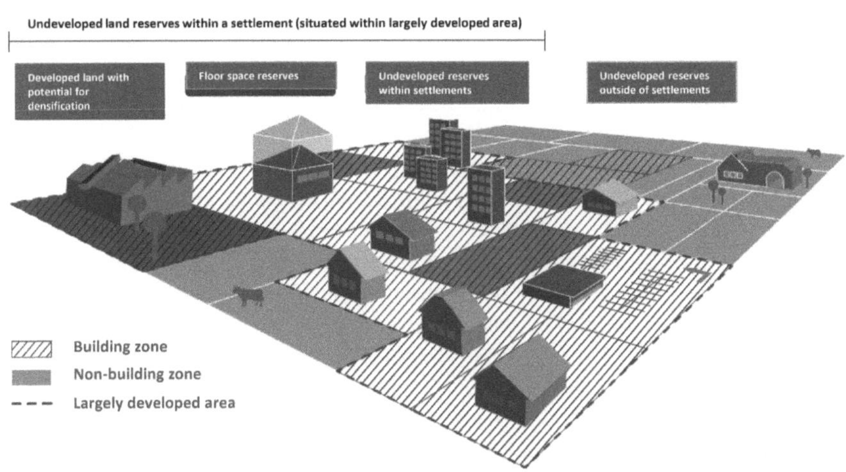

FIGURE 5.7 Urban densification. Concept of urban densification. *Modified from Nebel, R., Hollenstein, K., Carlo, Niedermaier, M., & Scholl, B. (2017). Schweizweite Abschätzung der Nutzungsreserven 2017: Bonriposi, vdf Hochschulverlag an der ETH Zürich.*

Urban densification can be achieved by increasing the density of already developed land with the potential for densification, by increasing floor space, by using undeveloped reserves within settlements, or by using undeveloped reserves outside of settlements (Nebel et al., 2017). Here, we considered only undeveloped building zone reserves within and outside of settlements. These are large reserves at the edge of settlement areas, gaps between existing buildings, and plots with the potential for inward development. For simplification, the following factors were not considered: Augmentation of floor space (postconstruction densification to make use of floor space reserves), developed land with potential for densification, designation, and rezoning of building land, building zone exchange, changes to the hazard map.

In order to identify these reserves, we overlapped the footprints of the building dataset with the building zones and applied the buffer method developed by Federal Office for Spatial Development (2007). Based on the model developed by Bonriposi et al. (2016), we additionally considered further spatially relevant distances to the 10-meter distance to buildings, namely the distance to forest (30 m), to motor- and expressways (20 m), to main roads (6 m), to secondary roads (4 m), to single-track railways (4 m), to multitrack railways (8 m), to tributary waters (5.5 m), and to main waters (15 m). These values represent nationwide averages of the distances stipulated in cantonal legislation. The resulting plots having a potential for densification were virtually filled with building footprints. We excluded implausible shapes of building footprints (e.g., long and narrow) and very small partial areas, as well as areas that do not fulfill the minimum requirements of plot width (12 m) or plot area (600 m^2), or footprint area (50 m^2) from the remaining developable areas. Following the selection of land parcels that are potentially developable, each plot was assigned to a building zone with the land use categories "residential," "work," "mixed," or "central zone." Depending on the building zone category, potential buildings with different sizes and distances can be built in each area. The typical sizes of building footprints were 225 m^2 for the residential zone, 900 m^2 for the work zone, 625 m^2 for the mixed zone, and 225 m^2 for the central zone. The minimum distances between these new buildings were 10, 20, 14, and 7 m for the respective zones. The minimum building widths were 15 m for the residential zone, 30 m for the work zone, 25 m for the mixed zone, and 15 m for the central zone. To assess the height of the potential buildings, we calculated the median of the heights of all existing buildings within a radius of 100 m around each potential new building's footprint. This method allowed the calculation of heights of about 97% of potential buildings. For a complete calculation of the building heights, the median of the heights of existing buildings was calculated per canton and divided into communal typologies. The height of a potential building from the remaining 3% was defined as the median of the combined cantonal and communal typologies at the location. Once all buildings' heights were virtually constructed, the volume was obtained by multiplying the building footprint area with the assigned height. Based on the specific cantonal cubic meter price, the building value (reconstruction value) could be obtained as described in the previous sections. The result of this procedure was a dataset of buildings within the identified land reserves used for urban densification. We assumed that these land reserves would be filled until the year 2040 with a growth rate of 1% per year. Thus, the urban densification scenario was also named the "2040 scenario." Fig. 5.8 shows the identified potential for urban densification in Bern, Switzerland.

Property-level flood protection measures

The second factor that is considered as a driver of flood risk change in the tool "Flood damage simulator" is the implementation of property-level flood protection measures. Structural property-level mitigation measures aim at preventing or minimizing flood damage to persons or property. They can be applied directly to or very close to the property (building or facility), with the aim of preventing or minimizing damage to persons and property. Structural property-level mitigation measures include (but are not limited to):

- waterproofing of doors and windows;
- adjusting the building's openings;
- raising the ground floor above the flood level;
- adjusting the terrain to optimize flow paths;
- permanent flood barriers such as dams, levees, berms, and walls;
- mobile flood barriers such as stop logs and metal or timber shields.

In the tool "Flood damage simulator," we considered the following types of measures: Elevating the property, positioning access points, or entrances outside the hazard area, creating corridors for controlled flow through the land

FIGURE 5.8 Urban densification. Example of the modeled potential for urban densification in Bern, Switzerland.

parcel, wet flood proofing, waterproofing, and shielding. Spatial planning measures and biological or organizational measures, mobile dams, or sandbags were not considered. As mentioned above, newly constructed buildings in blue hazard zones must be built in a way that they are not vulnerable against flooding up to a certain flood magnitude. Thus, new buildings are designed and planned with property-level flood protection measures. However, the older buildings, i.e., buildings constructed before this new legislation are not adapted to the flood hazard. Therefore, a high number of buildings remain located in yellow, blue, and red hazard zones that have no flood protection. The question was how high this share of buildings with property-level flood protection to all buildings in a region is. Fehlmann (2020) investigated this research question and conducted interviews to obtain a rough estimate of the share of buildings in flood hazard zones and nationwide that are fitted with structural property-level flood mitigation measures. Nine experts from cantonal administrations and insurance companies and four scientists and engineers working in this topic were interviewed. Moreover, available data from insurance companies were included in the estimation.

The experts estimated that 2% of the buildings in the yellow hazard zone (low hazard level) of the flood hazard zone map are protected by property-level mitigation measures, while the remaining 98% of buildings in this zone have no such measures in place. Within the blue hazard zone (medium hazard level), 11% of the buildings are fitted with flood mitigation measures; within the red hazard zone (high hazard level), the corresponding value is 10%. The figures represent a weighted average for whole Switzerland. However, the estimated shares of buildings fitted with property-level flood mitigation measures exhibit high regional variability. The values lie between 0.25% and 30% within the yellow hazard zone; between 2.5% and 35% in the blue hazard zone; and between 0.5% and 20% in the red hazard zone. This wide range of estimated values is largely due to cantonal variations in the handling of property-level flood mitigation measures. For example, some cantons also require buildings in the yellow zone to be fitted with property-level flood mitigation measures. However, apart from a few outliers, many of the estimated values are close together. The experts confirm that buildings in the yellow hazard zone (low hazard level) virtually have no property-level flood protection measures installed. The interviewed experts also discussed the reliability of the installed measures. The reliability of property-level flood mitigation measures is estimated at 50%, as many factors such as aging, lack of maintenance, incorrect handling of structural parts, incorrect dimensioning, and others reduce reliability. The reliability of temporary measures (e.g., sandbags) is estimated to be less than 50% effective. The experts estimate that the share of buildings with property-level flood mitigation measures will increase in the future because many buildings must be renovated. Besides the construction of a new building, a thorough renovation of an old building also requires the installation of flood protection measures.

Flood protection measures

In addition to the hitherto considered drivers of flood risk change, namely urban densification and property-level flood protection measures, we implemented a simplified method for considering the installation of flood protection measures in the river channel. In Switzerland, the hazard maps are updated after the implementation of flood protection measures, e.g., structural measures such as the construction of lateral levees. If a flood protection measure reduces the flood hazard in terms of flood intensity and frequency, the updated hazard map shows lower hazard levels than before. According to an analysis of the change of hazard maps after the implementation of several flood protection projects, we elaborated a simplified procedure to represent this kind of human adaptation in the tool. We assumed that red hazard zones have become blue zones and blue hazard zones became yellow zones after the implementation of a structural flood protection project. The yellow zones become white zones, meaning that there is no hazard anymore. The consequence of the assumed implementation of a flood protection measure is expressed in terms of a reduced potential damage. The overlay of the updated hazard maps with lower hazard levels and buildings results in a lower number of exposed buildings and in reduced vulnerabilities and therefore in a lower possible extent of damage.

Scenarios

The tool "Flood damage simulator" shows the possible extent of damage under current conditions, but also how it could change in the future for additional scenarios. The current state is called the scenario "Today" and is defined by the current state of hazard (represented by the hazard maps), by the current state of exposure (represented by the buildings dataset), and by the current state of vulnerability (represented by the current share of buildings with property-level flood protection measures). The reliability of all property-level mitigation measures is set at 0.5. In addition to the current conditions, there are two main future scenarios implemented. The scenario for the year 2040 (scenario "2040") shows changes in the possible extent of damage compared with scenario "Today." In an individualized scenario ("Own" scenario), users can personally configure the share of buildings with property-level mitigation, the share of building zone reserves developed, and the option of structural flood protection.

The scenario "2040" is based on estimated future shares of buildings with property-level mitigation in 2040 for each hazard level. In contrast to the shares of the current state (2%, 11%, 10% for the yellow, blue, and red hazard zones, respectively), the shares of buildings with property-level mitigation in 2040 are 2% for the yellow hazard zone, 22% for the blue hazard zone, and 26% for the red hazard zone on average for Switzerland. The values are adapted for each municipality according to their cantonal legislation. The reliability of all property-level mitigation measures was assumed to be 50%. In addition to the different implementation of property-level flood protection measures, the scenario "2040" has a higher number of buildings. The additional buildings in the exposure dataset were modeled by filling the potential areas for urban densification with buildings depending on the buildings' life cycle and the annual growth rate of the residential building stock. A building's life cycle depends on the service life of its structural components. The generally agreed standard service life of residential (apartment) buildings is 80–100 years. The service life of commercial buildings is considered to be shorter (50–80 years). Taking the maximum standard values (residential buildings: 100 years; commercial buildings: 80 years) and a share of commercial buildings of 15%, the average service life of all buildings is 95 years. Based on the information given in the buildings and dwellings statistics, the annual growth rate of the residential building stock is estimated at almost 1% of the building stock. We assumed an average share of building zone reserves developed of 36% by the year 2040 in Switzerland. The development of the building zone reserves by the year 2040 is estimated individually for each municipality and each hexagon.

The tool offers a possibility for the user to define their own scenario. The "Own" scenario of a user differs from the scenarios "Today" and "2040" by differentiating between the share of existing and new buildings with property-level flood protection measures. This aims at testing the hypothesis if and how flood risk can be reduced by adapting existing buildings with object-protection measures. Property-level flood mitigation on new buildings can only be defined by the yellow and blue hazard zones. It is assumed that no new buildings will be built in the red hazard zone. The reliability of all property-level flood mitigation measures is set at 0.5. Moreover, the user is free to assume the share of developed building zone reserves from the total of building zones. The share of building zone reserves developed can be set at four levels: current status (0%), minor (33%), substantial (67%), or complete development (100%). The "Own" scenario can also virtually implement structural flood protection measures in the river channel. This leads to a large-scale downgrading of the hazard level (red becomes blue, blue becomes yellow, and yellow becomes white). This is represented by changing the hazard maps as described above. Thus, the tool enables a user to conduct a sensitivity analysis of different flood risk management options and to test the effects of three different drivers of flood risk change on overall flood risk evolution. These options work at different scales and different societal hierarchies, from the property-level to the river reach level, and from the individual hierarchy to the municipal level. Fig. 5.9 summarizes the scenarios.

In addition to scenarios, the control panel enables a user to select the geographical aggregation unit, the hazard levels, and the share of damaged buildings in the hazard zone. Fig. 5.10 shows the graphical user interface of the web-based tool "Flood damage simulator."

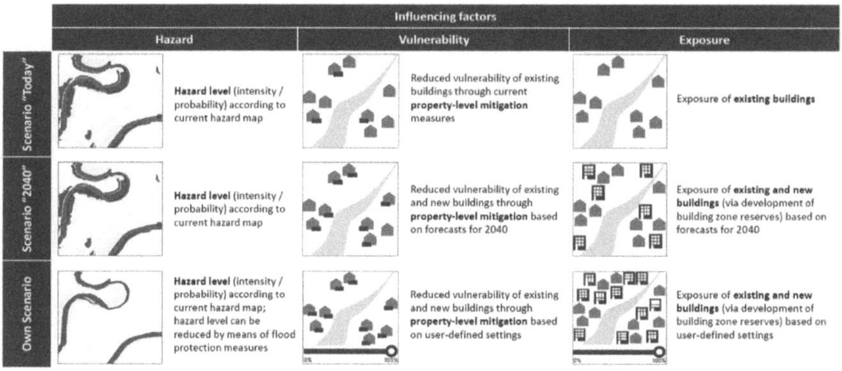

FIGURE 5.9 **Scenarios.** Overview of the scenarios of the tool "Flood damage simulator".

Results and discussions

In the following, the main results are summarized from a flood risk management perspective. The sensitivity analyses guided the distillation of a few conclusions for further developing the Swiss strategies.

First, the tool shows that the possible extent of damage in yellow zones is underestimated. The damage simulator shows that the so-called yellow hazard zones, i.e., zones exposed to a low level of flood hazard from a hazard perspective, are much more significant in terms of possible damage than previously assumed. From the damage perspective, therefore, there is a need for action not only in zones with moderate or major hazards, i.e., blue and red zones, but particularly also in the yellow zones. These areas have the highest number of buildings, and as a result, despite a lower extent of damage per

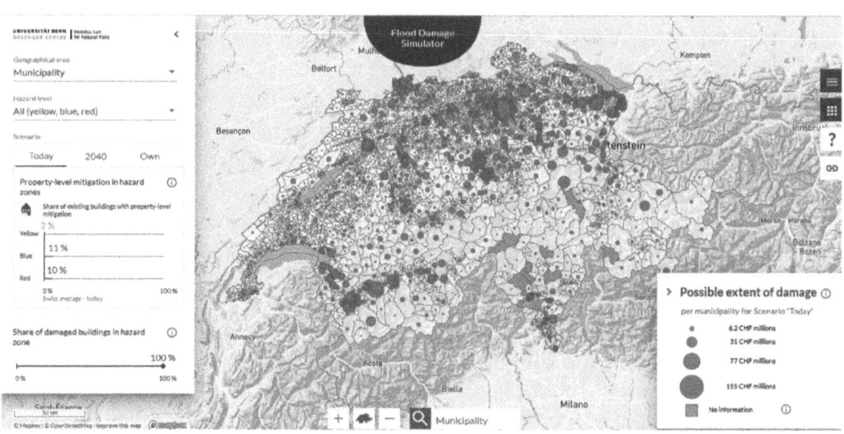

FIGURE 5.10 **Flood damage simulator.** Graphical user interface of the web-based tool "Flood damage simulator."

building, they have higher damage potential than zones with a higher hazard level. Precisely, these yellow hazard zones have often been neglected in flood risk management and land use planning. Unlike in blue zones, most cantons do not require specific flood protection measures in yellow zones, such as property-level flood mitigation measures. In view of the high possible extent of damage even with a low flood hazard level, it is advisable to promote or require property-level flood protection measures in yellow zones. Even small investments can have a big impact here: in the case of new buildings, the added costs of property-level mitigation measures are usually negligible.

Second, urban densification raises the damage potential by billions of Swiss Francs. The damage simulator also shows how densification or development of building zone reserves in Switzerland impacts the total value of property exposed to floods. Assuming that one-third of available building zone reserves were developed by 2040, the total value of buildings exposed to hazard could increase by CHF 5.3 billion. At 8000 ha, the newly built-up area would be about the size of 12,000 football pitches put together. Assuming that the available building zone reserves were built up entirely, the exposed property value could increase by CHF 16.6 billion. This breaks down into CHF 9 billion for residential buildings and CHF 7.6 billion for commercial, industrial, and service buildings. In this case, the total built-up area would be about 25,000 ha, which corresponds to 34,000 football pitches put together. If all of Switzerland's building zone reserves were developed, the building stock in the blue and yellow hazard zones would grow in similar proportions. This means that the ratio of buildings in the yellow zone to those in the blue zone would remain roughly the same as today. The possible extent of damage, in turn, would increase to about 3.4% in the case of a highly damaging flood. The reason for this rather modest increase is that the number of new buildings would be small in comparison with current existing buildings.

Third, the wider implementation of property-level flood mitigation measures can markedly reduce flood damage. It is not enough to apply property-level mitigation only to new buildings if the goal is to reduce the possible extent of damage over a large area. Due to the high number of existing buildings in hazard zones, these will continue to account for most of the damage potential in the future. They contribute to a higher possible extent of damage because property-level flood mitigation is mostly lacking or exists only for a small share of these buildings. The future development of the possible extent of damage thus greatly depends on whether property-level mitigation is also applied to existing buildings or not. To avoid increasing possible damage in the future, it is thus advisable to promote or require property-level flood mitigation measures for existing buildings. If, for instance, the number of buildings with property-level mitigation were doubled in the blue hazard zone, then about one in five buildings in this zone would be equipped with property-level mitigation measures. This would make it possible to maintain the possible extent of damage at the current level until

2040, despite the addition of new buildings. However, this assumes that climate change will not significantly alter the possible extent of damage from flood events between now and then. Fig. 5.11 shows the change in the possible extent of damage in the scenario "2040" compared to the current situation.

Conclusions from the complexity perspective

Flood risk assessment and the design of risk reduction strategies often neglect the influence of socioeconomic development on future exposure and vulnerability to floods and their development over time. Flood risk management options are assessed from the perspective of the current situation. The trajectory of flood risk evolution and the underlying drivers of change are not considered. Moreover, a thorough assessment of the effects of different flood risk management options at national scale with consideration of local characteristics is only rarely made. The tool gives the option to test the sensitivity of the three societal levers to govern flood risk evolution over time. It shows that flood risks will increase or decrease depending on proactive adaptations of both households and government. Property-level flood protection measures, for example, could achieve a significant risk reduction even at low cost. Moreover, the tool shows the path dependency of future flood risk management options. A fast-growing municipality with already a high share of exposed buildings and hitherto undeveloped building reserves will head toward increasing flood risks. However, the risk in this case will increase markedly in terms of absolute numbers because of the high number of new buildings. In contrast, a peripheral community with low growth rates is more likely to keep risks constant in terms of absolute numbers. In terms of relative numbers, a few new buildings constructed in flood hazard zones might increase flood risk relatively to the low number of existing buildings.

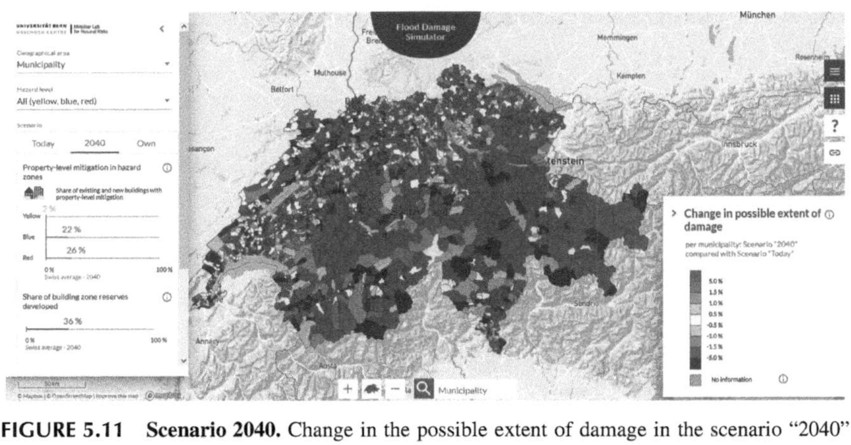

FIGURE 5.11 Scenario 2040. Change in the possible extent of damage in the scenario "2040" compared to the current situation.

The tool also shows that flood risk management options at different societal levels, at the individual level as well as at the community level, provide levers to govern flood risks. Thus, to achieve the goal of flood risk management to reduce flood risks or at least keep them constant over time, a cooperation between households and governments is therefore essential and may be reached by encouraging a constructive flood risk dialog. Moreover, a risk dialog is also needed to discuss the target conflicts between land use planning and flood risk management, e.g., by finding a trade-off between the target of urban densification as a measure to foster a sustainable use of the land resources and the targets of flood risk management.

As a communication and awareness-raising instrument for flood risk management, the "Flood damage simulator" tool is directed at various user groups such as policymakers, local authorities, spatial planners as well as researchers. The tool offers scenarios that represent the magnitude of flood damages to be expected today and show possible trends in the near future. Furthermore, it is possible to customize a user-defined scenario in which the key flood risk drivers hazard, exposure, and vulnerability can be individually adapted, thus breaking down a complex topic in more comprehensible subunits. Generated knowledge and awareness might promote proactive adaptation of both households and government leading to a reduction of flood risk. Last but not least, the damage simulator serves to raise awareness among the population and homeowners. The tool provides arguments for promoting or prescribing property protection measures on existing buildings as well—provided that the costs are in reasonable proportion to the benefits. Moreover, the damage simulator shows when property protection measures make sense even in areas with a low risk (yellow zone). The interactive tool is training the users in assessing the effects of societal levers to influence flood risks, either in their individual effects or in their combined effects on flood risk change.

The tool "Flood risk dynamics"—flood risk change at centennial scale

The tool "Flood damage simulator" offers to its user to test the effects of anthropogenic interventions to hazard, exposure, and vulnerability on the possible extent of present and future flood damage over Switzerland. This tool or model experiment focused on the sensitivity of flood risk evolution to different flood risk management options. The effects of these human interventions on flood damage can be analyzed in isolation but also in their combination. This simulation is based on assumptions of the future development of the drivers of flood risk. However, it remains an open question, how flood risk really evolved over time in the last two centuries. To answer this question with observed data, the Mobiliar Lab for Natural Risks developed another interactive web-based tool, namely the tool "Flood risk dynamics." This website interactively displays how the flood risk along the Emme River changed over the last 200 years in the river reach between Burgdorf and the border between the cantons of Bern and Solothurn in Switzerland. The tool was developed in the framework of an outreach project for flood risk communication. The interactive model experiment has been developed in a first version as software for a touch-screen application in an exhibition on climate change. In its final version, the model experiment has been implemented in an interactive web-based tool. The research on this topic that is behind the tool was described in detail in Zischg, Hofer, et al. (2018).

Method

We developed a model experiment setup that bases on the overlay of several timesteps of hazard maps and exposure. The main idea behind the model experiment was to reconstruct the state of the river morphology and the state of the settlements for selected timesteps between 1820 AD and 2016 AD. These two main drivers of flood risk change were manifested in space and therefore they were observable. We looked at how these two drivers of flood risk evolved over the past two centuries and how flood risk changed because of this coevolution of river morphology and urbanization.

River morphology and flood hazard

The first driver of flood risk change, the change in the river morphology, was dominated by two main processes. These were (i) the river correction including the construction of levees as a measure of flood protection, and (ii) the consecutive river incision triggered by the human intervention to the natural system. To represent the historic states of the river morphology in the model experiment, we reconstructed the historical morphology of the river with the help of historical maps. The procedure of reconstructing digital terrain models that represent historical river models was described in the previous chapter of this book. We selected three timesteps to represent the river morphology: 1820, 1910, and 2016. For these timesteps, enough reliable historical documents are available, e.g., topographic surveys. The topographic surveys of the river course were conducted in the late 19th century after two consecutive flood events in 1868 and 1876. These surveys represent the natural geomorphology of the Emme river, and thus we classified this state as of 1820. These flood events lead to the planning of the correction of the Emme river. From the georeferenced historic maps and the contemporary topographic surveys (Anonymous, n.d., 1820a, 1820b, 1898; Ritter, 1804, p. 151), we digitized the platform of the natural river course. In a subsequent step, we mapped all recent anthropogenic modifications of the terrain in the riverbed and in the floodplain and erased these areas from the present-day terrain model. In the resulting holes of the digital terrain model, we interpolated a mean surface based on the elevation values at the lateral borders of the erased areas (Zischg, 2016). This results in a terrain model free from artificial and hydraulically relevant landscape modifications (Ramirez et al., 2020). Afterward, we incised the digitized areas of the historic riverbeds into the historic digital terrain model. The incision depth was derived from the historic topographic surveys of cross section geometries. In 1820, the river course was on average 90 m wide and the riverbed was about 1−3 m below the surrounding floodplain. In contrast, the present-day digital terrain model shows an average river width of 42 m and the current riverbed on average 3 m below the natural riverbed. In 1884, the federal and cantonal authorities began with the construction of lateral levees. In 1910, the works were finished. We elaborated a digital terrain model for the river status of 1910 based on the maps of Goldschmied (1913) and Luder (1928). This status of 1910 represents an anthropogenically modified river channel but without the subsequent erosion. In the following decade, the narrowed riverbed was incised by up to 2.4 m. Consequently, soon after the first anthropogenic modifications the erosion had to be stopped by the construction of riverbed stabilization measures. In 1960, the process of riverbed erosion has stopped. Since then, the morphology of the river changed only locally. Thus, the current river morphology of 2016 can also be used to represent the river state of 1960. The river course in 2016 was reconstructed based on the LIDAR terrain model and the surveyed cross sections provided by the Canton of Bern. A digital terrain model of a historic river morphology is hereafter named a retromodel.

The reconstruction of the historical river courses (1820 and 1910) based on changing those parts in the landscape and in the digital terrain model, respectively, that have been anthropogenically modified. The two digital terrain retromodels of 1820 and 1910 and the digital terrain model of 2016 provided the basis for the creation of the computational mesh for a hydrodynamic model. The hydraulic simulations were conducted in the 2D BASEMENT modeling environment (Vetsch et al., 2017). The validation of the hydraulic model is described in Zischg, Mosimann, et al. (2018). This numerical model solves the shallow water equations on the basis of an irregular mesh, which was generated from a digital elevation model with a spatial resolution of 0.5 m. For each temporal status of the river morphology, we simulated several flood events ranging from the mean annual flood to an extreme flood. The water inflow fluxes at the upper boundary conditions were represented by synthetic design hydrographs. These hydrographs describe the typical flood event coming from the upstream river basin by defining the flood peak/flood volume relationship. With this relationship, the hydrographs can be scaled by increasing the peak discharge. The procedure was applied to generate synthetic design hydrographs for a continuous series of peak discharges with intervals of 50 m^3/s. The occurrence probability of each hydrograph was delineated from the extreme value statistics of the gauging station Emme Wiler (provided by the Federal Office for the Environment FOEN). Each of these synthetic hydrographs was fed into the hydraulic model and the resulting flood depths were calculated. For each flood model that represents one of the three states of the river morphology (1820, 1910, and 2016), a set of inundation maps therefore was elaborated. This set of flood maps provide the basis for calculating the number of exposed houses and the corresponding damages for each of the flood magnitudes. Fig. 5.12 compares the flood maps resulting from the river morphology in 1910 and 2016. The flood hazard maps show the peak discharge of a flood event that is needed to inundate a certain location. This might be a counterinitiative or at least an unusual way of representing flood hazards, but it directly synthesizes the relationship between the flood magnitude and the flood extent over a large number of scenarios. The comparison between the flood map of the current digital terrain model and the retromodel of 1910 shows that currently only flood events with a magnitude greater than a peak discharge of 450 m^3/s are needed to trigger inundation in the floodplain. In contrast, the 1910 retromodel shows that inundations in the floodplain can be triggered by flood magnitudes with peak discharges of 200 m^3/s.

Elements at risk

In this model experiment, we focused explicitly on buildings and houses. In a multitemporal exposure analysis, the building dataset was elaborated for selected temporal states. For the present-day situation, we used the building dataset elaborated by Röthlisberger et al. (2017). This dataset spatially represents the building's footprints, to which their building value and the number of inhabitants were attributed. This dataset is based on the digital terrain model of the Federal Office for Topography SWISSTOPO (Swisstopo, 2021) and the residential statistics of the Federal Office for Statistics (BFS, 2021a). As

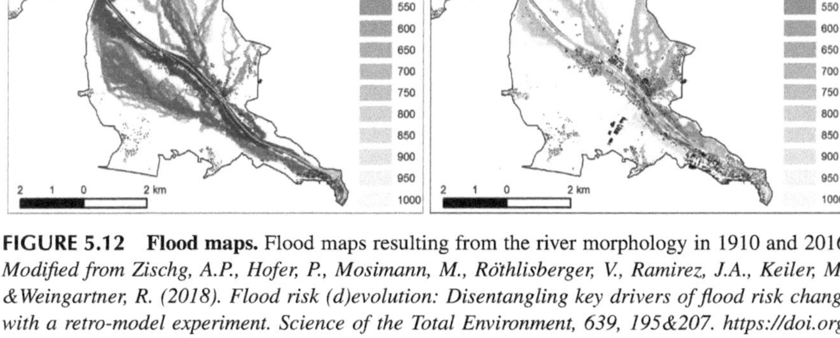

FIGURE 5.12 **Flood maps.** Flood maps resulting from the river morphology in 1910 and 2016. *Modified from Zischg, A.P., Hofer, P., Mosimann, M., Röthlisberger, V., Ramirez, J.A., Keiler, M., &Weingartner, R. (2018). Flood risk (d)evolution: Disentangling key drivers of flood risk change with a retro-model experiment. Science of the Total Environment, 639, 195&207. https://doi.org/10.1016/j.scitotenv.2018.05.056.*

described above, the building value was calculated based on each building's volume, which has been derived from LIDAR measurements. The monetary value was calculated based on average construction prices. This present-day situation provided the basis for the retromodels of the settlements. From a copy of this dataset, we removed all the buildings that were not present in the topographic maps of 1959 and 1960 of the Federal Office for Topography. This results in a dataset representing the settlement in 1960. The historical Siegfried maps (Swisstopo, 2020) of 1915, 1916, and 1917 lead to a dataset of the settlements in 1910. The historical maps of 1871, 1879, and 1880 lead to a dataset representing the year 1880. Finally, we used the historical Dufour map of the year 1845 (Dufour, 2021) and the maps of Finsler (1978) for reconstructing the status of settlements in 1820. The elements at risk, namely the dataset of the buildings, provide the basis for calculating the damage for each of the flood scenarios. Fig. 5.13 shows four timesteps of the historical maps used for attributing the construction years of the buildings.

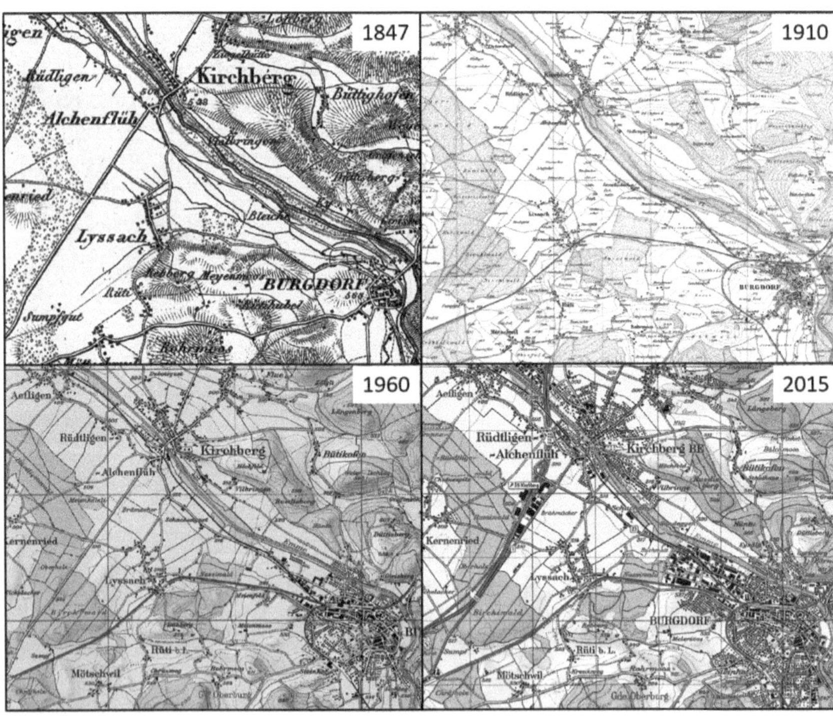

FIGURE 5.13 Timesteps of the historical maps. Four timesteps of the historical maps used for attributing the construction years of the buildings. *From Swisstopo: National Map 1:25,000, Swisstopo. (2020).*

Flood risk evolution

The set of hydrographs, the set of river morphology models and retromodels, and the set of settlement states were combined and overlaid. For each combination of one of the 19 hydrographs that represent different flood magnitudes and one of the river morphology models, a flood map was elaborated with the flood model of the respective river morphology. The resulting 57 maps of flood depths were combined with the set of five datasets of buildings. The flood depth at building scale was used to determine the vulnerability value for each building. This building-specific vulnerability value was then multiplied with the monetary value of the building. This resulted in the flood damage expressed in Swiss Francs for each building.

However, we let the reconstruction values, the type of functionality, and the vulnerability of the buildings constant over the whole period in the model experiment. This means that the flood damages vary only with the number of buildings exposed to floods and not with the inflation or extension of building values over time. This simplification was made because first, no information on past building's values and vulnerabilities was available, and second, we aimed to explicitly focus on the exposure as an element of the risk formula.

For each combination of flood map and settlement status, the flood damages were summarized for the study area. The procedure is shown in Fig. 5.14. Overall, this overlay resulted in 285 flood damage calculations. The integration of all damage calculations of a specific combination between all flood magnitudes of a certain river morphology and a certain settlement status

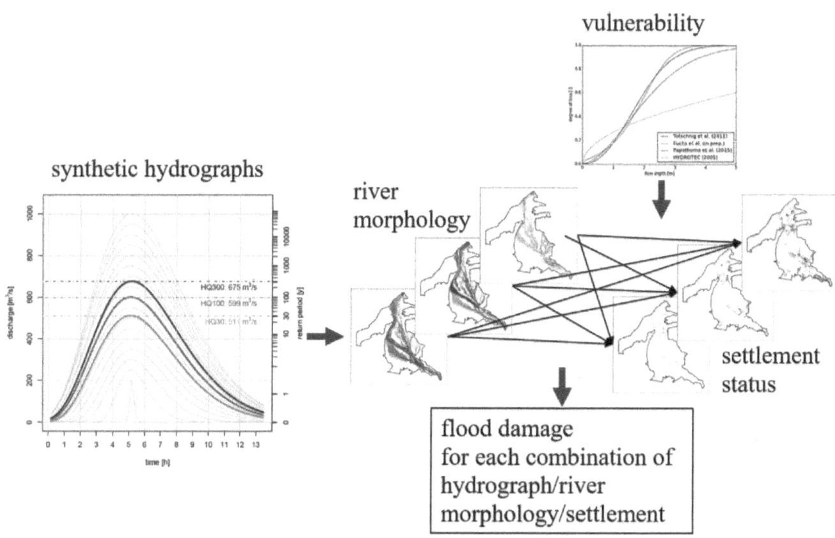

FIGURE 5.14 Model experiment. Schematics of the overlay between synthetic flood hydrographs, digital terrain model and retromodels, and settlement status.

results in a risk estimation at a certain point in time. The evolution of flood risk in the period 1820–2016 can be deducted by concatenating the combination between the river morphology and settlement status of the years 1820, 1910, and 2016. The settlement status of 1880 was overlaid with the river morphology of 1820 to quantify the flood risk of 1880, and the settlement status of 1960 was overlaid with the river morphology of 2016 to quantify the flood risk of 1960.

Results and discussion

The hydraulic and hydrological features of the river system have changed considerably since the beginning of the 19th century owing to river engineering activities. In the study area, the Emme river was transformed from a wide, shallow river into a much deeper body of water flowing in a narrow channel. The area of the water course declined from just under 2 km^2 to less than half after the massive river corrections of the early 19th century. Not only the course of the river but also the surrounding area and its settlements have altered significantly over the last 200 years. In the area being examined, the number of buildings increased from around 600 in 1820 to approximately 6500 today.

The annual cumulative flood risk (expected annual damage) exhibited huge changes between 1820 and 2016. From 1820 to 1880, flood risk increased because of settlement growth. From 1910 to 1960, it declined since flood protection measures and the successive riverbed erosion reduced the likelihood of a destructive flood. However, after 1960, flood risk increased again because of settlement growth and the associated rise in the potential for damage. The flood protection strategy adopted along the Emme is a special case. When the course of the river was corrected in 1910, it was simply made narrower and straighter.

It was thought that a smaller cross section would force the Emme to carve out a deeper riverbed for itself, through erosion. This process was expected to continue until a natural balance was achieved. In this example, the full impact of the flood protection measures did therefore not become evident directly after completion of the work in 1910, but with a certain delay. The minimum risk was achieved in 1960, when the erosion process and the associated deepening of the riverbed came to an end. Between 1960 and 2016, subsequent settlement growth caused the risk to increase again while river morphology remained more or less constant.

Fig. 5.15 shows the trajectory of flood risk in the floodplain of the Emme river near Burgdorf, Switzerland. According to the tool www.risikodynamik.ch, flood risk increased from 6.2 mio. CHF/year in 1820 to 21.6 mio. CHF/year in 1880 and decreased to 17.3 mio. CHF/year in 1910. After a 50-year period of riverbed incision, the flood risk decreased further to 0.6 mio. CHF/year. Due to the rebound effect of settlement growth after 1960, flood risk increased again by a factor of 5 .

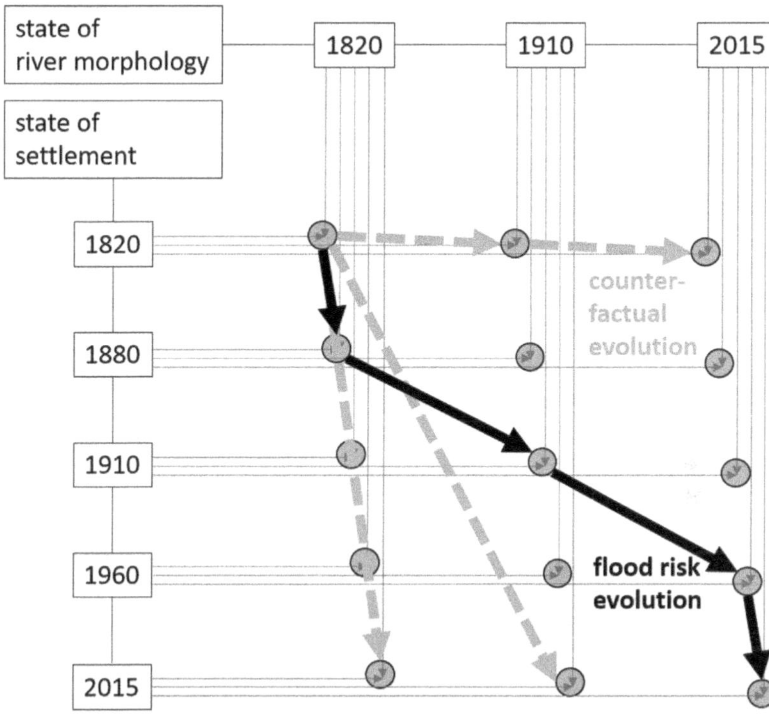

FIGURE 5.15 Flood risk change trajectories. Trajectory of flood risk evolution in the floodplain of the Emme river near Burgdorf, Switzerland. *Modified from Zischg, A. P., Hofer, P., Mosimann, M., Röthlisberger, V., Ramirez, J. A., Keiler, M. & Weingartner, R. (2018). Flood risk (d)evolution: Disentangling key drivers of flood risk change with a retro-model experiment.* Science of the Total Environment, 639, *195–207. https://doi.org/10.1016/j.scitotenv.2018.05.056*

In contrast to the interactive web-tool www.risikodynamik.ch, Zischg, Hofer, et al. (2018) focused in their model experiment on disentangling the isolated effects of the drivers of flood risk on flood risk evolution, namely the change of river morphology and settlement status. Moreover, their model experiment focused also on counterfactual pathways of flood risk evolution. For example, they stated that the most important driver of flood risk change in this case study was the effect of river incision. As shown in the previous chapter, the effect of riverbed incision on flood risk is highly nonlinear and is a negative feedback effect on flood risk evolution. The incision of the riverbed increases the river discharge capacity. Higher peak discharges are needed to flood the floodplain with the same extent as before. This in turn leads to a decrease of both the probability of inundation and the flood depths in the floodplain. Consequently, the vulnerability decreases and with it the flood damage. Urbanization plays a role after the end of the river corrections. Counterfactual trajectories of flood risk evolution would have been the result of different flood risk management options. Instead of investing in river corrections, an alternative option would have been to keep free the floodplain from settlement. This counterfactual scenario would have had a similar effect on flood risk evolution as the engineering method.

The interactive map partially allows to disentangle the factors affecting flood risk evolution and to consider them separately. The impact of the selected factors on flood damages and flood risk is impressively demonstrated when they are considered in isolation. On the one hand, the flood risk decreased considerably between 1820 and 2016 thanks to the flood protection work. Today, 1305 buildings are protected by flood defenses. On the other hand, settlement growth has led to a significant increase in the potential damages and therefore the flood risk. Overall, a simulation on the interactive map shows that the flood risk has roughly halved since 1820. Fig. 5.16 shows a screenshot of the graphical user interface of the tool "Flood risk dynamics" (www.risikodynamik.ch).

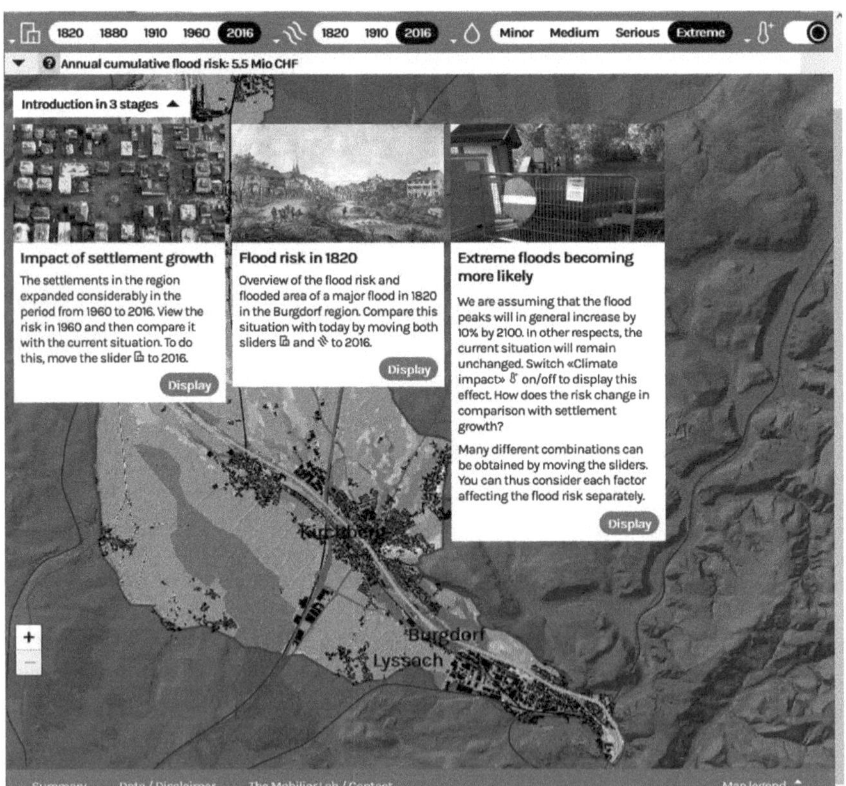

FIGURE 5.16 Flood risk dynamics. Screenshot of the graphical user interface of the tool risikodynamik.ch.

In contrast to the study of Zischg, Hofer, et al. (2018), the interactive web-tool “Flood risk dynamics” considered a third factor that influences flood risk evolution, namely the increase of peak discharge due to climatic changes. The “Climate impact” feature displays the likely effects of climate change on the flood risk evolution. Climate change is set to cause more frequent and more extensive flooding (Köplin et al., 2014). The simulated “climate impact” corresponds to a plausible climate scenario leading to peak discharges that are about 10% higher than today. In the web tool, this driver of change increases flood risk from 3 mio. CHF/year to 5.5 mio. CHF/year. Please note that this increase is based on the assumption that exposure is not increasing. Combined with the expected settlement growth in future, flood risk can be expected to increase further. Forward-looking spatial development may prevent a significant rise in flood risk, while additional protective structures may reduce it.

Another important finding is how much the form of the flood risk curves is changing over time. A flood risk curve shows the number of exposed elements or persons or the damage in relation to the flood magnitude. Fig. 5.17 shows the flood risk curves of 1820, 1880, 1910, 1960, and 2015. The flood risk curve of the Emme river floodplain in 1820 shows that buildings are affected by floods with a return period of 1–3 years. These buildings were hydropower and production facilities in the direct neighborhood of the river. They were built to be resilient for these very frequent floods or they suffered only small damages. The need for hydropower was higher than the risk to be damaged. In addition, the flood risk curve of 1820 is not increasing with the same slope as the flood risk curves of the more recent periods. This means that the exposure or the damage does not increase remarkably with the flood magnitude. This form of the flood risk curve thus does not show a sensitivity of the exposure to increasing flood magnitudes. This means that the floodplain in 1820 would not have been sensitive to the effects of climate changes to hydrology, as for example, the increase of peak discharges due to higher rainfall intensities in the river catchment. In contrast, the flood risk curve of 2015 shows a highly concave form. Buildings are not exposed to a flood event until a certain return period is reached. However, after crossing this threshold, exposure is heavily increasing with the flood magnitude. In relative terms, the slope is higher than 1:1, meaning that an increase in flood magnitude of 10% results in an increase of flood exposure by much more than 10%. Thus, the floodplain of the Emme river in Burgdorf today is highly sensitive to changes in peak discharge due to climatic changes. In terms of flood risk change, an increase of peak discharge by 10% will lead to an increase of flood risk by 185%. Such forms of flood risk curves can lead to so-called surprise effects of floods, which suddenly become catastrophes because of their extreme consequences (Kjeldsen & Prosdocimi, 2018; B. Merz et al., 2015). This surprise effect is an emergence of a complex system (Liu et al., 2007). The other flood risk curves are in between the two curves of 1820 and 2016. The multitemporal comparison shows how the sensitivity of the floodplain's exposure to increasing peak discharges changed

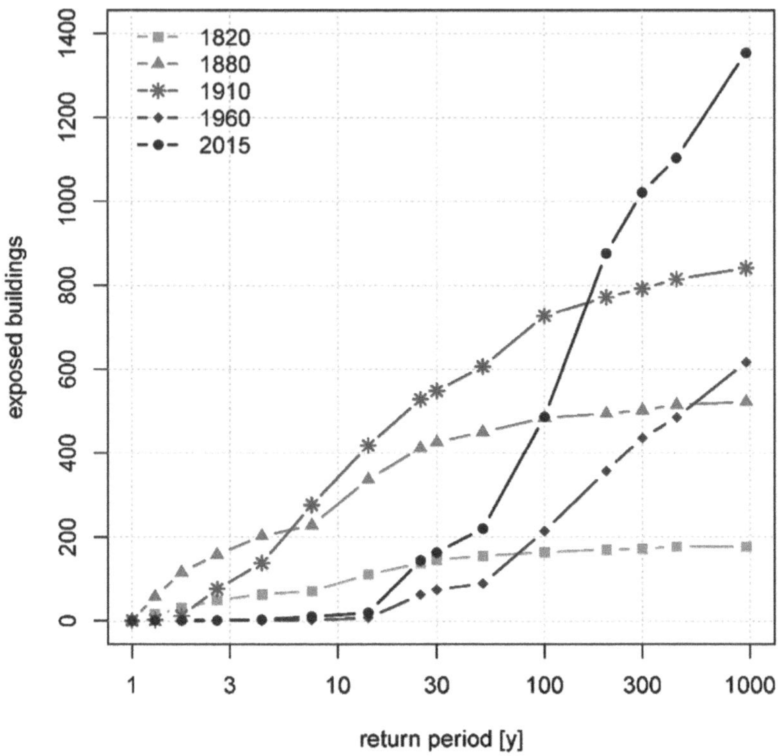

FIGURE 5.17 Flood risk curves. Flood risk curves of the Emme river floodplain in the period 1820–2015. *Modified from Zischg, A. P., Hofer, P., Mosimann, M., Röthlisberger, V., Ramirez, J. A., Keiler, M. & Weingartner, R. (2018). Flood risk (d)evolution: Disentangling key drivers of flood risk change with a retro-model experiment.* Science of the Total Environment, 639, *195–207. https://doi.org/10.1016/j.scitotenv.2018.05.056*

during the time period 1820–2015. It offers a visual explanation for the levee effect, exemplifying that in 2015 only rare events will be observed and thus the opportunities to learn from flood events are very rare. The case study is representative for many river reaches in Switzerland.

Conclusions from the complexity perspective

Flood risks are not constant. They increase with the increase of buildings and infrastructure in endangered areas and decrease when flood protection measures are implemented. In other words, the flood risk at a given location changes over time, it is dynamic in space and time. The tool "Flood risk dynamics" (www.risikodynamik.ch) allows to understand why this is the case and what role different influencing factors play in this process. The tool is like a model experiment that allows to analyze the effects of three selected risk drivers flood protection, settlement development, and climate change on the overall risk evolution, either considered in isolation or in combination. Another interesting aspect is how the flood risk curves change over time. These curves are representing the footprint of the spatial configuration between hazard, exposure, and vulnerability. The added multitemporal perspective exemplarily shows how the footprint is changing from a floodplain that is not sensitive to climatic changes to a highly sensitive floodplain. The flood risk curves change from a concave form to a convex form. This shows a shift of the sensitivity to increasing flood magnitudes toward extreme flood magnitudes. The evolution of these curves can be used to investigate the change of the behavior of a floodplain system and to detect if regime changes have been manifested, e.g., the change of the floodplain system from a quiet, robust, and nonsensitive behavior to a very high sensitivity to changing boundary conditions. The human interventions to the natural system of the river-floodplain system led to the introduction of tipping points or to a nonlinear system behavior. As long as the threshold of a certain flood magnitude is not exceeded, the system reaction in terms of flood damages is robust. However, above this threshold, the effects of a flood become quickly catastrophic.

The tool "Flood dynamics"—flood impact variability at hourly scales

The previous chapter described the tool "Flood risk dynamics." It was one of the first interactive web tools that allow to run model experiments with counterfactual scenarios of past urbanization and flood risk management policies. This allows to analyze flood risk change at a centennial scale. However, a question remains: how flood risks evolve in the short term?

Many risk factors vary in the short term, e.g., within hours or days. One example of the high temporal variability of risks is the risk of a car driver being hit by a natural hazard process. Especially in mountain areas, the occurrence probability of hazards varies with the meteorological conditions. After a heavy rainfall, soils are saturated and groundwater levels are high. These are preconditions for triggering of landslides or rockfall processes that might hit mountain roads. The moment of the release of the landslide is sometimes correlated with the rainfall peak, but is in most cases unknown. The occurrence probability remains high if the preconditions for a release are high. The same is valid for snow avalanches and glacier lake outburst floods. After a heavy snowfall period, the probability of avalanche release is high, but the exact moment of an avalanche release itself is unknown. The occurrence probability of glacier lake outburst floods depends on weather conditions that favor ice melting. However, also the trigger moment of this process is hardly to predict exactly. Therefore, risks for mountain roads vary in time with the weather conditions. Moreover, the risk of fatal accidents on roads due to natural hazards is varying also because of a highly variable traffic intensity.

Depending on the regional socioeconomic situation, traffic intensity can be high during rush hours or during weekends because of touristic attractivity. If both factors are high at the same time, a high probability of avalanche release and a high traffic density occur at the same time, and risk is peaking. Zischg, Fuchs, Keiler, and Meißl (2005), and Zischg, Fuchs, Keiler, and Stötter (2005) showed how avalanche release probability and traffic density on a high-alpine road vary during a winter season. Single days are responsible for the highest risks. The overall annual fatality risk in these areas is determined by a very few days with a high fatality risk. The road will be closed by the road managers in most of these days with a high hazard probability.

In flood risk management, the spatiotemporal evolution of a flood event plays an important role for the interventions of fire brigades and civil protection units before, during, and after a flood event. An extreme flood event affects large areas. However, the course of the event in space and time determines the availability of resources for the management of the event. If a heavy precipitation event passes over a large area in a short period of time, many communities might be affected by flooding at the same time. Therefore, supracommunal and regional intervention forces have to set priorities (i.e., triaging) in supporting the communal forces. Therefore, not all exposed areas will be managed ideally, e.g., some areas could not be evacuated in time. In contrast, if a rainfall event that is passing only over some parts of a river basin, the flood processes first will affect the subcatchments on which rain is falling and only in a second step it will affect the downstream floodplains as the flood waters from upstream will flow downstream. This setting is easier to be managed by civil protection units because there is enough time between the flooding in upstream and downstream floodplains and the available resources can be balanced. However, this makes evident that the spatial footprint of a flood event is much larger than the spatial footprint of the rainfall event. Moreover, the spatial footprint of a flood event depends on the considered types of flood impacts. If only the buildings are considered in impact analyses, the spatial footprint of the flood impacts maximally corresponds to the extent of the flooded areas. The footprint of the impacts results in this case from the intersection between settlement areas and flooded areas. If the impacts on the traffic networks are considered in the risk analysis, the spatial footprint of the flood impacts will be much larger than the extent of the flooded areas. Herein, systemic risks have to be considered. In the example of the traffic network, an interrupted road or railway section will lead to traffic deviations around the flooded parts of the traffic network. This in turn might lead to higher traffic loads and congestions in the redirected network segments. These network segments can be located far away from the flooded areas. Together with network segments that are directly affected by a flood event, the indirectly affected network segments remarkably increase the spatial footprint of the flood event. Moreover, the accessibility of service areas, workplaces, shopping malls, or hospitals is changing during flood events. The hospital in the nearest town might not be accessible for a citizen living in the suburb of this city. In case of need, this citizen might travel to another hospital in another town that is still accessible during the flood event. This is valid also for other attraction points. Thus, some towns will gain in attractivity during a flood because of the loss of accessibility of other towns because of traffic network interruptions (Loreti et al., 2022). Therefore, a flood event is disrupting the usual hierarchy of towns in a region and new patterns of attractivity of towns emerge out from the pattern of traffic network interruptions and subsequent losses of accessibility. Moreover, road and railway networks might be interconnected. If an important access road is blocked due to a flood event but the railway is still open, parts of the traffic amount might shift from the road network to the railway network. Vice versa, railway passenger might travel by car in case of interrupted railways. Especially the railway network is vulnerable to systemic interruptions because the possibility of redirections of trains to nonaffected railway segments is limited in comparison to the road network. The blocking of road segments due to a flood event leads also to difficulties in the maintenance of other network infrastructure. For example, if transformer stations of the electric current network are flooded, and the roads that are heading to the transformer station are flooded too, the maintenance staff of energy providers has no access to the transformer station and is not able to repair the damage. In contrast, the loss of electricity might disable the function of traffic infrastructure. Thus, the functioning of these two networks is not independent from each other in case of widespread flooding.

These emerging patterns of mobility must be foreseen and considered by the intervention forces. They must know which areas they cannot serve during a flood event because they are not accessible. This is the case when parts of the traffic network are detached from the whole network because the main access roads are interrupted. In turn, a part of the population might not have access to important critical infrastructure such as hospitals. In this case, helicopter flights have to be organized. This is frequent in mountain areas where villages are accessible only through one single access road. Moreover, the material for interventions to reduce flood losses is usually limited, e.g., personnel of civil protection units, pumps for pumping out cellars, sandbags, or mobile flood protection systems. Knowledge about typical patterns of how a flood event might regionally evolve in space and time is helpful for prioritizing and timing local interventions within a larger river basin. Communal intervention forces might collaborate instead of competing for limited available regional resources if their areas of responsibility are not flooded at the same time.

The intervention forces dealing with flood risk management, e.g., civil protection units and fire brigades, are preparing plans for their interventions during an emergency and are organizing trainings for their staff. In these trainings, the course of a flood event is described in the form of a storyboard. The storyline of a flood event is formulated based on past experiences (de Bruijn et al., 2016). This description of the course of a flood event in space and time within a delimited region is known to the planners of the training but not to the participants. With this, possible surprise effects and turns in the event management can be considered in the training. However, the formulation of storylines of flood events is limited to the knowledge of past flood events. Extraordinarily courses of flood events or flood events with very high magnitudes cannot be imagined and written down to story boards if the knowledge of past flood events is limited. In addition, the available information about flood events is very static. The available hazard maps are designed for supporting land use planning and therefore these maps contain a high number of possible flood scenarios. Therefore, the flood extent shown by these hazard maps is the spatial envelope of the superimposition of many different scenarios and of many different flood magnitudes. For the purpose of developing intervention plans or emergency plans, these static maps are useless. They do not show a single hazard event, and they do not show the consequences of the flood scenarios. Moreover, the hazard maps are usually showing local occurrence probabilities. This means that the inundation probability of river reach A in community B is independent from the inundation probability of river reach A in upstream community C and downstream community D. Therefore, the current hazard maps do not allow for preparing emergency plans that are coordinated between upstream and downstream municipalities. The hazard maps do not show the direct and indirect effects of flood events.

A paradigm change is needed in natural hazard and risk management. The mainly static perspective of assessing natural hazards is to be extended by a dynamic perspective. To support this paradigm change, new ways of cartography must be explored. Instead of mapping the cartographic sum of the spatial footprints of all flood scenarios, the spatiotemporal dynamics of flood scenarios must be shown. The new web-based technologies offer a potential for developing interactive tools. Based on computer-assisted and user-friendly web tools, the complexity of the spatiotemporal dynamics of flood scenarios and their impacts can be mapped. This should finally support training of intervention forces and the development of emergency plans (Stefanou et al., 2022).

In a project, the Mobiliar Lab for Natural Risks wanted to evaluate the feasibility of new web frameworks to create interactive tools for explicitly showing the spatiotemporal patterns of flood events and their impacts in Switzerland. The main goal of this feasibility study was to assess the requirements of intervention forces for such cartographic storylines and to present a prototype that meets their needs.

Method

The tool bases on a model chain that simulates hydrology, inundation hydraulics, and flood impacts based on weather scenarios as input data. The tool wants to exemplify the spatiotemporal pattern of flood impacts for selected extreme rainfall events. This aim requires physically plausible extreme weather scenarios. As the meteorological and climatological observations are relatively short, no—or only a few—extreme weather events could have been observed upon now. From historical documents we know that Switzerland was hit by severe events in previous centuries (e.g., Brönnimann et al., 2018). However, these historic events were not documented with enough detail because the meteorological monitoring system did not exist. Thus, the quantification of extreme rainfall events, including their magnitude, frequency, spatial extent, and temporal evolution is fundamentally constrained by the short length of the observational record (Thompson et al., 2017), resulting in a small flood sample, representative only for a limited number of extreme hydrometeorological configurations. We do not know about the maximal extent of heavy rainfall events over large areas. As Switzerland is divided by the Alpine Crest that in many cases acts as a meteorological divide, only very rarely an extreme rainfall event covers contemporarily both the southern and northern side of the Alpine arc. This complicates scenario definition for stress tests or for worst-case analyses.

Statistical or stochastic approaches to generate extreme rainfall events are limited in this complex topographic setting and do not guarantee physically plausible weather patterns. Therefore, we tested several approaches for generating extreme rainfall scenarios that can be used to develop storylines of extreme flood impact scenarios for Switzerland. The first approach was the development of a Monte-Carlo-based procedure to generate spatiotemporal rainfall patterns with a set of plausibility checks. This procedure was developed by Felder and Weingartner (2016) and was described in the previous chapter. In short, this approach generates a high number of rainfall patterns and a subsequent assessment module checks for plausibility criteria. These are consistency of rainfall intensity in space and time and maximum rainfall intensity. Rainfall patterns must focus on neighboring subcatchments and the area of most intense rainfall intensity must cover neighboring subcatchment at a similar timestep. Finally, the rainfall intensity in each subcatchment was checked for not exceeding the probable maximum precipitation according to WMO (2009). This method stochastically generates physically plausible rainfall scenarios. However, it is limited to single river basins. An application for whole Switzerland with its high climatological variability is questionable. Thus, we looked for another method to generate extreme rainfall scenarios.

A second approach for generating physically plausible rainfall events for stress tests was developed by Felder et al. (2018). This approach bases on the idea of "cherry-picking" extreme weather events from a long-running climate simulation. The principle is that a global climate model produces extreme events if the simulation period is long enough. From this global climate simulation, we selected timeseries of precipitation and temperature over 3−10 days where precipitation intensity was in the highest quantiles. The global climate model used in this study was the first version of the Community

Earth System Model (Hurrell et al., 2013). The model is a fully coupled global climate model considering components of the atmosphere, land, sea ice, ocean, and the carbon cycle. The model was run with a horizontal resolution of about 1° for all components. We extracted 400 years of an 850-year-long simulation to identify extreme precipitation events. The timeseries selected from the global climate model data were dynamically downscaled with a regional climate model. We used version 3.5 of the Weather Research and Forecasting Model (WRF) (Skamarock et al., 2008) for downscaling the data from the global climate model to the regional scale. This approach was successful in that it allowed to identify very extreme precipitation events. The magnitudes of these events are nearly in the order of a probable maximum precipitation event. However, this approach was computationally very intensive.

In the tool "Flood dynamics," we used a third method for generating nationwide and physically plausible precipitation events. We exploited a large set of weather hindcast archives for selecting extreme rainfall events that provide the basis for developing storylines of extreme flood events and their impacts. As shown by van den Brink et al. (2005), such archives of weather forecast simulation ensembles can provide a sound basis for describing the probability of weather patterns that trigger rare and extreme floods and their return periods. This data can be useful for the analysis of climate variability (Chen & Kumar, 2017; Weaver et al., 2014).

Hindcast archives of numerical (subseasonal, seasonal, or decadal) climate predictions are generally used to estimate forecast skill in a retrospective way and to statistically calibrate the forecasts (e.g., Robertson & Vitart, 2019). For providing medium-range weather forecasts, the weather services are running an ensemble of weather simulations for a certain period. After this period, namely 10 days, the ensembles markedly differ because the antecedent conditions or boundary conditions of the model do not play an important role anymore. The ensemble simulations that do not equal the observed weather can be defined as "near misses." They could also have been realized but have not because of the stochastic character of some processes considered in the model. Therefore, for each of the started simulation runs of the ensemble, we become different timeseries. The parts of these timeseries that are beyond the point in time where the forecasts are remarkably diverging can be concatenated. This results in a very long timeseries of weather simulations (see Fig. 5.18). All of the ensemble simulations have the same probability of occurrence although only one of the ensembles might later be similar as the observed weather. More generally, the model integrations are realizations of present-day atmospheric conditions, providing a large sample besides observational records. These artificially prolonged timeseries of weather simulations allow the analysis of a high number of extremes and the applications of extreme values statistics to describe flood events with a return period of precipitation sum over Switzerland of up to 1000 years.

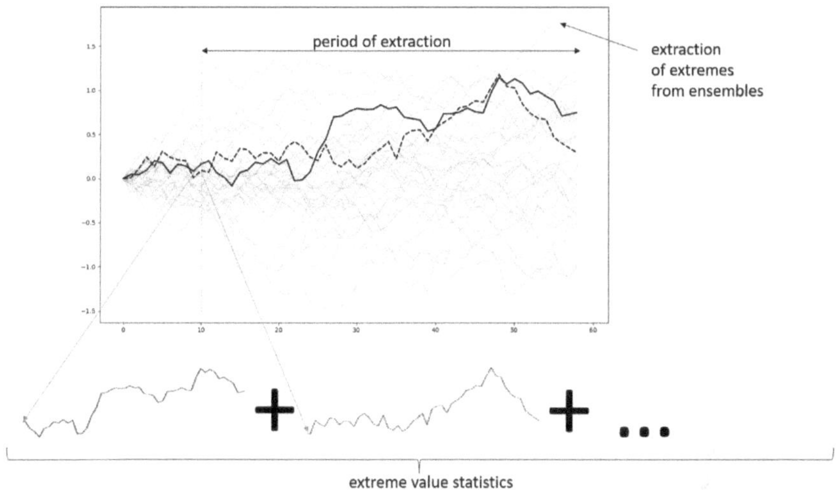

FIGURE 5.18 Analysis of hindcast archives. Schematical explanation of the extraction of extremes from ensemble hindcast archives and concatenating timeseries of ensemble simulations for extreme value statistics.

Rainfall scenarios

In this study, we used two (sub-)seasonal hindcasts provided by the European Center for Medium-Range Weather Forecasts ECMWF, the extended range forecasts (ENSext from 2018, ECMWF, 2019), and the seasonal forecasts (SEAS5, running since 2017, ECMWF, 2017). These hindcasts come from the integrated forecasting system IFS of the ECMWF. All data have a 6 h temporal resolution and a regular 0.4 degrees spatial resolution. ENSext is initialized every 2 weeks and runs for 46 days with 11 ensemble members. SEAS5 is initialized every month and runs for 215 days with 25 ensemble members.

We extracted the largest precipitation events from the ensembles by considering different spatiotemporal patterns and used these "near misses" events as input for a comprehensive modeling chain for assessing the impacts of the resulting flood simulations. Each ensemble member was considered as a physically consistent realization of meteorological patterns; thus, the ensembles can be regarded as independent sets that cumulate to over 2000 (ENSext) and more than 6000 (SEAS5) years of data. The rainfall data of these sets were used to calculate the return periods of precipitation events over Switzerland with a duration of 3, 5, 10 days. From these data, we extracted 245 precipitation events with different return periods (e.g., 100, 300, and 1000 years) for different durations. The selected events were downscaled with statistical methods (quantile mapping (Maraun et al., 2010)) to obtain a higher spatial and temporal resolution. The implementation of the quantile mapping used here closely followed the methods proposed by Ivanov and Kotlarski (2017) and (CH2018, 2018), i.e., by using a bilinear interpolation of the source fields to the target grid, followed by an empirical quantile mapping using the bilinearly interpolated values as input. The observation data used for the quantile mapping was the combined precipitation data product CPC (CombiPrecip) of Meteoswiss (MeteoSwiss, 2019; Sideris et al., 2014) that covers Switzerland with a horizontal grid spacing of 1 km. CPC is based on a statistical data fusion of radar fields and precipitation ground observations.

Out of the event set, we selected nine precipitation scenarios that differ in terms of the spatiotemporal pattern of the maximum rainfall intensity, the return period of the rainfall event, and the duration. Finally, we selected the events with the argument that all regions in Switzerland should be covered by at least one flood event. The nine selected rainfall events provided the basis for developing nine different storylines of flood scenarios and their impacts.

Hydrologic model

The nine selected rainfall scenarios were used as input to the hydrological model DECIPHeR (Coxon et al., 2019). The model was adapted to meet the requirements for modeling hydrological fluxes in a complex mountain topography. A module for modeling snow and ice melting was introduced. Specific routing and storage modules were introduced to consider the most relevant hydrological and hydraulic parts of the river network, e.g., regulated lakes, lake reservoirs, and floodplain retention capacity.

The hydrological model can write out hydrographs for selected points within the river basins and along the river network. These points have been selected in coordination with the hydraulic model. The hydraulic model covers all floodplains of the main river network in Switzerland. The floodplains were delimited under geomorphic criteria. This means that each river reach is delimited by a closed geometry of the adjacent floodplains where the water can flow in the floodplain. In most cases, the upper boundary conditions of the hydraulic models (of each floodplain and river reach) are locations where the river exits from a narrow valley without adjacent floodplains. The lower boundary conditions are points in the river network where, e.g., the river enters into a narrow valley or where the riverbed is deeply incised in the sediments and where flooding in the floodplains is no longer possible. Moreover, we considered only floodplains and river reaches with exposed elements located in the floodplains. In addition to the floodplains of the rivers, we added also the floodplains adjacent to lakes that are susceptible to lake flooding. Hence, the hydraulic models adjacent to lakes can simulate both dynamic river flooding and static lake flooding.

Hydraulic model

We set up more than 170 hydraulic models. These hydraulic models were nested into the hydrological model. Thus, the hydrological model was prepared to provide hydrographs at all upper boundary conditions of the hydraulic models. The overall routing and the simulation of the lake levels were done by the hydrological model. We followed the approach for nesting the hydraulic model into the hydrologic model as proposed by Zischg, Felder, Mosimann, et al. (2018). In contrast to this former study, the 1D hydraulic model for routing the water fluxes through the valley (Fig. 5.19, left) was exchanged by a direct implementation of the routing in the hydrological model (Fig. 5.19, right). Thus, the hydraulic models can directly be nested into the hydrological model.

The flood processes were simulated with the coupled 1D/2D model BASEMENT-ETH (Vetsch et al., 2017). The data for the river cross sections came from the Federal Office of the Environment (FOEN, 2018) and were harmonized. The hydraulic models were run in parallel and delivered outputs of flood extent, flood depths, and flow velocity in a coordinated hourly timescale.

In contrast to the previously presented tools and models, the tool "Flood dynamics" focuses on visualizing the spatiotemporal dynamics of a flood event in Switzerland. Therefore, we did not elaborate static hazard maps that show, for example, the highest flood depth per grid cell during the simulation time, but we elaborated flood maps and hazard maps for every hour in the simulation. The output data of every hydraulic model were merged into one output for each of the selected scenarios. This resulted in a scenario-specific and

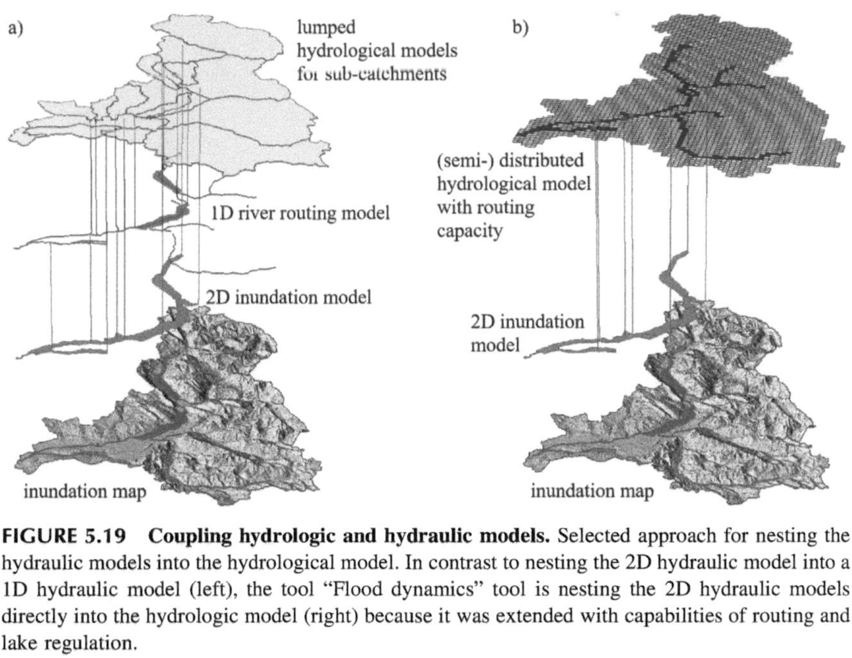

FIGURE 5.19 Coupling hydrologic and hydraulic models. Selected approach for nesting the hydraulic models into the hydrological model. In contrast to nesting the 2D hydraulic model into a 1D hydraulic model (left), the tool "Flood dynamics" tool is nesting the 2D hydraulic models directly into the hydrologic model (right) because it was extended with capabilities of routing and lake regulation.

dynamic map of flood depths and flood flow velocity with a hourly timestep. In the tool, two versions of dynamic flood hazard maps were visualized Fig. 5.20. The first map is a classified visualization of the flood depths. The second map is a derivative of the flood map and the flow velocity map.

The classification of the flood depths was derived according to the following thresholds:

- 1–30 cm
- >30–50 cm

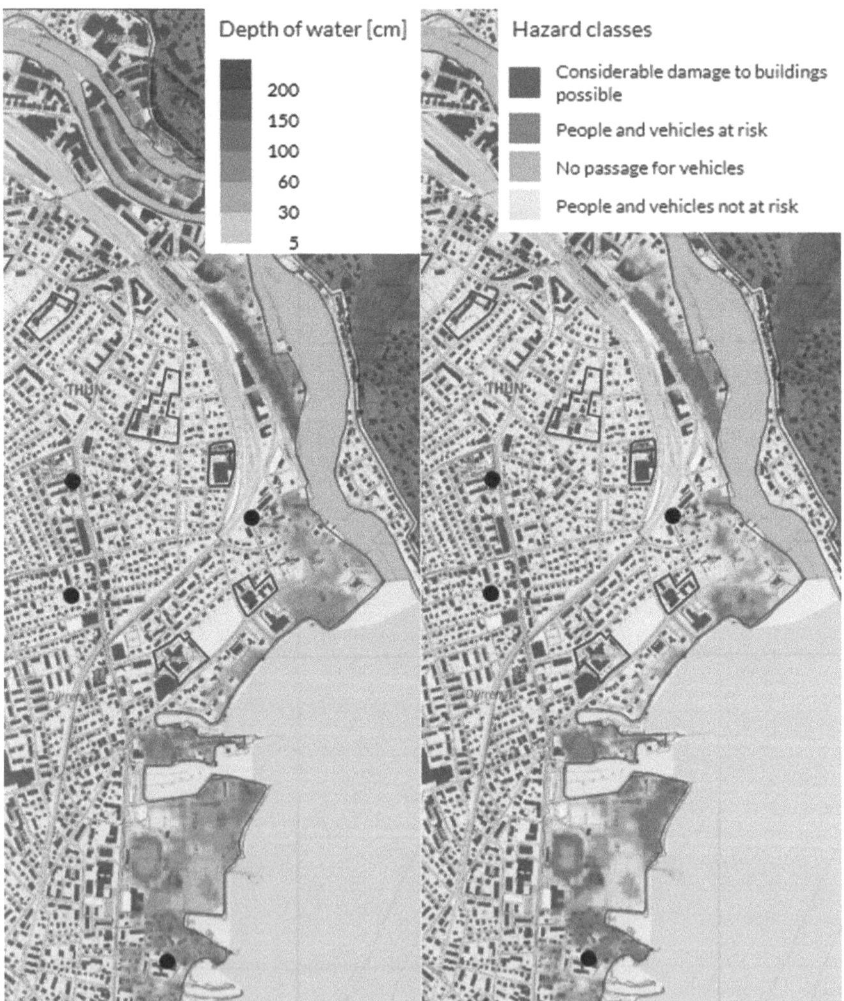

FIGURE 5.20 Hazard classification. Comparison of hazard classifications. The left map extract shows the classified flood depths, the right map shows the vulnerability-driven classification.

- >50–100 cm
- >100–150 cm
- >150–200 cm
- >200 cm

Areas with simulated flood depths below 0.01 m were not considered as flooded. The second map classifies the hydraulic impacts to persons, cars, and buildings. This is a cartographic representation of the hazard that is driven by the vulnerabilities of the different kinds of elements at risk considered. The hazard class related to people is driven by the physical stability of persons in a flood. The stability of people in flood waters depends on the combined effect of flood depth and flow velocity, expressed in the Froude number, and the height of a person (Arrighi et al., 2019). After these values are exceeding a threshold, a person will not be able to stand stable in the flood water and she will be washed away. The second vulnerability-driven classification of flood hazards is the classification of the stability of a car in flood waters. The stability of a car is dependent on the car type and on flood depth. We followed the procedure of Pregnolato et al. (2017) and assumed a medium car with a flood depth threshold of 0.3 m for being unstable. The third classification relates to the structural stability of buildings. A threshold of 1.5 m was used to describe the structural vulnerability of buildings. The classification is merged together similarly to the approach proposed by Costabile et al. (2021). This results in four hazard classes:

- Persons and cars are affected but not endangered
- Roads are not passable for cars
- Persons and cars are endangered
- Structural damages to buildings are possible

For the first class, flood depths and velocities must be below all three thresholds for persons, cars, and buildings. The second class must be below the threshold for persons but must exceed the threshold for cars. The third class describes flood values that exceed the thresholds for persons and cars, but not the threshold for houses.

Flood impact model

For each hour, we calculated the following flood impacts: the number of residents of flooded houses, the number of flooded buildings, the monetary damage at flooded houses, and the directly or indirectly affected roads. The number of residents affected by the floods result from the overlay of the hourly flood maps with the dataset of buildings that contain the number of residents per building. We used the buildings dataset of Röthlisberger et al. (2018). This dataset contains also an attribute for the reconstruction value of each building. The damages to buildings were calculated by attributing the flood depth of each hourly map to each building. We used a region-specific vulnerability function that was developed based on insurance claim data from Switzerland (Zischg et al., 2021). The degree of damage resulting from the flood depth and the vulnerability function was multiplied with the reconstruction value of the building. This resulted in the flood damage in monetary terms.

The impacts of a flood on the road network were calculated according to the following procedure. In a first step, the directly affected road segments were identified. We assumed that all road segments that are flooded with flow depths above 0.3 m were not trafficable. Thus, these road segments were considered as "closed" for the standard car. Special cars might still be able to drive along these roads. Bridges had to be considered differently. When flood depth maps are overlapped with a 2D road network without considering the third dimension, all road segments within the extent of the flood event would be identified as flooded. This does not play an important role for cases where roads are constructed on the ground. However, bridges would be considered as flooded too, although they might be located meters above the flood waters. Therefore, we checked the difference between the water surface elevation and the absolute height of the road surface at the bridge. Only if the water surface elevation was equal or higher than the lowest height of the road pavement at the top of the bridge plus 1 meter of freeboard, the bridge was considered as flooded and closed for the traffic. The indirect effects of flooding on the road network were calculated as follows. If a flood along a river valley affects a bridge that connects the two towns at both lateral sides of the valley, we assumed that residents of both towns would look for alternative routes to travel from one town to the other. Thus, we looked for the shortest paths from one of the two nodes of the flooded road segment at the outer border of the flood extent to the other. These two nodes would have been linked to each other by a short route and the bridge during a normal situation. During a flood situation, the length of the route between the two points increases remarkably. The road segments along the redirected route between the two nodes at the outer border of the flood were then classified as indirectly affected. For these redirections, we assumed that the traffic density is much higher during flood situations than during normal situations. This additional traffic density would lead to congestions. We mapped the directly and indirectly affected road segments for every hour. This results in a spatial extent of the flood impacts that is markedly enlarged in comparison to the spatial footprint of the flood itself.

Codesign of the graphical user interface

We invited several stakeholders from local and regional intervention forces and cantonal and national government authorities that are responsible for flood risk management. Before implementing the above described methods in a tool, we asked the stakeholders about the most important information content and about the way of cartographic visualization. Moreover, we collaborated with user experience and user interface designers to plan the graphical user interface.

During the development process, we iteratively showed the intermediate results to the stakeholders and asked for feedbacks if the presented information is meeting their needs. Also, the criteria for selecting the different rainfall events were coordinated with the stakeholders. Thus, the information content, the cartographic visualization, the background information, and the graphical user interface were coordinated and codesigned with the users' requirements.

Results

The tool "Flood dynamics", available at https://hochwasserdynamik.ch, starts with the principal description of the model chain. The main focus of the short introduction is laid on the process chain rainfall-hydrology-inundation-impact. It guides the user to a description of the storyline approach and the motivation of the developers. It asks the following question: "How does a flood event unfold in space and time?" The user is invited to discover the flood impacts based on possible precipitation scenarios. The tool offers the visualization of rainfall, flooded areas, and flood impacts at two different scales, namely at a national scale (overview) and at a local scale (zoom).

The first interactive choice has to be made by the user in the selection of the rainfall scenario. The user can choose from nine rainfall scenarios that vary in the rainfall magnitude, in the return period, and in the spatiotemporal pattern of the rainfall event. Users living in the Eastern part of Switzerland may prefer to select rainfall scenarios where the highest precipitation intensity and sum mainly covers the Eastern part of Switzerland. Residents and members of intervention forces of communities located at the downstream parts of the river network may prefer rainfall scenarios with a long duration while residents in the headwater catchments may select rainfall scenarios with a shorter duration but high rainfall intensities.

The selection of a rainfall scenario is the basis for all subsequent steps. It is the basis for the storyline of the flood event. Hence, the storylines of flood events are named after the description of the rainfall scenario. The user can look at the evolving of the rainfall scenarios with a time slider. She can choose to view rainfall intensity or rainfall sum over the event period. The time slider can be manipulated and the user can visualize either the dynamics of the rainfall event like in a movie or a static snapshot for a specific time within the event period. At the end of the time slider, the total sum of rainfall or the maximum intensity over Switzerland and over the event period can be visualized. The time slider to visualize the course of the event is similar to the time slider of the rainfall radar observation of the Federal Office for Meteorology Meteoswiss. This visualization of the rainfall over Switzerland is implemented in the MeteoSwiss-App for mobile phones. This is one of the most downloaded and used mobile phone apps in Switzerland. The map legend of the radar rainfall monitoring is well known in Switzerland, and therefore we used the same legend in terms of rainfall classification and colors (Fig. 5.21).

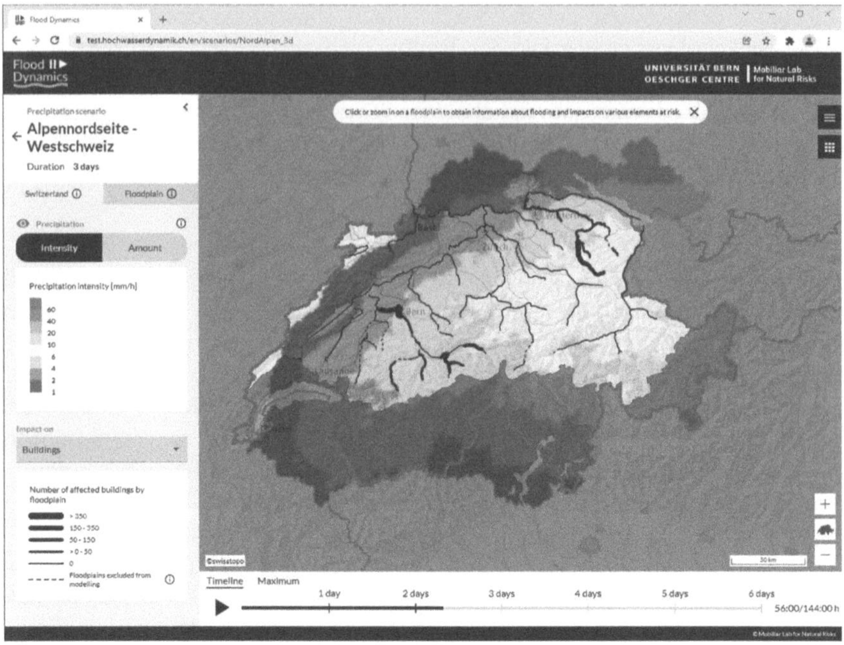

FIGURE 5.21 Flood dynamics. Screenshot of the tool "Flood dynamics" (hochwasserdynamik.ch), national overview of the rainfall scenario (rainfall intensity) and number of affected buildings.

At this national-scale overview, the user can navigate through the event with the time slider and look how the rainfall event unfolds in space and time. Moreover, the unfolding event can also be visualized in terms of flood impacts. To this aim, we aggregated the hourly values of exposure and damages to the floodplains. The floodplains are visualized at this scale as part of the river network (lines). The thickness of the line symbology represents the amount of flood impact. The user can choose to visualize either the number of buildings, the monetary damage to buildings, the number of flooded hospitals, nursing homes, and schools, the number of residents, nursing home places, workplaces affected by the flood, the length of impassable roads, as well as the length of redirected traffic routes. The user can look how the impact of the flood on the kind of selected exposure is unfolding in space and time with the rainfall and the flood event. This ability of the tool allows for delineating storylines of extreme floods. Time delays between one catchment and another or between upstream river reaches and downstream river reaches can be made explicit.

When zooming in to a selected floodplain, the cartography changes from the overview representation to the cartographic representation of the hazard map and the flood impacts at local scale. The hazard map varies with the unfolding of the flood event and with it the mapped flood impacts vary hourly.

Fig. 5.22 shows the flood hazard map at a selected timestep in the floodplain of the Thur River near Frauenfeld. In addition, the maps show also the element at risk that was selected by the user. In the example, the map shows the directly flooded roads and the roads that are indirectly affected by the flood event due to redirections and following traffic congestion.

With the overlay of the hourly hazard maps and the elements at risk, we can create timelines that show the unfolding of the flood impacts. Fig. 5.23 shows an example of such timelines for buildings, damage to buildings, people, and workplaces. The example shows the flood impacts of a 5-day rainfall event for the floodplain of the Thur River near Frauenfeld. The timelines show a sudden increase of the number of affected buildings as well as the calculated damage to buildings after a certain timestep. The first buildings are flooded after 90 h after the start of the rainfall events. The maximum of the affected buildings is reached at hour 114. The damages are increasing also after the maximum number of flooded buildings because the flood depths are still rising. The number of affected population evolves similarly as the number of the affected buildings, but the curve is steeper at the very beginning of the flooding. In contrast, the evolution of the number of affected workplaces follows a different form of the curve. Depending on the time of flooding of larger production facilities with a high number of workplaces, this indicator is increasing with a stepwise function.

The impacts of floods on roads during the unfolding of the flood event can also be visualized as a timeline (Fig. 5.24). The selected example of a flood event causes about 30 km of roads to be impassable and 15 km of roads that are flooded with a low intensity. These roads can be used with care or with special cars. The indirect effects of the flood event on the road network are remarkably higher. These comprehend road segments that are not flooded but function as alternative routes or redirections. About 100 km of roads are assumed to experience traffic congestions due to diversion traffic. This is more than twice the number of directly affected roads.

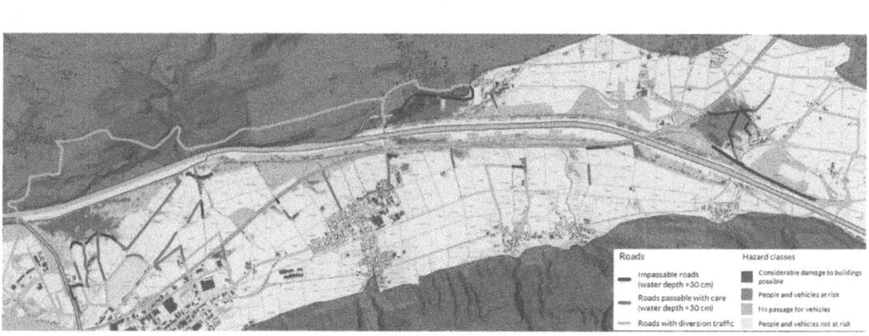

FIGURE 5.22 Flooded road segments. Mapping of road segments that are directly and indirectly affected by a flood event.

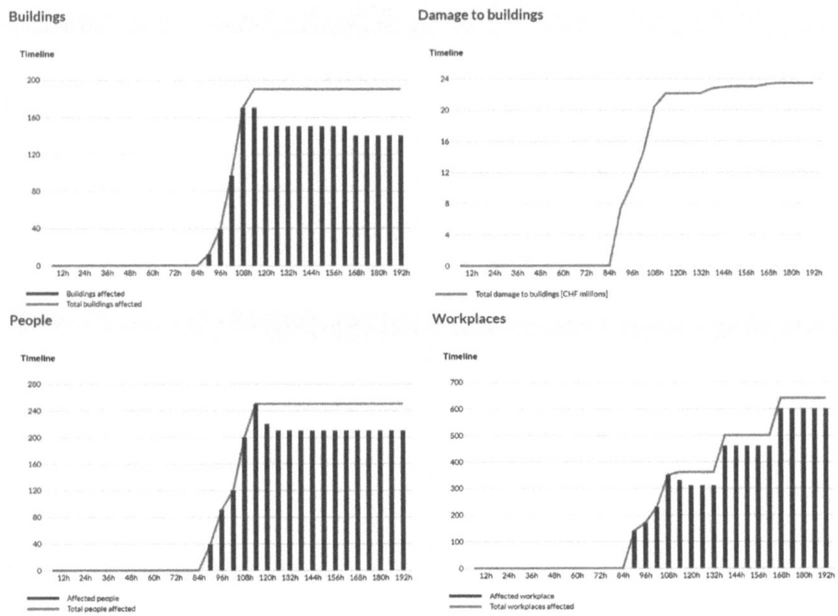

FIGURE 5.23 **Flood impacts.** Example of the temporal unfolding of the flood impacts on buildings, people, and workplaces in the floodplain of the Thur River near Frauenfeld during a 5-day event.

Conclusions from the complexity perspective

The flood impact simulations of the nine selected rainfall scenarios vary remarkably in space and time. This shows that the spatial and temporal pattern of flood impacts can be very different. Also rainfall events with the same amount of rainfall over the same period can result in remarkably different flood impacts. Thus, the rainfall event can be considered as very different storylines of unfolding flood events and their impacts. These storylines can be incorporated into storyboards of trainings of intervention forces. Moreover, the tool can also be used for flood risk communication.

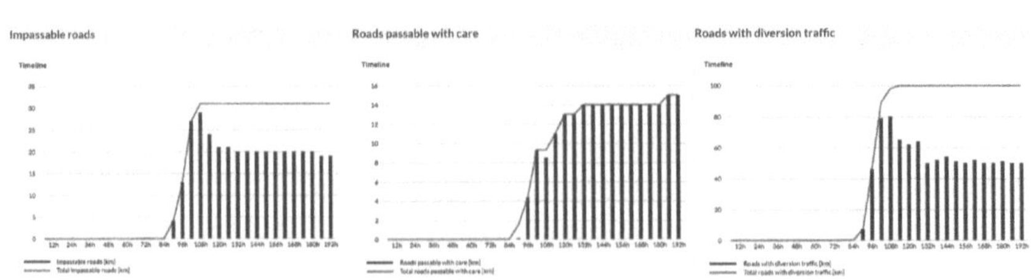

FIGURE 5.24 **Flood impacts.** Example of the temporal unfolding of the flood impacts on roads in the floodplain of the Thur River near Frauenfeld during a 5-day event.

The timelines of flood impacts are also varying remarkably between the floodplains and the rainfall scenarios. This shows how variable a flood event can locally unfold in space and time. This makes evident that some floodplains react quickly in terms of the number of exposed elements. The flood impacts are rising steeply after the overflowing of the lateral levees in some floodplains while other floodplains show a slow rise of exposure. This shows that some floodplains are supposed to the so called "surprise" effect while others do not. This relates to the topological relationship between the river with its characteristic morphology and the topography of the human environment. The floodplains with a fast increase of exposure after a certain flood peak discharge is exceeded are of special interest for the intervention forces and for the elaboration of their emergency plans. Knowledge on how the "human system" in terms of urban settlement and infrastructure reacts to disturbances from the "natural system" in terms of rivers and catchments is helpful for optimizing the preparedness. From the tool, the time available for preparing mobile flood protection measures during the onset of an extreme event can be delineated. This can be coordinated with the availability of flood forecasts and their lead times. It also allows to judge where preparatory works need more time to install than available. These areas can be prioritized for evacuation rather than for the installation of preventive temporary hazard management measures. Thus, knowledge of this variety of unfolding flood events contributes to preparedness.

The simulations represented by storylines show that rainfall events with a certain magnitude (e.g., return period of accumulated precipitation) can lead to highly diverse magnitudes at river reach levels. A 100-year rainfall event (e.g., 3, 5, 10 days) at national scale can lead to local return periods of more than 1000 years (e.g., return period of peak discharge). Contemporarily, some river reaches do not trigger flooding in floodplain.

Moreover, the timelines of the rainfall intensities and flood discharges at floodplain scale do not correspond exactly with the timelines of the flood impacts. The flood impacts can be delayed and can display a very different form of the curve. In some floodplains, the exposure is rising fast after the discharge exceeds a certain thresholds although the discharge is rising slowly. The simulations provide a lot of case studies to analyze the role of the spatial topology between the human and the natural system. This topology is culturally driven or is a result of the past trajectories of settlement growth or past human interventions to the river system. The timelines of flood impacts indicate also the difference between mountainous floodplains and floodplains in the Swiss Plateau. Flooding in narrow river valley results in steeply rising exposure curves that soon are reaching a plateau. In contrast, flood extent in flat areas is driven by flood volume. These floodplains exhibit steadily rising curve forms.

Last but not least, the cartographic representation of the indirect impacts extends the knowledge of the spatial footprint of the impacts of flood events.

The indirect impacts of floods on the road network are visualized by identifying the road segments that are not directly affected by the flood but are indirectly affected by diversion traffic. These road segments are providing alternative routes or routes with redirected traffic. The overall area affected by diversion traffic is much higher than the flooded areas and extend far beyond the floodplains itself.

In summary, the tool is contributing to the paradigm shift from a static mapping of flood hazards and risk to a dynamic mapping of flood events. The presented approach enhances the comprehension of the spatiotemporal dynamics of an extreme flood event over Switzerland. The storyline approach has proven to provide a useful framework for implementing modeling results in publicly accessible web tools for risk communication.

Decision-making leads to the emergence of spatial patterns of flood risk change

The previous modeling examples show mainly the effects of environmental changes or human interventions on flood risk change. In this last example, I will show a model experiment that aimed at demonstrating the effect of the philosophical basis that is underlying decision-making on long-term flood risk change. Thaler et al. (2018) conducted a model experiment that showed how the decisions of a central government in prioritizing investments in flood protection measures become manifest in space. The allocation of risks and benefits is evolving, depending on the principle of priority setting.

The experiment simulated the decision-making of a centralized government when following different philosophical schools. This was the Austrian government in the case study. The study tested the outcomes of different priority setting principles under the assumption that the strategy would have been followed strictly for a time period of 30 years.

In principle, the Austrian government is following a natural hazard funding policy that bases on a mixture of egalitarianism and utilitarianism (Thaler & Hartmann, 2016). In the early 2000s, the national government decided to prioritize the investment based on cost—benefit analysis, where the highest benefit-cost ratio gets implemented first. One reason for this decision was that the overall budget and the resource distribution over the federal states is mainly disaster-driven (Thaler, 2014), which means that after major strikes investments for mitigation and adaptation are repeatedly shifted toward these federal states (Raschky & Weck-Hannemann, 2007). This spontaneous reaction and shifting of budget raised questions of social justice. Regions that invested in prevention loose funding from central government in these cases of abrupt funding shifts while regions without previous investments in preventions but were hit by a flood event gained in funds. This inequality has focused the attention of policymakers and other stakeholders of how to approach the topic of natural hazard protection in the long term.

As we have learnt from the previous book chapters, the governance strategy and flood risk management policy are influencing flood risk change in the long run because they govern the spatial distribution of funds. The way how funds for investments in flood protections are allocated and distributed over the territory contributes to the emergence of spatial patterns of risk hotspots or cold spots. In an urbanized society in which the rural hinterland is only considered as a service and resource provider, public funds might be guided mainly to cities. In contrast, a cultural identification of citizens with the rural areas might favor keeping functioning rural infrastructure although it might have a low cost–benefit ratio. This is common in Alpine countries, which follow a strategy that avoids depopulation of remote areas. Every strategy has positive or negative effects on social inequality or social justice.

Thus, flood risk governance strategy must consider the topic of social justice. Basically, justice concerns questions on the allocation of wealth (resources), participation, and recognition across different members of a society (Schlosberg, 2007). The model experiment aimed to examine the potential impacts of changing the current natural hazard management policy in Austria on distributional effects in showing possible impacts of three philosophical schools (utilitarianism, Rawlsians, libertarianism). The focus lies on the question of the impact on the distributional effects when following strictly one of the three philosophical schools in natural hazard management. The study focused on the question of how distributional effects can be organized within the country based on three theoretical schools within the social justice debate. The selected schools focus their key concept and main arguments around distributional effects of new policy concepts and strategies.

The concept of utilitarianism, developed by classical economists, such as John Stuart Mill and Graham Bentham, understands justice and equity as the sum of individual benefits. The aim of utilitarianism is to "maximize the total potential happiness of society as a whole through the aggregation of individual happiness" (Johnson et al., 2007: 376). Thereby, the focus is on benefits of each individual (Bartel, 2002), and consequently, policy discussion should ensure maximal benefit of society ("greatest benefit to the greatest number" (Hunold & Young, 1998)). In flood risk management, the concept of utilitarianism is implemented in priority setting of investments in flood protection according to criteria, which should be chosen to secure the highest risk reduction per unit of resource input. Natural hazard risk management strategies are provided to those areas within country, where the benefits offer the greatest gain to the society (Johnson et al., 2007).

In contrast, Rawls (2005) defined justice as the equal distribution of basic rights and duties within a society. Justices will be defined by the society, who defines a common understanding of what social justice is, as well as which actions fulfill the basic needs for each individual within the society. Based on this understanding, Rawls tolerates injustices, if unequal developments in society increase the overall benefit (wealth) of a society. However, if the outcome reflects injustice, individuals will not be punished/discriminated in another aspect. Referring to natural hazard and risk management, an application of Rawls' concept of justice requests to distribute resources to most vulnerable people and objectives (Johnson et al., 2007). Therefore, the key objective of a just flood risk management policy is to select risk management measures not only and inevitably for high-value assets and areas of high-value aggregation.

The main focus of liberalism is in general a concept of free market thinking, with a particular focus on competition, availability of full information, equilibrium in market processes, and freedom of individual self-decision (economic freedom (Harrison, 2014; Hayek, 1991)). Key aspects include the availability of full information, which are mainly predefined rules equally to each individual.

Method

The model experiment bases on three steps. First of all, we overlaid the flood hazard maps of Austria (HORA (Merz et al., 2008; Stiefelmeyer & Hlatky, 2008)) of the Federal Ministry of the Environment with the building dataset of Fuchs et al. (2015). This dataset contains the building type, the reconstruction value of the building, and the number of residents. This intersection was than overlaid with the dataset of the Austrian municipalities. With these intersections, we can query the number and monetary value of all buildings exposed to floods per municipality, the number of affected residents, and the relative share between the number of affected buildings and residents to the total sum of buildings and residents in each municipality.

The model experiment based on the following assumptions: (a) the national government decides to allocate the available funds according to a ranked list of municipalities, (b) the ranking bases on one of the abovementioned philosophical schools, (c) the government has funds for implementing 60 flood protection projects per years, and (d) it follows the strategy for 30 years.

First, we prioritized the municipalities according to the Utilitarism approach by ranking the municipalities by the aggregated absolute monetary value of all exposed buildings. This simplifies a prioritization strategy that is based on a cost–benefit analysis. The available funds for protection projects were guided first to municipalities that have the highest monetary value exposed. This strategy was implemented in the model experiment as the 60 highest ranked communities will be eliminated from the list per year, for 30 years. At the end of this period, the flood risks were analyzed again.

Second, we prioritized the municipalities according to the Rawlsian approach by ranking the municipalities by their highest social vulnerability. Because there is no dataset for classifying the vulnerability and the coping capacity of the Austrian local governments, respectively, we elaborated simple indicators for a vulnerability classification. One simplified indicator was the share of the number of exposed buildings in comparison to the total number of buildings per local authority (Röthlisberger et al., 2017). The higher the number of exposed buildings or persons was in relation to the total number of buildings or persons within municipality, the higher was the vulnerability. In cities, the exposure ratio usually lays in the order of a few percent. In contrast, many municipalities in mountain areas have exposure rations of higher than 75%. A second indicator was based on the average monetary value of the exposed buildings compared to the average monetary value of all buildings in each local government. A value >1 means that exposed buildings are more costly than the average of all buildings in the local government. A value <1 means that the exposed buildings are less worth than the average of the total building stock in the local authority. The highest ranking in terms of vulnerability had the local government with the lowest value for this indicator. It is assumed that municipalities where the exposed buildings have a lower monetary value than the average are more vulnerable than others. Hence, this indicator represents an important factor to classify social vulnerability of each local government (Cutter, 2003). The local government was classified and ranked by each of these three indicators in a descending order. This priority list simplifies a prioritization strategy with a focus on social vulnerability. Again, the available funds for protection projects were guided first to municipalities that have the highest social vulnerability. This strategy was implemented in the model experiment as the 60 highest ranked communities will be eliminated from the list per year, for 30 years. At the end of this period, the flood risks were analyzed again.

Under the third philosophical school, the liberalism, we assumed that the Austrian natural hazards system change toward a self-oriented risk management policy; including no public funding from the national government. The citizens living in the hazard-prone areas have to organize their individual protection scheme. Thus, no allocation of public funds was modeled.

Results

The modeling results show some emerging spatial patterns. Depending on the respective political direction, various local governments gain and others loose within the actual distributional system of mitigation strategies. This manifests in a spatial pattern at national scale. The implementation of a pure utilitarian policy approach would cause that high-income communities in hazard-prone areas mainly benefit. These are cities and touristic hotspots. Those areas experienced a long period of economic growth that resulted in a remarkable urban spreading and finally in a high flood exposure in absolute numbers. Consequently, this policy direction would encourage the public administration to ignore their own failure in the past natural hazards management and prevention. On the other hand, following a Rawlsians approach, mainly peripheral communities would gain from a new policy direction. These peripheral and rural areas often show large socioeconomic challenges besides the natural hazards problem. However, investments in flood protection works in these areas are limiting further depopulation and the implementation of protection projects might bring in an economic stimulus. This may not necessarily lead to an increase of exposure because these communities often have a larger potential of land that is free for development. In these cases, avoiding risky areas for development is easier than in urban areas. Finally, the most radical change would include the implementation of a liberalism policy, whereabouts the state only provides hazard information, but no further mitigation measures. These findings highlight the distributional consequences of future mountain hazard management strategies and point to the crucial selection of policy direction in navigating the selection of various adaptation schemes.

Conclusions from the complexity perspective

This model experiment examined how different social justice traditions might change the spatial distribution of funding for natural hazard protection schemes. It shows how philosophical schools are underlaying decision making in flood risk management. With it, philosophical thinking becomes manifest in space and contributes to the emergence of spatial patterns in the flood risk landscape. Thus, it contributes to spatial complexity. The study showed the emergence of spatial patterns from a chosen prioritization strategy for public investments in flood protection. Depending on the selection of philosophical tradition, the winners and losers of public funds remarkably differ in space. For instance, prioritization of structural protection measures based on utilitarianism would mainly support the tourism hotspots as well as industrial complexes. The consequences would be that the public administration would acknowledge and encourage constructing new and more high-value buildings in hazard-prone areas. This may lead to positive feedback effects because it would result in a further increase of potential losses across the country according to the levee effect. On the other hand, a focus on the most vulnerable groups would mainly support local governments in peripheral areas and mountainous regions. As such, this policy would support the main losers from a utilitarianism economic thinking. Yet, the number of individuals who gain from this policy would be much lower in comparison to the other strategy. The municipalities that gain from this strategy will not contribute remarkably to further increase of flood risk because they anyway have low growth rates. In addition, the economically strong and fast-growing urban areas will grow further under this strategy but without investments in flood prevention. In this regard, this policy will not contribute to keep flood risk constant but may lead to increased residual flood risk in the long term.

Monitoring flood risk change

Instead of analyzing the sensitivity of human–environmental systems to external changes, or to analyze future flood risk change, the presented modeling framework can be used also to monitoring flood risk change. In this case, the model framework is used to model the flood hazard (hydrological and hydraulic models) and the damages (exposure and vulnerability models) at selected timesteps and the risk analysis is repeated periodically.

As seen in the previous chapters, the factors that determine flood risk are changing periodically. Hazard maps are updated regularly, e.g., after 5 or 10 years. After the implementation of flood protection measures, the hazard maps are also updated. Extreme value statistics are repeated when the timeseries become longer. Regarding exposure and vulnerability, the data is also dynamic. The national or regional topographic or cartographic services are regularly updating all the geodata used for risk analyses. For example, the geodata of buildings or streets, as well as of other infrastructures, are updated every 1–5 years. Moreover, the vulnerability of exposed infrastructure is evolving. The dates of the updates of each of the datasets that are needed for flood risk analysis are not coordinated. Therefore, every year, one of all these datasets is updated. This is influencing overall flood risk that is changing over time. Thus, monitoring this change in flood risk can be achieved by repeating regularly the risk analysis. All the updates of geodatasets result in a continuously changing data stream. This data stream can be used to continuously updating the risk analysis. This results in a monitoring system that observes flood risk change.

This concept of a system for monitoring flood risk change was implemented by the regional government of Carinthia (Zischg et al., 2013). The information system dynamically combines the results of the flood hazard modeling with all other geodata that is needed for the risk analysis. If one of all the underlying data is updated, the system is recalculating the flood risk over the region and adds a record and a date to the timeseries of risk analyses. The entries in the database are historized and attributed with timestamps. This allows to analyze the overall trend of flood risk change and to identify the municipalities in which flood risks are increasing. Based on this risk monitoring system, decisions can be drawn to initiate the planning of flood protection projects to avoid the further increase of flood risks. In the long term, this monitoring system should be able to monitor the signal of climatic changes in the evolution of flood risk (Fig. 5.25).

Modeling framework for considering complexity in the analysis of flood risk change

In this chapter, as well as in the previous chapters, we have seen many different model experiments for analyzing flood risk change. Here, I am

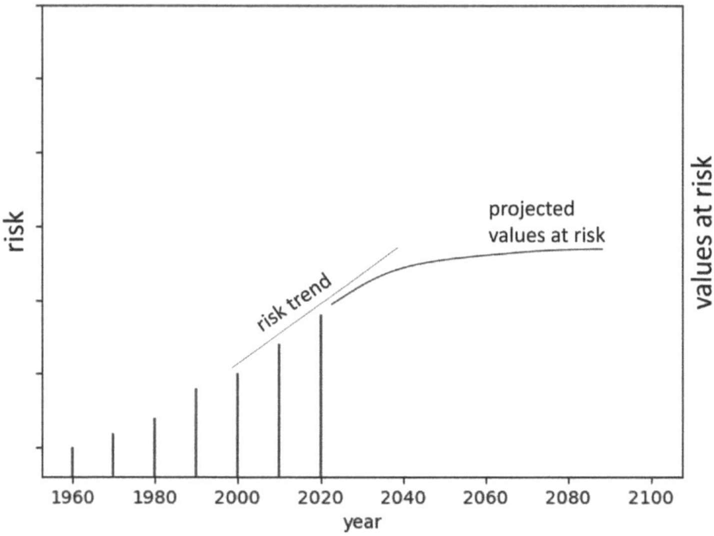

FIGURE 5.25 Flood risk monitoring. Simplified concept for monitoring the temporal development of the economic values at risk and the related risks. The dark bar shows the situation of 2010, the light bars show hypothetical values of future and historic trends. *Modified from Zischg, A., Schober, S., Sereinig, N., Rauter, M., Seymann, C., Goldschmidt, F., Bäk, R. & Schleicher, E. (2013). Monitoring the temporal development of natural hazard risks as a basis indicator for climate change adaptation.* Natural Hazards, 67*(3), 1045–1058. https://doi.org/10.1007/s11069-011-9927-0*

outlining what is common to all of these model experiments. From the examples, the main principles are distilled and described.

First of all, all the examples base on the framework of model coupling. All model experiments were conducted with coupled component models (CCMs). These are composed of specialized disciplinary models representing parts or processes of a studied system. Coupled models integrate submodels to form a model chain that represents a whole system (Blair & Buytaert, 2016). Synonymously, this type of model is often defined as an integrated environmental model (Laniak et al., 2013; Welsh et al., 2013). The examples proved that CCMs are flexible regarding the level of integration and are relatively transparent because the submodels have mostly been validated in their specific discipline. The coupling of models facilitates the representation of the full chain of natural processes from atmospheric and meteorological processes to hydrologic and hydraulic processes and finally flood risk change. Coupled models integrate these specialized models for each of the processes in the process chain to form a model chain that represents a whole system. Thus, model coupling allows to synthesize relevant scientific progress in the different disciplines of flood risk research. Under a reductionist philosophy, all the subdisciplines of hydrology and flood risk research have made relevant progress in developing reliable models for specific and narrow purposes. From a Universalist view, this fragmented knowledge has to be compiled and incorporated to meet the requirements of holistic thinking in solving our actual problems, which are often interconnected, complex, and interdisciplinary. The described examples of model experiment proved that coupled component modeling frameworks have the potential to unify specialist knowledge and to bring together knowledge from different disciplines for holistically analyzing coupled human and natural systems.

Second, the model experiments have shown that the submodels and model components can be coupled across different spatial and temporal scales. Thus, a CCM framework allows to bridge different spatial and temporal scales. Another advantage that has been demonstrated by the model experiments is that CCM can potentially combine both lumped and spatially explicit models. Geographical information systems provide key functionalities to bridge spatial scales and to combine lumped and spatially explicit models by spatial disaggregation. Moreover, we have seen examples of model experiments where the information of global climate models with a spatial resolution of hundreds of kilometers and a daily temporal resolution can be downscaled to a submeter spatial resolution and hourly temporal resolution (Felder et al., 2018). The coupled model for simulating large wood recruitment, transport, and deposition (Zischg, Galatioto, et al., 2018) is an example for coupling models across different timescales. The first submodel is analyzing LiDAR data and forest inventories, identifies individual trees, and assesses the volume of wood that can be recruited by flood processes. The second submodel is a hydraulic model that is simulating the flood process. This model is working at seconds scale and writes the output at selected timesteps, e.g., every 10 min. The third submodule is nested into the hydraulic simulation and runs at a higher spatial or temporal resolution. For each timestep of the model output of the hydraulic model, this procedure assesses the hydraulic impact on each of the individual trees and decides if the tree is eroded and recruited by the flood process for transport. The fourth module is checking for each timestep if the conditions for transporting the recruited large wood log further downstream are still met. If the hydraulic conditions do not support the transportation of the wood logs, the process stops and the wood log is deposited. Moreover, coupled component models are able to couple raster-based and vector-based modeling approaches. These two main benefits, coupling across scales and coupling of different modeling approaches into hybrid modeling frameworks, together with the multimodel and multitemporal approach at component level, allow new study designs for analyzing flood risk change.

Third, coupled component models can be used to analyze the uncertainty of model chains. One submodule that simulates one specific physical process can be exchanged by another modeling approach and the effect of the different model approaches to the overall model outcome can be analyzed. This can be done in a multimodel approach. One presented example was a coupled component model that represents physical processes in the atmosphere and hydrosphere as well as processes leading to damages to the elements at risk. This coupled component model covers the Aare River basin upstream of Bern, Switzerland. This model chain (Keller et al., 2019; Zischg, Felder, Weingartner, et al., 2018) was set up in a multimodel framework at the component level, enabling us to analyze uncertainties of the full model chain and disentangle the effects of selected submodules to the overall uncertainties of the full model chain. With variations of this coupled component model, we analyzed the effects of the spatiotemporal rainfall pattern, the effects of downscaling methods, the effects of the model coupling setup, the effects of the vulnerability functions, the effects of scale, and the effects of river engineering measures on the estimation of flood losses. The multimodel approach at component level leads to a higher interpretability of the results in that it highlights the most sensitive component of a model chain.

Fourth, model coupling allows to simulate features of complexity (Zischg, 2018). One of the presented examples showed how the feedback effect between riverbed erosion, river conveyance capacity, and flood risk in the floodplain can be modeled. The hydraulic model that assesses flood hazard and inundation probability in the floodplain in two dimensions can be coupled with a morphological model that simulates riverbed erosion and the evolution of the river morphology over time. This coupling allows to compute the changing risk because of river morphology changes. The model experiment of how flood risk is distributed in space and time and how it depends on the philosophical foundation of the flood risk management policy showed us that coupled models can simulate the emergence of spatial patterns. This is especially the case, if spatially explicit agent-based models are coupled with process models. The theoretical concept and the underlying assumptions of decision-making in flood risk management can result in very different outcomes after a long period of implementation. The decisions of agents become manifest in space. Moreover, the presented examples of model experiments showed that the effects of selected drivers of change on overall flood risk change can be analyzed in isolated way or in a combination. This opens opportunities to study coevolution and self-reinforcing or dampening effects of flood risk factors changing in space and time.

Last but not least, coupled component models allow the simulation of the effects of anthropogenic modifications of the landscape on flood risk change. The model experiments showed that human impacts on hydrology and river morphology increasingly dominate over natural causes of flood risk change over centuries. In the presented cases, this is mainly the human dimension of hydrological and geomorphological change.

In summary, CCMs are able to model flood risk change in a spatially explicit way compared to system dynamics models that are in most cases spatially lumped models. In contrast to integrated assessment models and earth system models, CCMs do not require bringing all submodules and processes to common spatial and temporal resolution and can be validated at the component level. Already validated models can be coupled together. As such, coupled component models, even if highly complex, offer a higher level of transparency and reproducibility than other modeling frameworks. CCMs provide an interesting approach for analyzing flood risk change, for modeling feedback mechanisms between human activities and the natural environment, and for the regionalization of global environmental and socioeconomic changes. The examples of this book give an outlook on how coupled modeling frameworks can predict and evaluate the effects of different adaptation strategies on flood risk evolution. I showed how model experiments can test alternative pathways of flood risk evolvement by assuming specific management strategies. This is supporting the search for robust and sustainable risk management strategies in the context of adaptation to global changes.

Development of coupled component models

Theoretically, CCMs offer many perspectives for modeling flood risk change and for considering features of complexity in models. Practically, the development of such model frameworks might be a technically complicated task. As described above, model coupling requires to bridge scales, to combine lumped data with raster and vector data, and to deal with different model specifications. In the following, I am summarizing my experiences in the development of CCMs that have been used to elaborate the model experiments shown in this book.

Model coupler

The core of a model chain is the model coupler. It is a specific software or a programming code that connects the submodules of the model chain. The coupler coordinates temporal and spatial scales of the submodules. Moreover, the coupler is "listening" to the submodules, it is starting or ending the simulations of submodules if specific conditions are met. After Laganier et al. (2014), the model coupling strategy can either be of unidirectional or of bidirectional type. The first case is also called external coupling or a model cascade; the information is exchanged in one direction only. In the second case, the coupled submodels interact between each other. An overview of common coupled or integrated modeling approaches is given by Kelly et al. (2013) and Laganier et al. (2014).

In a unidirectional coupling, the models are coupled in a cascading framework with harmonized simulation periods. The timesteps of the submodules must not necessarily be harmonized. An example of a model cascade is the model chain "regional climate model-hydrologic model-hydraulic model-flood impact model." The regional climate model runs at an hourly scale. The same is valid for the hydrological model. The latter writes the output at hourly scale. In contrast, the hydraulic model is running at seconds or subseconds scale, depending on the spatial resolution. This model interpolates the boundary condition given by the hydrologic model at hourly scale. At the end, the flood impact model might only use the maximum values of flow depths or velocities per mesh cell and over the whole simulation period. Fig. 5.26 is exemplarily showing the bridging of spatial and temporal scales.

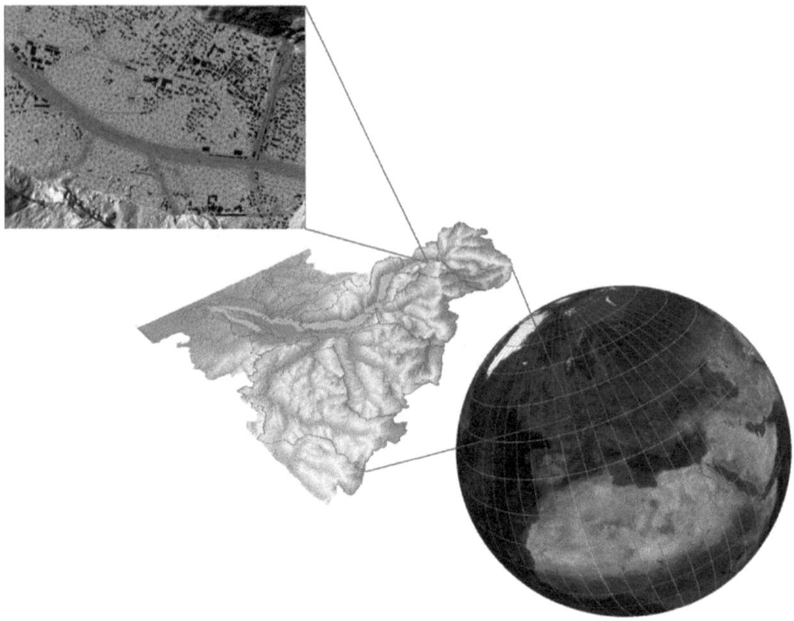

FIGURE 5.26 Bridging scales. A coupled component model framework across spatial and temporal scales.

In this example of a cascading model chain, the model coupler fixes the simulation period. In a first step, it reads the data from global climate simulations, and it runs the regional climate model. The coupler triggers the execution of the regional climate model and delivers the input data and the parameter file to this submodule. After this simulation is finished, the coupler takes the output data from the first submodule and converts it into a format that is required as input data for the second submodule. Then, it triggers the execution of the second submodule, in this case the hydrological model, by providing the required input data and the parameter file. The third module is again triggered when the simulation of the second submodule is finished. Again, the coupler converts the output data from one model into the required format of the other model. After the hydraulic simulation, the model coupler might query about the maximum flow depth per mesh unit over the whole simulation unit if the hydraulic model does not provide it. The model coupler then is attributing the outputs of the hydraulic model to the elements at risk. For example, if the elements at risk are houses, the model coupler is attributing the summary statistics of the flow depth of all mesh elements around a building to the building itself. With this, the next submodule, the flood impact module can calculate the damages. If the elements at risk are roads, the model coupler is analyzing the flood conditions along the road network segments that are flooded with flow depths above a critical threshold. The subsequent impact module then identifies the road segments or bridges that are not viable. In this case of loosely coupling disciplinary simulation models, the submodels can be independent executables. The only requirement to the selection of suitable programs that can act as submodules and can be coupled together is that the selected simulation models must be automatically executable by the model coupler. The programming language of the model coupler must be allowed to run command lines at operating system level. In all of the examples shown in this chapter and in previous chapters, the model coupler was programmed with Python. The "os" library of Python enables a Python program to run any executable from the command line of the operating system and provides notices from the operating system to the Python program when the simulation is finished. Of course, the program must provide an executable that can be run from the command line of the operating system. The executable must also be able to read the input data and a parameter file from the command line. This requirement favors open source models. These can be changed if needed so that they meet these requirements. The coupling of commercial models that can only be executed by clicking on the graphical user interface is not possible in this case. If such software must be used, the model chain must be run in a cascade by running the programs manually. As a consequence, this exhibits the simulation of large ensembles, a robust sensitivity analysis, and an uncertainty analysis. It prohibits also the development of a multimodel environment, which allows to analyze uncertainties of a submodule by running different models in parallel for a specific natural hazard process.

A multimodel framework allows to simulate system changes or regime changes. An example of this was the model experiment in the Emme River floodplain near Burgdorf (Switzerland) that aimed at disentangling the drivers of flood risk change. In this example, three different flood models were implemented in the model chain. One flood model simulates flooding of the current situation. This model is based on the current digital elevation model. The other two models were set up in the same way and with the same parameters, but they based on different digital elevation models and a different land use. All three models are valid from a certain time in the timeline of the simulation. The first model represented the geomorphological situation at around 1820 and was valid for the period 1820–1910. The second model was valid for the period 1910–60, and the third flood model was valid for the period after 1960. A long-run hydrologic simulation, for example, would feed the correct hydraulic model, depending on the simulation time. The model coupler decides which flood model to start and for how long (Fig. 5.27). The same is valid for the flood impact model.

Of course, this is only possible in a retrospective way, namely when the geomorphology change of the river channel is known before starting the simulations. This way comes to a limit when the aim is to simulate future evolvements. This requires a direct coupling between the morphological

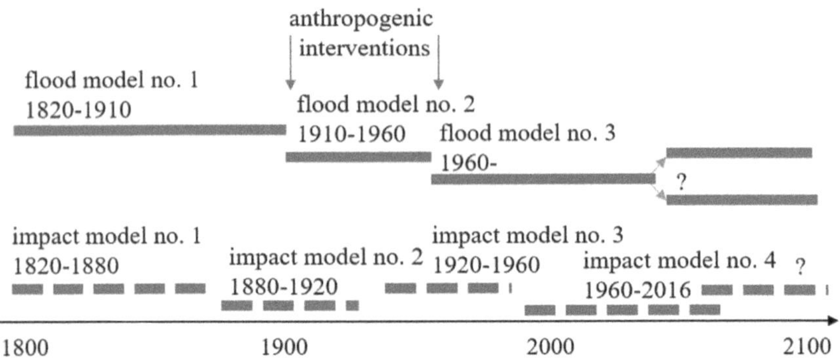

FIGURE 5.27 Multimodel approach. Schematic of modeling system change with a multimodel approach and continuous exchange of models that are representing different system states.

model and the flood model. In this case, the morphological model is simulating river channel changes that influence the flood model. Ideally, these two simulation models are already combined in one software, as this is the case for, e.g., CAESAR-Lisflood (Coulthard et al., 2013) or BASEMENT (Vanzo et al., 2021). If the two processes of morphological channel adjustments and floods are simulated in one software, the model coupler does not need to coordinate between these two simulations. Different from this bidirectional coupling between two physical processes is the reaction of the submodels that represent the human system to the outcomes of the submodels that represent the natural system, and vice versa. One example for such dependency is the construction of flood protection measures by human agents after a flood event. Here, the flood event resulting from the inundation model triggers a human intervention. To simulate this human intervention in a coupled human and natural system, a submodule must be developed that modifies the flood model in a way that it can represent the adaptation (Hossain et al., 2020). This can be implemented by a model coupler that reacts on the signal of an agent based model and triggers a model that implements flood protection works into the model chain. In the simplest cases, the heights of the lateral dikes of a river are virtually heightened after a flood event that exceeds a certain magnitude. This can be simulated by an agent-based model that has one simple solution for flood risk reduction in the toolbox. This human intervention intends to reduce flood hazards and thus completely changes the flood model. From the point in time when a flood event with a certain magnitude occurs, the model coupler must be able to exchange the flood model with a new flood model. It becomes more complex, if more than only one adaptation strategy must be considered in the decision-making of the agents. An example is the reaction of households and the implementation of property-level flood risk reduction measures as exemplarily shown by Haer et al. (2017, 2019). This must be dynamic in time. Therefore, the model coupler must coordinate between the flood model and the submodel that simulates human adaptation to floods. In this way, the implementation of climate adaptation measures along the timeline from now to the end of the 21st century can be modeled (Haer et al., 2020). However, this coupling between models for simulating physical processes and models for simulating human adaptation processes in space and time is one of the biggest challenges for modelers. There is still a lack of integrating process-based models with socioenvironmental models. It is a research frontier in the research on climate change impacts and climate change adaptation. Adaptation and mitigation are in many cases coupled. If more resources are invested in mitigation, fewer resources have to be invested in adaptation, and vice versa. Ideally, also flood risk management decision rules or principles of social justice are incorporated into such models (Thaler, 2021). Therefore, a model coupler for modeling the spatiotemporal evolution of climate change impacts and adaptation or mitigation measures must be able to consider these connections. Some examples of sociohydrological models, e.g., the models shown by Di Baldassarre et al. (2015, 2013), Viglione et al. (2014), are demonstrating the ability of modeling feedbacks between human and natural systems. However, these models are stylized and lumped models only and do not consider the spatial evolution and change of flood risk.

Development of model components

Coupled component models have an advantage in that they are flexible regarding the level of integration and are relatively transparent because the submodels are in most cases validated in their specific discipline. This is one of my favorite arguments for using coupled component modeling frameworks. Submodules of model chains or the so-called components of coupled component models can be developed and validated independently from the other modules. This is easening the overall model development. Clearly, new submodules must fit into the model chain, and they must not require input data that the other components do not provide. Except from this requirement, the development of the components must provide interfaces for the model coupler. The model coupler must be able to "communicate" with the newly developed component. The program must be enabled to start or stop a simulation of a model component or to influence the simulation of one component by the simulation of another component. This is absolutely required to model feedback processes and to simulate a change in the system's behavior. Moreover, the model coupler must transfer information of changes in the boundary conditions to the model components.

The development of model components that are coupled to a model chain offers the possibility of combining top-down modeling approaches with bottom-up analysis approaches. In many cases, top-down model chains represent a system behavior that is relatively constant in time. One example of such a shortage is to study future flood risks without implementing the future system status of floodplains with their values at risk and the adaptation of flood risk management strategies over time. Moreover, the development of the studied system over time is influenced by its sensitivity to changes in the boundary conditions. This means that both changes in the boundary conditions and internal changes in the system determine the development path of a changing system behavior. The drivers of change are interwoven, and a sound analysis of changes in coupled human and natural systems needs to consider them. Thus, a second main approach in the development of model components is to study the sensitivity of process simulations to changing conditions. In this bottom-up approach, the focus is laid more on the internal behavior and change of the system rather than on the boundary conditions. In the coupled models, this is studied on the one hand by sensitivity analyses of the sub-modules in an isolated way and on the other hand by sensitivity analyses of the whole model chain.

Therefore, a bottom-up development of model components is focused on representing the sensitivity of a process to changing conditions. If the sensitivity of a component is known, than thresholds of change or tipping points can be deducted where the whole system is changing its structure and behavior. An example of such an approach can be an analysis of the magnitudes of past flood events that triggered either the implementation of local flood protection works or a regional/national adaptation of the legislation or funding policy. A historical analysis can lead to estimate the order of magnitude of flood events that are potentially leading to paradigm changes in flood risk management and decision making. After the experience of such extreme flood events, the further adaptation of the society to floods will follow a different pathway than it would have been followed without this experience. Therefore, a coupled component model that aims at simulating future adaptation pathways should incorporate such tipping points. If a tipping point is reached, the model coupler or a specified agent-based model that simulates adaptation pathways can select from different possible development pathways. This flood event therefore will influence the future development pathways. This shows how a bottom-up sensitivity analysis can be combined with the models of natural processes. Especially the coupling of process models with agent-based models that simulate the interactions of individuals and institutions with the changing environment, offer a huge potential for extending the capabilities for simulating complex adaptive systems such as floodplains. Thus, the inclusion of the bottom-up modeling approach leads to a more holistic application for prediction purposes than process models alone. The combination of physics-based process models and agent-based models offers a thorough simulation of the spatiotemporal dynamics in floodplains.

While both top-down and bottom-up modeling approaches offer a high potential for the development of methods and tools for the analysis of changes in complex environmental systems, a research gap is identified in bridging both approaches. Therefore, the main aim of future research in modeling floodplains as complex adaptive systems should be to integrate both approaches. This should lead to an extension of the capabilities of coupled component modeling. If the sensitivity of a hydrogeomorphic system is analyzed in detail and a model of adaptive behavior is developed (bottom-up approach), a subsequent analysis of the impacts of changing the boundary conditions (top-down), and consequently, a prediction of future development paths can be done more satisfyingly. This means that changes over time in the boundary conditions meet system-specific sensitivities and adaptive capabilities. Fig. 5.28 schematizes a possible combination of top-down and bottom-up approaches in modeling floodplains as complex adaptive systems with a coupled component model that potentially allows to simulate emergent phenomena and system changes.

Top-down and bottom-up modeling strategies can be integrated when model developers are collaborating with stakeholders in stress-test analyses or in developing storylines of flood scenarios or adaptation scenarios. With the involvement of stakeholders, the sensitivity of the system that is studied can be assessed. An example is a food logistics company that is distributing goods to all grocery stores in a certain region. This stakeholder depends on the

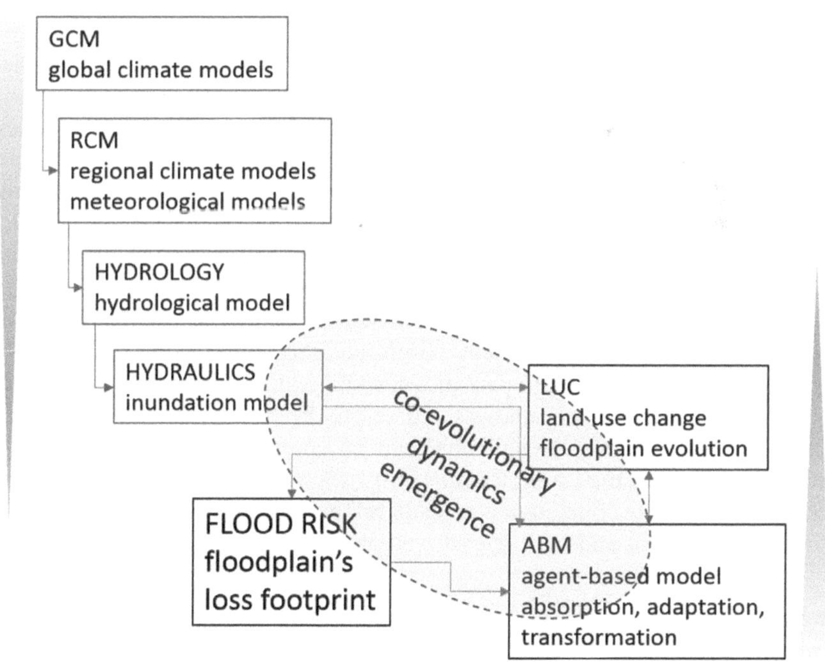

FIGURE 5.28 **Model coupling.** The proposed schema for merging top-down and bottom-up approaches in the framework of coupled component models. *From Zischg, A. P. (2018). Floodplains and complex adaptive systems— perspectives on connecting the dots in flood risk assessment with coupled component models.* Systems, 6*(2). https://doi.org/10.3390/systems6020009*

functioning of the road network. If some remote valleys or villages cannot be accessed by the trucks because parts of the road network are flooded, these villages must make use of their stocks. A top-down modeling approach can assess these indirect consequences only vaguely. Without the specific knowledge of which logistic company is transporting goods from where to where and without the knowledge of the current stock of goods in the grocery stores, it cannot be assessed from which day on after the flood event, the situation becomes critical. The involvement of the persons that are managing this logistic network and the implementation of this knowledge in the modeling chain remarkably increase the reliability of the flood impact assessment. Moreover, these persons or experts can help in selecting the extreme weather events or flood events as the basis for the flood impact assessments. The coinvolved experts know the vulnerability of their logistics networks and can advice the modelers in the identification of flood scenarios that are crucial for storylines of flood impacts. Around these selected events, a narration of the spatiotemporal evolution of the flood impacts and of the weak points in the current situation can be developed. The model chain is then integrating the top-down cascade of model components with the bottom-up codeveloped vulnerability components. The storyline approach is a way to integrate both.

In contrast to the codevelopment approach or the transdisciplinary research approach, a bottom-up approach can also be followed by a data-driven approach. If enough and reliable data about the system behavior under external disturbances are available, a systems vulnerability function can be distilled from these data. Moreover, a pure data science or statistical approach can be combined with expert's elicitation approaches. An example of such a development of a flood vulnerability function for buildings based on both approaches can be found in Zischg et al. (2021). In this study, we developed a flood vulnerability function by quantifying expert heuristics and deriving an impact model out of this expert knowledge and a flood vulnerability function by analyzing insurance claim data. Both functions provide relatively similar results of flood damage assessments. Both functions describe the sensitivity of a core element of the flood impact model change, the buildings as elements at risk considered in the risk analysis. Therefore, both functions are bottom-up components of the model chain that can be connected with the other top-down components. Fig. 5.29 schematically shows how the top-down model components that are coupled in a cascade can be combined with the bottom-up model development and with data-driven analyses.

Data sources

The development of models depends on reliable data. In all of the model experiments shown in this chapter and in the previous chapter, we worked in a region where many data sources are available that are needed for risk analysis. However, in many regions of the world, the provision of accurate data and

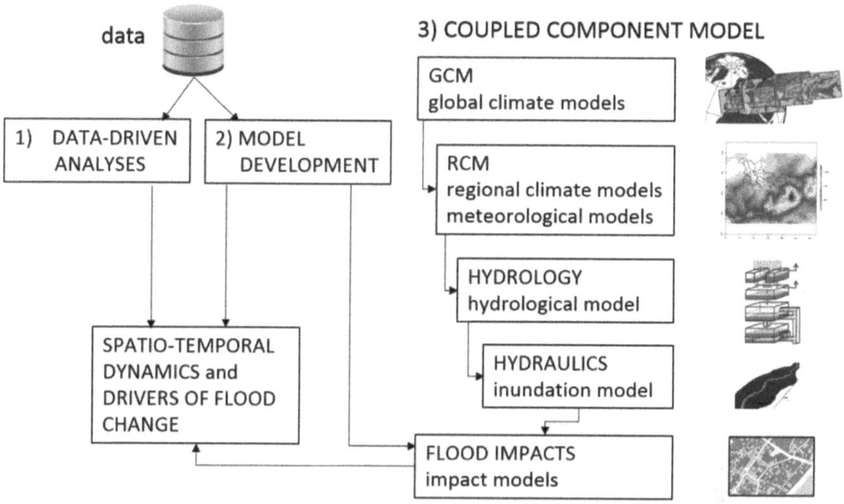

FIGURE 5.29 Model coupling. The proposed schema for integration of process models with data-driven analyses in coupled component models.

geodata by the state is not guaranteed. Therefore, I am fully aware that the presented model experiments and tools cannot be easily transferred to data-scarce regions (Malgwi et al., 2021). However, many data sources are currently in production by research institutes that cover the whole globe. Datasets are growing where citizen can contribute, e.g., Open Street Map. Nevertheless, in the following I am summarizing the data needed for sound risk analyses and for the analysis of flood risk change. All of the presented examples were conducted at the scale of a single building, respectively at meter scale. Modeling and analyzing flood risk at this scales absolutely require object-based and high-resolution data for both, flood modeling and impact modeling. Disaggregation of low-resolution data to high-resolution data is possible for selected data but it is limited. If the focus is laid on analyzing flood risk change, all the data used in the flood risk analysis must be available for historic periods, namely continuously updated data must be historized.

Data for modeling flood hazards

Flood modeling requires either hydrological or meteorological data. If the timeseries of river discharges are available at many locations in the river network and are long enough to derive extreme value statistics, flood models can base on these date. In this case, the hydrographs of flood events can provide the boundary conditions for the hydraulic model. If hydrological data are not available or if the flood risk analyses should base on specific scenarios, we need meteorological data. Thus, data of spatiotemporal patterns of rainfall must be available or generated. In this book, I described several methods to elaborate physically plausible precipitation scenarios, namely the "cherry picking" of extreme events from long-running climate simulations or from hindcast archives or the statistical simulation of precipitation events.

Besides precipitation, the hydrological model requires additional data. These are, for example, temperature, glaciers, evapotranspiration, or wind. Moreover, the hydrological model incorporates land use and topography. Thus, a land use map and a digital terrain model must be available. Of course, timeseries of discharge must be available to calibrate and validate the hydrological model at sample points in the river network.

The hydraulic model requires digital terrain models with a high spatial resolution. In recent years, a remarkable progress can be observed in the availability and accuracy of digital elevation models, also at global scale (Hawker et al., 2018, 2019; Liu et al., 2021). At regional and local scale, digital elevation model based on laser scanning technologies becomes increasingly available. In all presented examples, we used digital elevation models that base on this technology. For Switzerland, such digital elevation models are available with a spatial resolution of 0.5-2 m. However, working with flood models at this scale requires to eliminate artifacts in these data. For example, bridges are often represented as flow-blocking features because of the 2.5D character of the available digital elevation models. In this case, bridges and similar anthropogenic artifacts must be corrected prior to their use for flood modeling. Another problem is that the laser signal cannot go through water. Therefore, the digital elevation models are showing the water surface elevation in the river channels and not the z-coordinates of the river bed. Thus, the river channels must be considered separately and the channel geometry must be interpolated from surveyed river cross sections. The availability of river cross sections is therefore crucial for a reliable representation of the river discharge capacity. Without an exact representation of the river channel, the extent of the simulated flood processes as well as the flood depths is remarkably overestimated. Ideally, the data of river cross sections are available in a historized data model. The comparison of old surveys with newer cross section geometries allows to calibrate morphological models that simulate bed erosion and aggradation. The same is valid for the digital elevation models. A comparison between an older digital elevation model and a newer one allows to analyze the effects of newly implemented flood protection measures. These are in most cases visible as topographic features in the digital elevation model.

Data for modeling elements at risk

The absolute requirement for analyzing flood risk change is data on changes of the elements at risk. Settlement growth and the development of production facilities and infrastructure are one of the main drivers of flood risk change. Thus, historized data of the elements at risk is indispensable besides data on changing flood hazard. This requirement might be a bottleneck for dynamic flood risk analysis, especially when flood risk analysis should be conducted at single building scale. Therefore, the buildings datasets might limit the analysis of the trajectories of flood risk change. In some countries (e.g., Switzerland and Austria (Fuchs et al., 2017)), the public residence register is available for research. This dataset provides the best data for flood risk analyses. This data is continuously updated and therefore it can be analyzed at different timesteps, e.g., at annual timesteps. However, the availability of such data is limited due to the data protection and privacy legislation. It is absolutely necessary that the data has to be anonymized and it should not be joined with other data. Therefore, the use of these data needs a thorough legal advice and consideration of the privacy and data protection laws. Many other elements at risk can be derived from topographic maps or digital terrain models. For example, the footprints of buildings can be extracted from the digital terrain model and the high-resolution vector maps, which are provided by some national topographic services. Also the road and railway networks or other infrastructure networks can be extracted from digital terrain models or vector maps. These are continuously updated and therefore historized datasets of these elements at risk could be available in principle. However, the historized data are not made available in countries without an open government and open data policy. The pros and cons of open data and open government strategies should not be discussed here, but it is clear that a policy that exhibits the use of government geodata for research exhibits the development of new methods for flood risk analysis and especially for flood risk change analysis. A perspective on this topic on the value of open data is given by Wildman and Lewis (2022).

Data for modeling vulnerability

This statement on data availability leads to the next topic in analyzing flood risk change. Besides the analysis of hazard and exposure, flood risk analysis must base on the analysis of vulnerability. Vulnerability is a further characterization of the elements at risk and a vulnerability function is needed to quantify flood damages from hazard and exposure data. However, the development of flood vulnerability functions requires localized data of flood damages. Only georeferenced damage data can be overlaid with hazard data, e.g., the flood depths during the flood event that caused the damage. For example, many vulnerability functions are being developed out from the relationship between a parameter of flood magnitude, in many cases flood depth, and the degree of damage (Molinari et al., 2020). The degree of damage is the relationship between the damage of a building and the reconstruction value of the building. Therefore, this relationship between flood depth and degree of damage can only be deduced if the flood depth can be attributed to the building or the point in space with the attribute of the flood damage. This can only be done if insurance claim data are available and can be geolocalized. These data includes the value of an insured object, in this case a house, and the monetary damage out at a specific day that was paid out from the insurance company. The data record of the flood damage might not necessarily be geolocalized but the contract between the insurer and the client contains an address of the insured object. This address can be used to geolocalize the insured building. With it, the data record of the insurance value and the damage can be geolocalized too. However, this data combination can only be made internally by the insurance company itself due to privacy regulations. The insurance company can give out an anonymized record of the building value and/or the degree of damage at a specific date. With these data, flood modelers can overlay the flood depth data with the geolocalized and anonymized insurance claims and data scientists thereafter can develop statistical vulnerability functions out of the data. If no data of flood damages are available, expert's elicitation might be an option to develop vulnerability functions (Neubert et al., 2016). Another necessary information needed for exposure and vulnerability assessment is the monetary value of each building. In the previous examples, we used different models to estimate the reconstruction values of the buildings (Röthlisberger et al., 2018). These models that characterize the elements at risk can only be developed in such a detailed level if insurance data is available for research. The insurance contracts contain the information of the building value. This value is normally estimated by a regional property assessment expert and is comparable across a region. Especially in regions with a monopoly public insurance institution, the building's value is estimated also for the purpose of wealth taxation and therefore it is made with expertise. These data provide unique opportunities to derive exposure models. Additional data such as the characteristics of the regional economic situation and the land use can be used to optimize the calibration of the exposure models. The availability of insurance claim data for research is one of the main bottlenecks that avoid the optimization of exposure and vulnerability models. Also the use of these data is regulated by privacy protection laws. However, a cooperation between the regional insurance companies and the research group of the author proved to be successful and both sides benefitted from the collaboration. Sensible data can remain within the insurers and data can be anonymized. Moreover, the overlay procedure of hazard data and the insurance claim data can also be conducted intra muros by the insurer and the data records can be given to researchers without the geolocalization. This approach is absolutely avoiding any conflicts with privacy legislations. However, it needs the willingness for collaboration from the insurer. The data-driven analyses clearly show that newly available data such as the residential register and insurance claim data proved to be a valuable basis for analyzing the effects of socioeconomic change on flood risk. However, data-driven approaches alone do not allow the disentanglement of coevolutionary dynamics in both environmental and socioeconomic changes.

Perspectives of coupled component models in hydrology

Human activities related to water are challenged by both climatic changes and anthropogenic influences to the water systems. Hence, sustainable adaptation in the water sector requires an understanding of hydrology in its full complexity. In the examples of model experiments, I showed how changing human and natural systems are shaping hydrology and how changing hydrology is framing human activities and ecosystem services. Extending process models, such as climate models and weather forecast models, coupled with hydrologic/hydraulic models with socioeconomic impact and loss models, is a way to integrate modeling frameworks. This coupled component modeling framework enables considering the impacts that a technological society could have on natural systems when reacting to extreme events. This requires an analysis of driving forces that shape water systems and the analysis of feedback processes between hydrology and socioeconomy. Thus, coupled component models provide an approach for studying both the context for hydrology and hydrology as a context.

Context for hydrology

In studying the changing context for hydrology, often a top-down approach of analyzing the effects of global environmental changes to the hydrology of rivers is followed. This includes changes in the climate, in the natural environment, and in the human footprints on the river systems (hydromorphology). The analysis of changing contexts for hydrology is based on the coupling of process models in combination with data-driven analyses. The main focus herein lays on the regionalization of the impacts, i.e., predicting the spatiotemporal dynamics of global changes on selected river basins at regional to local scale. This requires bridging spatial scales from global climate models to local impacts and temporal scales from centennial to subhourly flood dynamics.

Hydrology as a context

In contrast, the analysis of hydrological conditions as context for society must follow a bottom-up approach of studying the vulnerability of human activities and socioeconomic systems related to water. In this approach, hydrology is a main context for their future evolvement and changes in hydrology affect their functioning. The main focus in this research topic must be laid on changes in the river systems and their special role for society. Rivers are remarkably modified by flood protection works, dams, and water/energy infrastructure. Thus, heavily modified river systems act as a crucial framework for settlement development or even as lifelines in the Anthropocene. Changes in hydrology markedly drive future pathways of socioeconomic development. The research question in studying hydrology as context is the sensitivity of human water systems to hydrological changes. Furthermore, studying future pathways of evolving water systems in the Anthropocene requires a comparison between natural and engineered river systems.

Feedbacks: unifying top-down and bottom-up modeling approaches

Finally, top-down and bottom-up approaches have to be brought together and bidirectional feedback mechanisms between rivers and society have to be implemented in integrated modeling frameworks. This means that coupled component models for studying the impacts of hydrometeorological extremes and global changes on water systems have to be extended with capabilities for capturing the effects of adaptation measures. The adaptation strategies of governments and individuals must be implemented into impact models. This book presented approaches for enabling coupled model frameworks to predict and evaluate the effects of different adaptation strategies. This is supporting the search for robust and sustainable risk management strategies in the context of global changes. It requires first to study feedback mechanisms empirically or theoretically. One main foundation for this is model experiments. Herein, different states of the environment and the human system are represented in multitemporal models. With this, model experiments can test alternative pathways of flood risk evolvement by assuming specific management strategies.

Acknowledgments

I am thanking the team of the Mobiliar Lab for Natural Risks of the Oeschger Center for Climate Change Research at the University of Bern. The following people have contributed to the development of the presented examples: M. Kauzlaric, M. Mosimann, S. Schick, V. Röthlisberger, G. Felder, D. Bernet, S. Trefalt, M. Keiler, O. Martius-Romppainen, L. Thomi, R. Sturny, R. Weingartner, A. Fehlmann, L. Munz, D. Reichenbach, J. Lang, G. Meier, G. Tyedmers, N. Weber, S. Loreti, M. Mertin, M. Brughelli, S. Schürmann, J. Ryser, N. Rindsfüsser, T. Papilloud, T. Bertschinger, N. Galatioto, J.J. Gómez-Navarro, C. Raible, M. Lang, D. Vogt, B. Stocker, C. von Matt.

References

Anonymous. (1820a). *Querprofile Emme. StAB AA V 189e-h.*

Anonymous. (1820b). Plan des Emmen-Bettes von Kirchberg bis Zuchwyl-Brücke. *Massstab, 1*, 5000 (in Bern-Fuss). Aus: StAB AA V 189a-b.

Anonymous. (1898). Emme, Mittellauf: Querprofile der beiden Ufer von km 15-34. Aus: StAB AA Vol. 199a.

Anonymous. (n.d.). In Plan der Güter, die zwischen der Wynigenbrücke und der Kirchbergbrücke an der Emme liegen.

Arrighi, C., Pregnolato, M., Dawson, R. J., & Castelli, F. (2019). Preparedness against mobility disruption by floods. *Science of the Total Environment, 654*, 1010–1022. https://doi.org/10.1016/j.scitotenv.2018.11.191

Bartel, R. (2002). Neo-/Liberalismus und öffentliche Finanzen. *Das Öffentliche Haushaltswesen Österreich, 43*.

Bernet, D. B., Prasuhn, V., & Weingartner, R. (2017). Surface water floods in Switzerland: What insurance claim records tell us about the damage in space and time. *Natural Hazards and Earth System Sciences, 17*(9), 1659–1682. https://doi.org/10.5194/nhess-17-1659-2017

Bernet, D. B., Trefalt, S., Martius, O., Weingartner, R., Mosimann, M., Röthlisberger, V., & Zischg, A. P. (2019). Characterizing precipitation events leading to surface water flood damage over large regions of complex terrain. *Environmental Research Letters, 14*(6). https://doi.org/10.1088/1748-9326/ab127c

Bezzola, G. R., & Loat, R. (2016). Integrated Risk Management – identify, Evaluate and Manage Natural Risks. Integrales Risikomanagement – naturrisiken erfassen, bewerten und steuern. In *INTERPRAEVENT 2016 conference proceedings* (pp. 18–26).

BFS. (2021a). *Eidg. Gebäude- und Wohnungsregister GWR.*

BFS. (2021b). *Statent - Statistik der Unternehmensstruktur.*

Blair, P., & Buytaert, W. (2016). Socio-hydrological modelling: A review asking "why, what and how?". *Hydrology and Earth System Sciences, 20*(1), 443–478. https://doi.org/10.5194/hess-20-443-2016

Bonriposi, M., Bettex, L., & Chenal, J. (2016). *Renforcement du Monitoring du Plan Directeur Cantonal Genève 2030 - Phase 3: Estimation des réserves à bâtir - Commune de Lancy.* EPFL.

van den Brink, H. W., Können, G. P., Opsteegh, J. D., van Oldenborgh, G. J., & Burgers, G. (2005). Estimating return periods of extreme events from ECMWF seasonal forecast ensembles. *International Journal of Climatology, 25*(10), 1345–1354. https://doi.org/10.1002/joc.1155

Brönnimann, S., Rohr, C., Stucki, P., Summermatter, S., Bandhauer, M., Barton, Y., Fischer, A., Froidevaux, P., Germann, U., Grosjean, M., Hupfer, F., Ingold, K., Isotta, F., Keiler, M., Martius, O., Messmer, M., Mülchi, R., Panziera, L., Pfister, L., … Zischg, A. (2018). *1868 – the flood that changed Switzerland: Causes, consequences and lessons for the future* (Vol. 94).

de Bruijn, K. M., Lips, N., Gersonius, B., & Middelkoop, H. (2016). The storyline approach: A new way to analyse and improve flood event management. *Natural Hazards, 81*(1), 99–121. https://doi.org/10.1007/s11069-015-2074-2

CH2018: Climate scenarios for Switzerland. (2018). National Centre for Climate Services Zurich.

Chen, M., & Kumar, A. (2017). The utility of seasonal hindcast database for the analysis of climate variability: An example. *Climate Dynamics, 48*(1–2), 265–279. https://doi.org/10.1007/s00382-016-3073-z

Collenteur, R. A., de Moel, H., Jongman, B., & Di Baldassarre, G. (2015). The failed-levee effect: Do societies learn from flood disasters? *Natural Hazards, 76*(1), 373–388. https://doi.org/10.1007/s11069-014-1496-6

Costabile, P., Costanzo, C., De Lorenzo, G., De Santis, R., Penna, N., & Macchione, F. (2021). Terrestrial and airborne laser scanning and 2-D modelling for 3-D flood hazard maps in urban areas: New opportunities and perspectives. *Environmental Modelling and Software, 135*, 104889. https://doi.org/10.1016/j.envsoft.2020.104889

Coulthard, T. J., Neal, J. C., Bates, P. D., Ramirez, J., de Almeida, G. A. M., & Hancock, G. R. (2013). Integrating the LISFLOOD-FP 2D hydrodynamic model with the CAESAR model: Implications for modelling landscape evolution. *Earth Surface Processes and Landforms, 38*(15), 1897–1906. https://doi.org/10.1002/esp.3478

Coxon, G., Freer, J., Lane, R., Dunne, T., Knoben, W. J. M., Howden, N. J. K., Quinn, N., Wagener, T., & Woods, R. (2019). DECIPHeR v1: Dynamic fluxEs and connectivity for predictions of HydRology. *Geoscientific Model Development, 12*(6), 2285–2306. https://doi.org/10.5194/gmd-12-2285-2019

Cutter, S. L. (2003). The vulnerability of science and the science of vulnerability. *Annals of the Association of American Geographers, 93*(1), 1–12. https://doi.org/10.1111/1467-8306.93101

Di Baldassarre, G., Viglione, A., Carr, G., Kuil, L., Salinas, J. L., & Blöschl, G. (2013). Socio-hydrology: Conceptualising human-flood interactions. *Hydrology and Earth System Sciences, 17*(8), 3295–3303. https://doi.org/10.5194/hess-17-3295-2013

Di Baldassarre, G., Viglione, A., Carr, G., Kuil, L., Yan, K., Brandimarte, L., & Blöschl, G. (2015). Debates—perspectives on socio-hydrology: Capturing feedbacks between physical and social processes. *Water Resources Research, 51*(6), 4770–4781. https://doi.org/10.1002/2014WR016416

Di Carlo, G., Hollenstein, G., Nebel, K., & Niedermaier, R. (2017). *Schweizweite abschätzung der Nutzungsreserven 2017, vdf Hochschulverlag AG.*

Dufour, G. (2021). *Background information on the Dufour map.*

ECMWF. (2017). *SEAS5 user guide: European centre for medium-range weather forecasts.*

ECMWF. (2019). *IFS documentation.* European Centre for Medium-Range Weather Forecasts.

Fanta, V., Šálek, M., & Sklenicka, P. (2019). How long do floods throughout the millennium remain in the collective memory? *Nature Communications, 10*(1). https://doi.org/10.1038/s41467-019-09102-3

Federal office for spatial development.(2007).

Fehlmann, A. (2020). *Estimating the share of buildings in flood hazard zones fitted with structural property level flood mitigation measures.*

Felder, G., Gómez-Navarro, J. J., Zischg, A. P., Raible, C. C., Röthlisberger, V., Bozhinova, D., Martius, O., & Weingartner, R. (2018). From global circulation to local flood loss: Coupling models across the scales. *Science of the Total Environment, 635*, 1225–1239. https://doi.org/10.1016/j.scitotenv.2018.04.170

Felder, G., & Weingartner, R. (2016). An approach for the determination of precipitation input for worst-case flood modelling. *Hydrological Sciences Journal, 61*(14), 2600–2609. https://doi.org/10.1080/02626667.2016.1151980

Finsler, H. C. (1978). Landschaftswandel und wirtschaftliche Umstrukturierung seit dem frühen 19. *Jahrhundert, 3.*

FOEN. (2018). *Cross-sections and river bathymetrie for the main river network in Switzerland.* Federal Office for the Environment.

FOEN and SwissRe. (2008). *Aquaprotect.* Bern: Federal Office for the Environment FOEN.

Fuchs, S., Keiler, M., & Zischg, A. (2015). A spatiotemporal multi-hazard exposure assessment based on property data. *Natural Hazards and Earth System Sciences, 15*(9), 2127–2142. https://doi.org/10.5194/nhess-15-2127-2015

Fuchs, S., Röthlisberger, V., Thaler, T., Zischg, A., & Keiler, M. (2017). Natural hazard management from a coevolutionary perspective: Exposure and policy response in the European Alps. *Annals of the American Association of Geographers, 107*(2), 382–392. https://doi.org/10.1080/24694452.2016.1235494

Goldschmied, H. (1913). *Emmekorrektion 1. Sektion Kantonsgrenze Bern/Solothurn bis Grenze Burgdorf.*

Haer, T., Botzen, W. J. W., & Aerts, J. C. J. H. (2019). Advancing disaster policies by integrating dynamic adaptive behaviour in risk assessments using an agent-based modelling approach. *Environmental Research Letters, 14*(4). https://doi.org/10.1088/1748-9326/ab0770, 044022.

Haer, T., Botzen, W. J. W., de Moel, H., & Aerts, J. C. J. H. (2017). Integrating household risk mitigation behavior in flood risk analysis: An agent-based model approach. *Risk Analysis, 37*(10), 1977–1992. https://doi.org/10.1111/risa.12740

Haer, T., Husby, T. G., Botzen, W. J. W., & Aerts, J. C. J. H. (2020). The safe development paradox: An agent-based model for flood risk under climate change in the European Union. *Global Environmental Change, 60*, 102009. https://doi.org/10.1016/j.gloenvcha.2019.102009

Harrison, J. L. (2014). Neoliberal environmental justice: Mainstream ideas of justice in political conflict over agricultural pesticides in the United States. *Environmental Politics, 23*(4), 650–669. https://doi.org/10.1080/09644016.2013.877558

Hawker, L., Bates, P., Neal, J., & Rougier, J. (2018). Perspectives on digital elevation model (DEM) simulation for flood modeling in the absence of a high-accuracy open access global DEM. *Frontiers in Earth Science, 6*. https://doi.org/10.3389/feart.2018.00233

Hawker, L., Neal, J., & Bates, P. (2019). Accuracy assessment of the TanDEM-X 90 digital elevation model for selected floodplain sites. *Remote Sensing of Environment, 232*, 111319. https://doi.org/10.1016/j.rse.2019.111319

Hayek, F. A. (1991). *The road to serfdom.*

Hossain, M. S., Ramirez, J. A., Haisch, T., Speranza, C. I., Martius, O., Mayer, H., & Keiler, M. (2020). A coupled human and landscape conceptual model of risk and resilience in Swiss Alpine communities. *Science of the Total Environment, 730*. https://doi.org/10.1016/j.scitotenv.2020.138322

Hunold, C., & Young, I. M. (1998). Justice, democracy, and hazardous siting. *Political Studies, 46*(1), 82–95. https://doi.org/10.1111/1467-9248.00131

Hurrell, J. W., Holland, M. M., Gent, P. R., Ghan, S., Kay, J. E., Kushner, P. J., Lamarque, J. F., Large, W. G., Lawrence, D., Lindsay, K., Lipscomb, W. H., Long, M. C., Mahowald, N., Marsh, D. R., Neale, R. B., Rasch, P., Vavrus, S., Vertenstein, M., Bader, D., … Marshall, S. (2013). The community earth system model: A framework for collaborative research. *Bulletin of the American Meteorological Society, 94*(9), 1339–1360. https://doi.org/10.1175/BAMS-D-12-00121.1.

Ivanov, M. A., & Kotlarski, S. (2017). Assessing distribution-based climate model bias correction methods over an alpine domain: Added value and limitations. *International Journal of Climatology, 37*(5), 2633–2653. https://doi.org/10.1002/joc.4870

Johnson, C., Penning-Rowsell, E., & Parker, D. (2007). Natural and imposed injustices: The challenges in implementing "fair" flood risk management policy in England. *Geographical Journal, 173*(4), 374–390. https://doi.org/10.1111/j.1475-4959.2007.00256.x

Keller, L., Zischg, A. P., Mosimann, M., Rössler, O., Weingartner, R., & Martius, O. (2019). Large ensemble flood loss modelling and uncertainty assessment for future climate conditions for a Swiss pre-alpine catchment. *Science of the Total Environment, 693*. https://doi.org/10.1016/j.scitotenv.2019.07.206

Kelly, R. A., Jakeman, A. J., Barreteau, O., Borsuk, M. E., ElSawah, S., Hamilton, S. H., Henriksen, H. J., Kuikka, S., Maier, H. R., Rizzoli, A. E., van Delden, H., & Voinov, A. A. (2013). Selecting among five common modelling approaches for integrated environmental assessment and management. *Environmental Modelling and Software, 47*, 159–181. https://doi.org/10.1016/j.envsoft.2013.05.005

Kjeldsen, T. R., & Prosdocimi, I. (2018). Assessing the element of surprise of record-breaking flood events. *Journal of Flood Risk Management, 11*, S541–S553. https://doi.org/10.1111/jfr3.12260

Köplin, N., Schädler, B., Viviroli, D., & Weingartner, R. (2014). Seasonality and magnitude of floods in Switzerland under future climate change. *Hydrological Processes, 28*(4), 2567–2578. https://doi.org/10.1002/hyp.9757

Laganier, O., Ayral, P. A., Salze, D., & Sauvagnargues, S. (2014). A coupling of hydrologic and hydraulic models appropriate for the fast floods of the Gardon River basin (France). *Natural Hazards and Earth System Sciences, 14*(11), 2899–2920. https://doi.org/10.5194/nhess-14-2899-2014

Laniak, G. F., Olchin, G., Goodall, J., Voinov, A., Hill, M., Glynn, P., Whelan, G., Geller, G., Quinn, N., Blind, M., Peckham, S., Reaney, S., Gaber, N., Kennedy, R., & Hughes, A. (2013). Integrated environmental modeling: A vision and roadmap for the future. *Environmental Modelling and Software, 39*, 3–23. https://doi.org/10.1016/j.envsoft.2012.09.006

Liu, Y., Bates, P. D., Neal, J. C., & Yamazaki, D. (2021). Bare-earth DEM generation in urban areas for flood inundation simulation using global digital elevation models. *Water Resources Research, 57*(4). https://doi.org/10.1029/2020WR028516

Liu, J., Dietz, T., Carpenter, S. R., Folke, C., Alberti, M., Redman, C. L., Schneider, S. H., Ostrom, E., Pell, A. N., Lubchenco, J., Taylor, W. W., Ouyang, Z., Deadman, P., Kratz, T., & Provencher, W. (2007). Coupled human and natural systems. *Ambio, 36*(8), 639–649. https://doi.org/10.1579/0044-7447(2007)36[639:CHANS]2.0.CO;2

Loreti, S., Ser-Giacomi, E., Zischg, A., Keiler, M., & Barthelemy, M. (2022). Local impacts on road networks and access to critical locations during extreme floods. *Scientific Reports, 12*(1). https://doi.org/10.1038/s41598-022-04927-3

Luder, F. (1928). *Situationsplan von Burgdorf, erstellt nach dem Kataster (Druck).*

Malgwi, M. B., Schlögl, M., & Keiler, M. (2021). Expert-based versus data-driven flood damage models: A comparative evaluation for data-scarce regions. *International Journal of Disaster Risk Reduction, 57*. https://doi.org/10.1016/j.ijdrr.2021.102148

Maraun, D., Wetterhall, F., Ireson, A. M., Chandler, R. E., Kendon, E. J., Widmann, M., Brienen, S., Rust, H. W., Sauter, T., Themel, M., Venema, V. K. C., Chun, K. P., Goodess, C. M., Jones, R. G., Onof, C., Vrac, M., & Thiele-Eich, I. (2010). Precipitation downscaling under climate change: Recent developments to bridge the gap between dynamical models and the end user. *Reviews of Geophysics, 48*(3). https://doi.org/10.1029/2009RG000314

Merz, R., Blöschl, G., & Humer, G. (2008). National flood discharge mapping in Austria. *Natural Hazards, 46*(1), 53–72. https://doi.org/10.1007/s11069-007-9181-7

Merz, B., Vorogushyn, S., Lall, U., Viglione, A., & Blöschl, G. (2015). Charting unknown waters—On the role of surprise in flood risk assessment and management. *Water Resources Research, 51*(8), 6399–6416. https://doi.org/10.1002/2015WR017464

MeteoSwiss. (2019). *Hourly precipitation estimation through rain-Gauge and radar: CombiPrecip, Federal Office of Meteorology and Climatology Meteo Swiss.*

Molinari, D., Scorzini, A. R., Arrighi, C., Carisi, F., Castelli, F., Domeneghetti, A., Gallazzi, A., Galliani, M., Grelot, F., Kellermann, P., Kreibich, H., Mohor, G. S., Mosimann, M., Natho, S., Richert, C., Schroeter, K., Thieken, A. H., Zischg, A. P., & Ballio, F. (2020). Are flood damage models converging to "reality"? Lessons learnt from a blind test. *Natural Hazards and Earth System Sciences, 20*(11), 2997–3017. https://doi.org/10.5194/nhess-20-2997-2020

Nebel, R., Hollenstein, K., Carlo, Niedermaier, M., & Scholl, B. (2017). *Schweizweite Abschätzung der Nutzungsreserven 2017: Bonriposi, vdf Hochschulverlag an der ETH Zürich.*

Neubert, M., Naumann, T., Hennersdorf, J., & Nikolowski, J. (2016). The Geographic Information System-based flood damage simulation model HOWAD. *Journal of Flood Risk Management, 9*(1), 36–49. https://doi.org/10.1111/jfr3.12109

Pregnolato, M., Ford, A., Wilkinson, S. M., & Dawson, R. J. (2017). The impact of flooding on road transport: A depth-disruption function. *Transportation Research D: Transport and Environment, 55*, 67–81. https://doi.org/10.1016/j.trd.2017.06.020

Ramirez, J. A., Zischg, A. P., Schürmann, S., Zimmermann, M., Weingartner, R., Coulthard, T., & Keiler, M. (2020). Modeling the geomorphic response to early river engineering works using CAESAR-Lisflood. *Anthropocene, 32*. https://doi.org/10.1016/j.ancene.2020.100266

Raschky, P. A., & Weck-Hannemann, H. (2007). Charity hazard-A real hazard to natural disaster insurance? *Environmental Hazards, 7*(4), 321–329. https://doi.org/10.1016/j.envhaz.2007.09.002

Rawls, J. (2005). *A theory of justice*. Harvard University Press.

Ritter, J. (1804). *Plan der Emmenkorrektion von Joseph Ritter: Abschnitt der Emme zwischen Ziegelbrücke und Eysteg (nach dem Emmenplan von. J. A. Riediger)* (p. 151). Massstab k.A. Aus: BAB PP.

Robertson, A. W., & Vitart, F. (2019). *Sub-seasonal to seasonal prediction: The gap between weather and climate forecasting*. Elsevier.

Röthlisberger, V., Zischg, A. P., & Keiler, M. (2017). Identifying spatial clusters of flood exposure to support decision making in risk management. *Science of the Total Environment, 598*, 593–603. https://doi.org/10.1016/j.scitotenv.2017.03.216

Röthlisberger, V., Zischg, A. P., & Keiler, M. (2018). A comparison of building value models for flood risk analysis. *Natural Hazards and Earth System Sciences, 18*(9), 2431–2453. https://doi.org/10.5194/nhess-18-2431-2018

Schlosberg, D. (2007). Defining environmental justice: Theories, movements, and nature. In *Defining environmental justice: Theories, movements, and nature*Oxford University Press. https://doi.org/10.1093/acprof:oso/9780199286294.001.0001. Vol. 9780199286294, pp. 1–256).

Sideris, I. V., Gabella, M., Erdin, R., & Germann, U. (2014). Real-time radar-rain-gauge merging using spatio-temporal co-kriging with external drift in the alpine terrain of Switzerland. *Quarterly Journal of the Royal Meteorological Society, 140*(680), 1097–1111. https://doi.org/10.1002/qj.2188

Skamarock, W. C., Klemp, J. B., Dudhia, J., Gill, D. O., Barker, D., Duda, M., Huang, X., Wang, W., & Powers, J. (2008).

Stefanou, N., Kazantzidou – Firtinidou, D., Sakkas, G., Theodoridis, G., & Rousakis, V. (2022). Training and exercises for Critical Infrastructure–a Hellenic computer-assisted exercise use case analysis. *International Journal of Disaster Risk Reduction, 69*. https://doi.org/10.1016/j.ijdrr.2021.102729, 102729.

Stiefelmeyer, H., & Hlatky, T. (2008). Hora – an Austrian platform for natural hazards as a new way in risk communication. In *INTERPRAEVENT 2008 conference proceedings* (pp. 229–236).

Swisstopo. (2014). *Swissalti3d, Swisstopo, Bern*.

Swisstopo. (2020). *Siegfriedkarte* (pp. 1870–1926).

Swisstopo. (2021). *The topographic landscape model TLM*.

Thaler, T. (2014). Developing partnership approaches for flood risk management: Implementation of inter-local co-operations in Austria. *Water International, 39*(7), 1018–1029. https://doi.org/10.1080/02508060.2014.992720

Thaler, T. (2021). Social justice in socio-hydrology—how we can integrate the two different perspectives. *Hydrological Sciences Journal, 66*(10), 1503–1512. https://doi.org/10.1080/02626667.2021.1950916

Thaler, T., & Hartmann, T. (2016). Justice and flood risk management: Reflecting on different approaches to distribute and allocate flood risk management in Europe. *Natural Hazards, 83*(1), 129–147. https://doi.org/10.1007/s11069-016-2305-1

Thaler, T., Zischg, A., Keiler, M., & Fuchs, S. (2018). Allocation of risk and benefits—distributional justices in mountain hazard management. *Regional Environmental Change, 18*(2), 353–365. https://doi.org/10.1007/s10113-017-1229-y

Thompson, V., Dunstone, N. J., Scaife, A. A., Smith, D. M., Slingo, J. M., Brown, S., & Belcher, S. E. (2017). High risk of unprecedented UK rainfall in the current climate. *Nature Communications, 8*(1). https://doi.org/10.1038/s41467-017-00275-3

Tobin, G. A. (1995). The levee love affair: A stormy relationship? *JAWRA Journal of the American Water Resources Association, 31*(3), 359–367. https://doi.org/10.1111/j.1752-1688.1995.tb04025.x

Tobin, G. A., & Montz, B. E. (1994). The flood hazard and dynamics of the urban residential land market. *JAWRA Journal of the American Water Resources Association, 30*(4), 673–685. https://doi.org/10.1111/j.1752-1688.1994.tb03322.x

Vanzo, D., Peter, S., Vonwiller, L., Bürgler, M., Weberndorfer, M., Siviglia, A., Conde, D., & Vetsch, D. F. (2021). BASEMENT v3: A modular freeware for river process modelling over multiple computational backends. *Environmental Modelling and Software, 143*. https://doi.org/10.1016/j.envsoft.2021.105102

Vetsch, D., Siviglia, A., Ehrbar, D., Facchini, M., Gerber, M., Kammerer, S., Peter, S., Vonwiler, L., Volz, C., Farshi, D., Mueller, R., Rousselot, P., Veprek, R., & Faeh, R. (2017).

Viglione, A., Di Baldassarre, G., Brandimarte, L., Kuil, L., Carr, G., Salinas, J. L., Scolobig, A., & Blöschl, G. (2014). Insights from socio-hydrology modelling on dealing with flood risk—roles of collective memory, risk-taking attitude and trust. *Journal of Hydrology, 518*, 71–82. https://doi.org/10.1016/j.jhydrol.2014.01.018

Weaver, S. J., Kumar, A., & Chen, M. (2014). Recent increases in extreme temperature occurrence over land. *Geophysical Research Letters, 41*(13), 4669–4675. https://doi.org/10.1002/2014GL060300

Welsh, W. D., Vaze, J., Dutta, D., Rassam, D., Rahman, J. M., Jolly, I. D., Wallbrink, P., Podger, G. M., Bethune, M., Hardy, M. J., Teng, J., & Lerat, J. (2013). An integrated modelling framework for regulated river systems. *Environmental Modelling and Software, 39*, 81–102. https://doi.org/10.1016/j.envsoft.2012.02.022

White, G. F., Kates, R. W., & Burton, I. (2001). Knowing better and losing even more: The use of knowledge in hazards management. *Environmental Hazards, 3*(3–4), 81–92. https://doi.org/10.1016/S1464-2867(01)00021-3

Wildman, G., & Lewis, E. (2022). Value of open data: A geoscience perspective. *Geoscience Data Journal*. https://doi.org/10.1002/gdj3.138

WMO. (2009). *Manual on estimation of probable maximum precipitation (PMP)* (3rd ed., Vol. 1045). World Meteorological Organization.

Zischg, A. (2016). River corrections and long-term changes in flood risk in the Aare valley, Switzerland. *E3S Web of Conferences, 7*. https://doi.org/10.1051/e3sconf/20160711010, 11010.

Zischg, A. P. (2018). Floodplains and complex adaptive systems—perspectives on connecting the dots in flood risk assessment with coupled component models. *Systems, 6*(2). https://doi.org/10.3390/systems6020009

Zischg, A. P., Felder, G., Mosimann, M., Röthlisberger, V., & Weingartner, R. (2018). Extending coupled hydrological-hydraulic model chains with a surrogate model for the estimation of flood losses. *Environmental Modelling and Software, 108*, 174–185. https://doi.org/10.1016/j.envsoft.2018.08.009

Zischg, A. P., Felder, G., Weingartner, R., Quinn, N., Coxon, G., Neal, J., Freer, J., & Bates, P. (2018). Effects of variability in probable maximum precipitation patterns on flood losses.

Hydrology and Earth System Sciences, 22(5), 2759–2773. https://doi.org/10.5194/hess-22-2759-2018

Zischg, A., Fuchs, S., Keiler, M., & Meißl, G. (2005). Modelling the system behaviour of wet snow avalanches using an expert system approach for risk management on high alpine traffic roads. *Natural Hazards and Earth System Science, 5*(6), 821–832. https://doi.org/10.5194/nhess-5-821-2005

Zischg, A., Fuchs, S., Keiler, M., & Stötter, J. (2005). Temporal variability of damage potential on roads as a conceptual contribution towards a short-term avalanche risk simulation. *Natural Hazards and Earth System Science, 5*(2), 235–242. https://doi.org/10.5194/nhess-5-235-2005

Zischg, A., Galatioto, N., Deplazes, S., Weingartner, R., & Mazzorana, B. (2018). Modelling spatiotemporal dynamics of large wood recruitment, transport, and deposition at the river reach scale during extreme floods. *Water, 10*(9). https://doi.org/10.3390/w10091134, 1134.

Zischg, A. P., Hofer, P., Mosimann, M., Röthlisberger, V., Ramirez, J. A., Keiler, M., & Weingartner, R. (2018). Flood risk (d)evolution: Disentangling key drivers of flood risk change with a retro-model experiment. *Science of the Total Environment, 639*, 195–207. https://doi.org/10.1016/j.scitotenv.2018.05.056

Zischg, A. P., Mosimann, M., Bernet, D. B., & Röthlisberger, V. (2018). Validation of 2D flood models with insurance claims. *Journal of Hydrology, 557*, 350–361. https://doi.org/10.1016/j.jhydrol.2017.12.042

Zischg, A. P., Röthlisberger, V., Mosimann, M., Profico-Kaltenrieder, R., Bresch, D., Fuchs, S., Kauzlaric, M., & Keiler, M. (2021). Evaluating targeted heuristics for vulnerability assessment in flood impact model chains. *Journal of Flood Risk Management, 14*(4). https://doi.org/10.1111/jfr3.12736

Zischg, A., Schober, S., Sereinig, N., Rauter, M., Seymann, C., Goldschmidt, F., Bäk, R., & Schleicher, E. (2013). Monitoring the temporal development of natural hazard risks as a basis indicator for climate change adaptation. *Natural Hazards, 67*(3), 1045–1058. https://doi.org/10.1007/s11069-011-9927-0

Chapter 6

Confronting complexity in flood risk management

The analysis of flood risks gives us an overview of the flood hazards, the elements at risk, and their vulnerability in a certain region, namely in floodplains within a river basin. In these landscape elements, the determinants of risk—hazard, exposure, and vulnerability—spatially overlap. Floodplains provide living space, room for socioeconomic activities, recreation, ecosystem services, biodiversity, and connect terrestrial with aquatic ecosystems. Hence, floodplains are dynamically shaped by a variety of processes and forces. In this book, floodplains are therefore considered as coupled human and natural systems. The systems of floodplains adapt to changing environmental conditions or to human interventions. The processes and forces that shape the evolution and adaptation of floodplains are numerous, diverse, interdependent, influence each other, coevolve, balance each other out, or amplify each other. The dynamics of processes in floodplains can lead to the emergence of new system characteristics and to the adaptation of the floodplain systems to the changing conditions. Thus, floodplains can also be defined as complex adaptive systems.

In the previous chapters, I summarized the literature of the forces and drivers of change that shape floodplains and flood risks in floodplain. I described model experiments for analyzing the effects of the individual drivers of flood risk change on overall flood risk evolution. Moreover, I summarized and described the behavior of floodplain systems when they adapt to changing external conditions and how feedback effects work in the adaptation process. But, what is the significance of these findings? Does the consideration of floodplains as complex adaptive systems matter for flood risk management? Do feedback effects between natural processes and human actions matter? Are they relevant for climate adaptation? In the following sections, I want to draw the conclusions from the examples shown in this book and to answer these questions.

A new perspective on flood risks

The book sheds light on the dynamic nature of flood risks, namely on flood risk change. We have seen how dynamically the factors or determinants of

Flood Risk Change. https://doi.org/10.1016/B978-0-12-822011-5.00001-6

flood risk change in space and time and how flood risk itself is changing with the drivers of change. This **dynamic perspective** extends the static perspective in that it adds further information to the classical risk analysis. To the static value of risk of a current risk situation, the value of the rate of change is added. This has some practical implications. If risk analyses are used for priority setting in the distribution of public funds for investments in flood protection, this additional information allows for fine-tuning the decisions. Hot-spot areas of flood risks in a certain area can be weighted with the rate of change. Hot-spot areas with a high positive rate of flood risk change will become more riskier in the future. In contrast, flood risk in cities where the hazard zones are already developed and new developments only find place outside of hazard zones will not increase remarkably because of low rates of change. Flood risk in these areas will only grow due to changes in vulnerability or hazard. Moreover, places with a relatively lower current flood risk but with very high rates of change will soon outpace the current hot spots. Thus, a priority setting that bases on the current situation only might lead to an underestimation of hot-spot areas after a certain period in this case. With the consideration of the rate of change, we can anticipate the development of future hot-spot areas to a certain extent. This enhances the ability to imagine development trajectories and to anticipate future flood risks.

Another aim of flood risk analysis is to prepare the information needed for cost—benefit analyses. If flood risk is to be reduced in a certain floodplain or city, different management options must be compared and the optimal solution must be selected from the set of alternatives. In many cases, the optimal solution is selected by comparing the ratio between costs and benefits of each protection measure. In this case, risk analysis is made for the reference scenario, namely the current situation, and once again for each situation with a planned flood risk reduction measure. The benefits of the protection measure result from the comparison between these risk analyses. The benefit is measured in monetary units per time. The costs of the protection measure are the effort needed to implement the measure plus their financial costs and their maintenance costs over a certain period. It is also measured in monetary units per time. Interestingly, such cost—benefit analyses consider the costs of the maintenance and the interest rates to calculate the financial costs for the measure's lifetime but neglect the future situation of hazard, exposure, and vulnerability. This is inconsequent and needs to be revised. At minimum, a thought about the functionality and effectivity of the planned measure under changing factors of flood risk is necessary. A pure static basis for decision-making in a dynamically changing environment might lead to a selection of the wrong flood protection measures and to unsustainable solutions. Moreover, the implementation of the flood risk protection measure that is optimal for the current situation might reduce the degree of freedom for alternative flood risk reduction measures in the future.

A dynamic perspective on flood risks requires the development of an information system that is **monitoring flood risk change**. Flood risk monitoring systems are observing the changes in the factors of risk and are continuously recalculating flood risk. Such a monitoring system allows to detect areas with fast-growing risks and to anticipate emerging flood risk hot-spot areas. Thus, a flood risk monitoring system supports the development of adaptive flood risk management strategies. Because we will never be able to predict changes in nature and society over long periods, we need to continuously look at how flood risks are changing and adapt risk management strategy accordingly. The monitoring of changing flood risks is a precondition for adaptive risk management strategies under climate and socioeconomic change. An interesting concept to confront deep uncertainties and flood risk change is the development of dynamic adaptive pathways (Haasnoot et al., 2013). In contrast to the current planning paradigm that bases on the assumption of a static situation, this concept guides planners to create a strategic vision of the future and to develop short-term actions for achieving the vision but also to establish a framework to guide future actions. The future actions comprise many different actions that will be selected depending on the future development of the system. According to the changes of the system along the timeline, one or another action will be selected.

The concept of dynamic adaptive policy pathways bases on the two concepts of adaptive policymaking and adaptation pathways. Adaptive policymaking bases on assumption-based planning of actions and on making markers and signposts explicit in the timeline that describe critical system changes and therefore must be monitored. The crossing of such markers triggers new actions for achieving the goals of the policy. Thus, adaptive policymaking requires a contingency plan to adapt the actions to new system conditions or information over time. In contrast, adaptation pathways base on exploring possible actions for achieving the goal of governing risks that are sequenced in time. Depending on the trajectories of possible system changes, alternative pathways of adaptation options are designed. The adaptation pathways are often visualized as decision trees. A route along one of the tree branches is an adaptation pathway. This approach uses transient scenarios. Thus, also this policy requires a monitoring system that allows identifying forks, junctions, and branching along the pathways of development. The combination of both approaches with an information system that monitors system changes provides a method to confront a dynamically changing environment. This is absolutely needed in the era of the Great Acceleration. For flood risk management, this method has a potential to systematically confront flood risk change. The approach of dynamic adaptive policy pathways bases on these steps: After the description of the current system characteristics, the uncertainties of future developments and the objective of the flood risk management strategy, a problem analysis is conducted. Herein, the sensitivity and vulnerability of the system to changing environmental conditions are analyzed. Moreover, the full range of possible actions is analyzed based on transient scenarios of changing risk factors. This leads to the identification and determination of actions that allow achieving the goals. These actions are assessed and iteratively reassessed in terms of efficacy, point in time of implementation, and according to the system vulnerabilities. The selected actions are concatenated in a timeline that describes adaptation pathways. The adaptation pathways are visualized in a map and along a time axis. This map enables to identify adaptation pathways that exclude future actions, e.g., by occupying solution space that can never be used by subsequent actions. From this set of adaptation pathways, a preferred pathway is selected that optimally meets the decision criteria under current system conditions and under the foreseen future changes. The selected pathway determines contingency actions and points in time where a further decision is needed, e.g., triggers, tipping points, or critical system vulnerabilities. The contingency actions aim at keeping the selected adaptation pathway viable. This leads to the definition of a dynamic adaptive plan. The implementation of this adaptive plan bases on continuously monitoring the system conditions and by iteratively implementing the contingency actions. Examples of dynamic adaptive pathways can be the decision between the implementation of flood risk protection measures in the river or the implementation of property-level flood protection measures at the level of individual buildings. The future decision along the adaptation pathway depends on either changes in the hazard, in exposure, or in vulnerability. The map of the dynamic adaptive pathways provides the basis for designing the monitoring system. It determines which factors should be monitored that describe critical system states and transitions.

In the previous chapter, we have seen how a coupled component model can be used for monitoring flood risk change. The concept of coupled component modeling potentially provides the possibility for monitoring the changes in the system components, i.e., the risk factors, as well as change in the system characteristics. In the examples, we used flood risk as the main quantitative description of the characteristics of floodplain systems. Moreover, coupled component models can simulate the range of future development trajectories. If the future urban development can be modeled transient and assumed scenarios of future infrastructure types and vulnerabilities can be modeled transient, the hazard modules of coupled component models can model the future development of floods in transient climate change scenarios. Furthermore, the effects of adaptation actions on the system behavior can be modeled. With its capabilities of modeling the determinants of flood risk transient in space and time as well as adaptation options, a coupled component modeling framework can provide all the information needed to develop dynamic adaptation pathways. It furthermore provides a method to quantify the solution space. This is the space where the future development of the system meets with possible adaptation actions. It is limited by the constraints and opportunities that determine why, how, when, and who adapts to risks resulting from climatic spaces (Haasnoot et al., 2020). Coupled component models can help to draw the solution space and therefore fully support to follow a dynamic perspective on flood risk management.

In addition to a dynamic perspective on flood risks, a **complexity perspective** on flood risks helps to understand the dynamics of flood risk change and to approach several problems at the same time in a multidisciplinary framework. Dealing with flood risks in our times means dealing with wicked problems that arise from solutions for problems of the past. In many cases, the unintended and perhaps negative effects of monodisciplinary flood risk management strategies show up only in the long term. Examples are the unintended negative effects of river engineering measures for the aquatic ecosystems or the generation of lock-in situations by the restoration activities after a historic flood event or by land use planning strategies. These legacy effects of past solutions narrow the degree of freedom for problem-solving of today. It is time to rethink our nowadays risk management strategies and screen their possible outcomes for long-term unintended effects (Baldassarre et al., 2018). The effects of climatic changes will further increase uncertainty and complexity in this topic.

Moreover, decision-making in flood risk management is embedded in or touches a high number of other policies, strategies, plans, actions, and measures. The implementation of flood protection works in rivers always has to be coordinated with environmental protection issues. Flood risk protection measures have to be compliant with environmental protection laws. In an increasing number of cases, flood risk protection measures are designed to enable or to foster environmental protection. River restoration actions are embedded in flood risk protection measures and vice versa. Thus, flood risk management has to consider all aspects of river ecology. It has to consider also the landscape perspective. Rivers should be laterally connected with their floodplains and should longitudinally not be disconnected by anthropogenic structures (Delong & Thoms, 2016; Gilvear et al., 2016). Moreover, river management and flood risk management are connected with the energy sector. Hydropower plays an important role for the energy supply in mountainous regions and the adjacent lowlands. The energy market and the laws for energy generation and provision determine flood risk management practice as well as the evolution of flood risk over large parts of the river network. Therefore, stakeholders of this sector are usually part of the planning team of flood protection measures. Many synergies can result out from this collaboration (Zischg, 2012). In urban areas, city planners that look at flood protection measures from their perspective as well as traffic engineers are participating in the planning process. Moreover, infrastructure operators will bring in their ideas and constraints into the process. Flood protection measures also provide ecosystem services (Barth & Döll, 2016). Therefore, the stakeholders dealing with this topic will be involved. Such measures for flood protection are often defined as nature-based solutions that provide cobenefits for climate adaptation, heat protection, biodiversity, and local recreation. Last but not least, flood protection measures usually need land and thus agriculture becomes a participating stakeholder in the planning process. Moreover, agriculture can provide levers for attenuating or dampening flood risks (Zischg et al., 2012). Therefore, the number of participants of a round table for participatory planning processes can become large. From the perspective of a hydraulic engineer that is responsible for planning the flood protection measure, this can be a pity. From this monodisciplinary perspective, the many different views on the topic might narrow the solution space of hydraulic engineering methods. However, a multiplicity of perspectives on one topic radically increases the number of possible solutions by making use of the higher creativity. The multiple views guarantee a very important argument for sustainable solutions, namely the search for low regret measures and the identification of possible unintended consequences of a solution in flood risk management. The higher the diversity of views on a project, the higher is the probability that the project team is avoiding unintended effects of a narrowly one-topic-designed project on other topics. Therefore, a variety of different perspectives increases the ability to understand the complexity in flood risk management projects.

A complexity perspective therefore is a mindset that allows embracing complexity as suggested by (Boulton et al., 2015). Challenges arising from complex interdependences between different components of flood risk, or river ecology, or floodplains that are understood as coupled human and natural systems can be confronted with a theoretical bases. Complexity in flood risk management must not be reduced, but its consideration in the analysis and planning process offers an evolutionary leap in decision-making practice. Note that the selection of an optimal solution in terms of cost benefits from a set of solutions for flood risk management is leading to a econimically efficient solution in the long term, but the extension of the solution space by a variety of perspectives and the selection of a solution that meets the most requirements of the stakeholders will still leave space for future adaptation. This necessarily comes with higher costs. The redundancy of such measures is higher. However, the redundancy of measures is increasing the resilience of floodplain systems. Systems with higher redundancy exhibit more possibilities to absorb external shocks and are therefore more resilient than highly efficient systems. In the previous examples, we have seen how other (mono-)disciplinary strategies can have unintended effects on flood risk change. One example was the successful implementation of the clear water acts or water framework directives. With the increased water quality, the water bodies become attractive recreation areas and thus the shorelines of water bodies are becoming attractive living spaces with a certain time lag. The industrial transformation, i.e., the abandonment of old industrial sites along the water bodies and their transformation into residential areas, is increasing flood risk. Such unintended consequences of one policy to the domain of another policy can be avoided by looking at the whole system with a complexity perspective. In summary, a complexity perspective markedly enhances robustness of decision-making in flood risk management. Decisions are better supported and the possibilities for finding unsustainable solutions can be limited.

Lessons learnt from the Kander river deviation

The example of the Kander River deviation in 1714 AD in the Canton of Bern in Switzerland allows reflecting on the meanings of deep uncertainty and complexity in flood risk management. The pioneer project was one of the earliest large-scale geoengineering projects. It aimed at using the retention capacity of the Lake Thun in the Bernese Alps to dampen the flood peaks of the Kander River in Bern. In the end, this should reduce the risk of flooding in Bern, the capital of the Canton of Bern. The reduction of flood peaks should also reduce the braiding of the Aare River downstream of Thun and to stabilize the shipping business from Thun to Bern along the Aare River. In these times, the knowledge of fluvial erosion was poor. Thus, the planners of this river deviation were not aware of this epistemic uncertainty. This lack of knowledge was a so-called "unknown unknown". The initial plan to deviate only flood waters to the Lake Thun and to keep low flows and average flows in the former river bed failed because of this lack of knowledge. This uncertainty led to the first unintended consequence of a cascade of unintended consequences, namely the erosion of a 27 m deep gorge that suddenly made the river deviation total and irreversible. After this initial human intervention, the coupled human and natural system of the river-floodplain network began to adapt to the new circumstances. Here, I considered the floodplain of the old Kander River, the floodplain of the Aare and Gürbe River between Thun and Bern and the floodplains along the shorelines of Lake Thun as the Aare and Kander River system. The adaptation of this system was driven by natural processes and anthropogenic interventions. Both processes coevolved and influenced each other. The drying out of the old riverbed of the Kander River led to the direct loss of water provision and thus to a loss of hydropower and indirectly, with a time delay, to the drying out of all groundwater sources and thus to a loss of freshwater provision. Society reacted by deviating a second river, a former tributary of the Kander River, toward the villages along the old and now dry riverbed of the Kander River. The lake system reacted to the doubling of the contributing catchment area by a rising lake level. This natural adaptation led to the adaptation of the lake outflow system. The higher lake level led to a higher lake outflow and consequently the channel morphology of the lake outflow, namely the Aare River, began to adapt to the new flow regime by lateral erosion. This in turn triggered human adaptation. Society adapted their infrastructure along the shorelines and installed one flood protection measure after another. Moreover, the higher lake level frequently flooded some areas of the city of Thun and made them unusable for human activities. Both the initial natural adaptation process and the subsequent cascade of human interventions led to a final adaptation of the lake system that lasted for 300 years and ended with the construction of a relief tunnel in 2009. The Aare River floodplain downstream of Thun adapted to the lack of sediment input from the Kander River by the incision of the riverbed. The river channel stabilized but the lateral erosion made the shipping difficult. Thus, the downstream adaptation triggered a further adaptation of the lake outflow system by installing weirs that can regulate the lake level.

All these cascading processes could not have been foreseen by the planners of the Kander River deviation. Only a minority of engineers warned and suggested to adapt the lake outflow before deviating the Kander River into the Lake Thun. These early warners have been ignored and the government decided to finance only the main intervention and not the integral modification of the whole system. The warning engineer had finally to flee from the region because he feared the angry of the population of Thun that had to live with the unintended consequences of "his" project for many years. Only after 300 years, the reputation of the lead engineer that imagined an integral solution instead of a partial intervention was restored by historians. In addition, many others raised the hand for other possible unintended consequences. For example, fishermen's feared negative effects of the colder and turbid water of the Kander for the lake ecosystem. Citizens that were resident of low-lying areas adjacent to the lake feared higher lake levels. They were not heart and the government followed "the science", i.e., the newly arising river engineering discipline. Nevertheless, the government had to react to the cascading effects and to continuously adapt their policy.

Thus, in retrospect we can learn from this narrative of a complex adaptive system. We can observe that there are **feedback** effects between the natural and the human systems. The coevolving risk components can result in **nonlinear** relationships between one and another driver of change. We can see the effects of **time lags** in the cascade of unintended processes. We can learn from this example, how society of these times confronted **deep uncertainty**. The **epistemic uncertainty** was confronted with societal learning. The experience in observing the massive and very unexpected erosion processes led to the foundations of the concepts of modern river geomorphology. The adaptation of the human system, namely the businesses that related on the river and lake system, was stepwise. They and the government adapted their policies after each new phenomenon arises. Hence, they reacted to **emergent phenomena**. Both the natural processes and the human adaptation together led to a **self-organization** and **self-regulation** of the river-lake-floodplain system. In a first period, these reactions were ill-defined and targeted to a single problem, but in the course of time the reactions became more **system-oriented** or integrated. Erosion protection and the enlarging of the lake outflow were planned hand in hand after the first lessons learnt. The example allows learning lessons from a **systemic perspective**. The later phase of human intervention that aimed at dealing with the unintended consequences of the previous interventions showed that interventions can have **synergistic effects**. The adaptation of the lake outflow by both the installation of a new lake outflow and the installation of weirs for lake level regulation targeted many problems at the same time, ranging from erosion protection to ecology, drinking water provision, and flood protection to enabling the shipping business. We can also learn from the example that a **dynamic perspective** is important. Intervening into a coupled human and natural system and assuming a static behavior might misguide the planning process. We can observe that the decisions were taken at **multiple scales** and **hierarchies**. The financing and thus the main authority for decision-making came from Bern. Nevertheless, local stakeholders influenced the decisions in the details. The general public was not involved in the decision-making and did not create huge opposition. From the historical documents, we can learn that a narrow elite was responsible for decision-making and a **diversity of opinions** was suppressed. The warnings of possible failures were ignored. This lack of diversity hampered the integrated approach. The described cascade of unintended consequences of an early human intervention to a natural river system explains the concept of **path dependency**. For example, once the initial river deviation was made and once the subsequent erosion made the deviation total and irreversible, the solution space for solving the subsequent problems shrunk remarkably. Since a restoration of the initial situation was becoming impossible, all subsequent flood risk management solutions can only deal with the consequences but not with the root causes of the problem. Therefore, the path of the river deviation determines all subsequent pathways of adaptation. The future evolution of the floodplain system is therefore contingent on historical and contextual factors. The order or sequence in which subsequent or cascading processes occur or phenomena arise is a consequence of the context that is shaped by the first intervention. We can also observe a **regime shift**. As shown by the model experiments that analyzed the effect of all human interventions to the river system on the hydrologic regime, the characteristics and probabilities of low and high flows at Bern changed drastically. The Aare River changed from a wild mountain river to a lowland river. The initial point of the regime change was the Kander River deviation. Therefore, this human intervention can be considered as a **tipping point** for a 300-year long and irreversible transition of the Aare River system from one system behavior to another. This transition process is superimposed by natural **fluctuations** and **episodic changes** in the hydrologic regime. For example, the Kander River deviation took place during the Little Ice Age and the transition phase after this human intervention was also influenced by the warming after the Little Ice Age in the European Alps. Moreover, the system change was also superimposed with a societal change. The society in this region transitioned from agricultural and trading society toward an industrialized technocratic society. This transformation could not have been foreseen by the planners of the Kander River deviation. The changes of societal values influenced decision-making over time. Only 100 years after the Kander River deviation, the correction of the Aare River became relevant. The industrialization shaped the imaginative force of the society to modify the environment at large scales. The arising of the railway and the construction of the railway line between Bern and Thun made the shipping business obsolete. The highly vulnerably railway had to be protected from flooding. This required to stabilize the Aare River and to correct its river course. The river channel morphology was harmonized and the river course was straightened. To protect the new railway, the riverbanks had to be stabilized with huge boulders and lateral dikes were constructed. Thus, the new technology of traffic and the new technology for shaping landscapes at large scales increased the technical ability to modify rivers. Thus, the Aare River between Thun and Bern soon after the construction of the railway became an anthropogenically modified river. Together with the long-lasting effects of the Kander River deviation, e.g., the sediment deficit in the Aare River downstream of Lake Thun, the new human interventions triggered other cascades of unintended consequences. The lateral ecological connections between the river and the floodplains were lost and the riparian forests began to transform. The groundwater level sank with the fluvial erosion that followed after the straightening and narrowing of the river. After the industrial revolution, society transformed into a service-oriented economy. This influenced urbanization, traffic, and communication. All these further societal changes could not have been foreseen in 1714 AD. Thus, we can learn from this example how massive future changes in society and societal values might influence flood risk change in the future. Therefore, a static planning perspective becomes definitively obsolete or hampering. We can learn from this example that we cannot foresee the future evolutions, but we can learn from it by considering path dependency. Our decisions in flood risk management of today make space for future decisions or might limit the future solution space. With a complexity perspective, we are at least able to approach future developments up to a certain limit. Moreover, the complexity perspective can guide us to the most relevant levers with which we can influence the rate of change of flood risk. Finally, this identification of the most relevant parameters of levers for influencing flood risk change can lead to the ability of thinking outside the box. We can read the overall long-term trends, for example, the trend toward fossil-free cities, and break this trend down to the problem of flood risk management. Nowadays, altering one of the most influencing factors for flood risk, namely the discharge capacity of rivers, touches in many places a sacred cow because the adjacent lateral spaces of the rivers are used as streets or parking spaces in many cities around the world. If fossil-free cities are becoming car-free cities, we might identify new drivers of flood risk change. The arising of potential future use of space that is currently deserved for traffic might open completely new possibilities to adapt current flood protection to climate changes by giving rivers more room. In the following, a summary of the components of risks is given.

Levers for controlling flood risk in the long term

Flood risk is composed of many factors, variables, determinants, and processes. Many of these are more or less interdependent from others while some are dependent only from a few others. A factor that is changing over time might trigger changes of another factor. The previous chapters described these dependencies and feedback in detail. Here, I try to summarize these interdependent factors of risk in the form of a network. I am not strictly distinguishing between the formal characteristics of factors, determinants, drivers of change, variables, or processes, but I am focusing on drawing the network of influences. The nodes of this network are factors or processes that describe characteristics, status, or change of subsystems. I call the nodes of the network components of flood risk. The edges between these nodes draw the influences between the risk components. These influences are visualized as directed edges. Therefore, the network of influencing risk components is a directed network. The directions are set from one node that influences another node. The influence can be positive or negative. The change of one risk component influences the change of the other node at the end of the directed edge. The network contains only the main components of flood risk that were described in the previous chapters. I am fully aware that the social components of the flood risk management might be underrepresented from a sociologists' perspective, and the ecological components are missing from an environmentalist's perspective. Moreover, many risk management options, for example, insurance, are not considered in this network of risk components. A more comprehensive causal-loop diagram is described by (Mai et al., 2020). Although a more biased view on the geomorphological/hydrological variables of the big picture on flood risks, the network of interdependent risk components can be used to determine the levers for controlling flood risk in the long term. Flood risk is a component of the network and results from the probability of a flood impact. From this node in the network, namely flood risk, we can follow all the paths that influence flood risk directly or indirectly.

Thus, the graph supports the selection of levers or drivers of change that influence flood risk. The network is shown in Fig. 6.1. The sizes of the nodes represent the out degree. This is the number of outgoing edges from a node. The higher the number of outgoing edges from a node, the higher is the out degree. The color of the nodes represents the in degree. This is the number of incoming edges to a node. The higher the number of incoming edges, the darker is the color of the node. The font size represents the betweenness centrality. The betweenness centrality measures the importance of nodes in the network. Nodes that are located on many shortest paths between other nodes in the graph have a high betweenness centrality score.

The nodes with the highest betweenness centrality are the risk components discharge capacity, flood extent, roads, and extreme weather event. The nodes with the highest out degree are surface runoff, river discharge, discharge

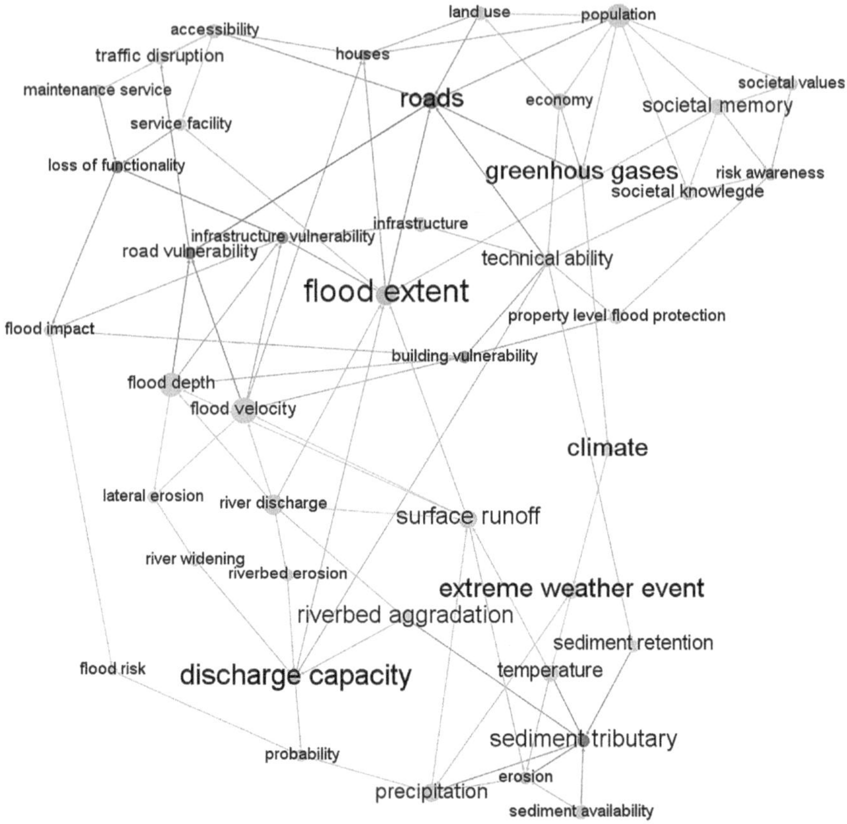

FIGURE 6.1 Network of risk components. Network of risk components. The sizes of the nodes represent the out degree. The higher the number of outgoing edges from a node, the higher is the out degree. The color of the nodes represents the in degree. The higher the number of incoming edges, the darker is the color of the node. The font size represents the betweenness centrality.

capacity, flood extent, flood velocity, flood depth, population, societal memory, and climate. The nodes with the highest in degree are roads, road vulnerability, infrastructure vulnerability, loss of functionality, and discharge capacity.

Flood risk is determined by the probability and the impact of flood events or flood scenarios. These two nodes are influenced by many others. Probability is influenced by the process chain climate/extreme weather event/precipitation. This link shows that climate has a relatively short path to flood risk. This path from climate to flood risk crosses only three nodes. Thus, climatic changes that lead to a change in the probability of extreme rainfall events have a direct influence on flood risk. Another relatively direct link to flood risk is the path from population to houses and their vulnerability to flood impacts, if the houses are exposed to floods. However, this path from population and houses to flood risk can encounter some bypasses or detours. With growing population, societal knowledge is increasing and with it the technological ability to change the human environment according to the societal values. Technical ability thus can lead to a large-scale implementation of flood protection measures. This manifests in increased river discharge, and via the reduced probability of flooding, it influences flood risk. Technological ability also might be used to implement property-level flood protection measures. This in turn reduces building vulnerability, and thus it influences flood impacts and flood risks. The network shows other feedback described in the examples of the previous chapters. A flood event, namely a realization of an extreme weather event on the ground, affects a share of the houses and with it a large number of people. A flood event thus increases societal memory and thus societal knowledge. This might lead to an alteration of the societal values. Via this societal learning, society becomes risk aware. Moreover, the increased societal knowledge leads to an increase in the technical ability of flood protection. Consequently, society might invest in property-level flood protection or in structural flood protection measures. This uses two different levers that influence flood risk from two sides. Property-level flood protection decreases the vulnerability of buildings and thus reduces flood risk via flood impacts. Feedback loops on the physical parts of the network of risk components are related to sediments. Climate change is altering precipitation and temperature. The increase in precipitation intensity and probability will have effects on surface runoff and on soil erosion in the river catchments. This increases sediment load and sediment transport in the tributary rivers. Moreover, increasing temperature leads to melting of permafrost in high alpine areas. This increases sediment availability. Together with increased surface runoff and erosion, this will increase sediment load in the tributaries. Increased surface runoff in the catchments leads to an increasing discharge in the tributary, and consequently, this will increase riverbed erosion and lateral erosion. This is an additional contribution to an increase in sediment input from the tributaries to the main rivers. The additional sediment input from the tributaries might lead to riverbed aggradation and therefore to a reduced discharge capacity of the main river downstream of the tributary. This increases the probability of inundation in the floodplain, and finally, it contributes to an increase in flood risk. Another feedback loop works in the opposite direction. The installation of sediment retention basins for protecting settlements from sediment transport and debris flows coming from the tributaries reduce the sediment input to the main river. This influences sediment connectivity and sediment transport. The sediment balance is also influenced by hydropower dams that break the sediment connectivity between upstream and downstream river reaches. A reduced sediment input from the tributaries plus a higher river discharge lead to an increase in riverbed erosion and lateral erosion. This increases discharge capacity and therefore reduces probability of inundation in the floodplain and thus reduces flood risk. These examples of a feedback loop between climate change and flood risk are one of the visible clusters in the network of flood risk components. Another cluster is visible around societal memory, societal knowledge, societal values, and risk awareness. A third cluster is visible around the characteristics of flood processes. River discharge during a flood event and the river discharge capacity determine the inundation of the floodplain. Depending on discharge capacity, a higher or lower volume of water is exiting the river channel and flooding the adjacent floodplain. Flood extent, flood depths, and flood velocity quantify the inundation process in the floodplain. These three metrics influence flood impacts. Flood extent determines the amount of exposed houses, the length of flooded roads, and the number of affected infrastructure elements. Flood depth and flood velocity determine the vulnerability of the elements at risk that are exposed to the flood extent. These three metrics therefore are connected and result in the connected exposure and vulnerability. Both determine flood impacts and therefore flood risk. Another feedback loop lays in components that influence flood impacts on infrastructure. In multilayer networks such as the traffic network and the energy network, disturbances spread from one network to the other. A power outage in the electricity networks results in disturbances in the traffic network. Railways are directly affected by power outages and roads are indirectly affected, e.g., by the failure of traffic control systems and subsequent traffic congestion. Vice versa, disturbances of the road network by floods lead to a lack of accessibility of maintenance staff of electricity providers to important nodes in the electricity network. The flood event itself or other circumstances can lead to a failure of a node element, e.g., a transformer station. The maintenance staff cannot reach this node, and therefore they cannot repair it within useful time. Thus, the power grid might fail with a certain time lag. This leads to a loss of functionality that is an indirect flood impact. These detours in the network of risk components can be seen as a positive feedback or a link between two systems.

Implications for flood risk research and management

The model experiments and analyses were a first attempt to approach the dynamics and complexity of flood risk change. This is relevant for **flood risk research**. Several case studies described the influences of changing risk components to other components or to the overall evolution of flood risk. The findings base on experiments in a virtual world, namely in models that represent reality. After (Sterman, 2006), learning about complex systems from evidence is hampered by several factors. Complexity hinders both the learning from evidence and the generation of evidence. Replication is difficult or even impossible in studying complex systems. Long-term evolutions or time delays in feedback might hide important drivers of change. Feedback effects are blurring cause–effect relationships. Thus, counterintuitive system behavior can arise from the interactions of the system components and eventually from time delays. This creates many confounding variables for deriving simple cause–effect relationships, and thus it challenges the reductionist or mechanistic perspective. Nonlinearity is challenging our minds and the emergence of new patterns or system behavior often cannot be imagined before being observed in modeling. A multiplicity of factors and actors in a complex system results in dampening or self-amplifying actions. Resulting system behavior cannot be explained with a narrow focus on selected parts of the system. Hence, a systemic perspective is essential for generating knowledge in an interconnected and dynamic world.

To learn more about complex systems in the topic of flood risk research, we can make use of modeling and simulation, a branch of science. Mitchell (2009) describes several approaches of how computation and simulation lead to gaining new knowledge of complex systems. For this approach, we create virtual worlds, i.e., models or simulations with which scientists or decision-makers can conduct experiments and test hypotheses or strategies, respectively. Such virtual worlds provide low-cost laboratories for learning (Sterman, 2006). They allow analyzing processes in a time lapse or in slow motion. The virtual implementation of strategies, actions, or measures in the virtual world allows testing their outcome. The experiments can be repeated under different boundary conditions. Moreover, even experiments that test the outcomes of human decision-making can be conducted. In contrast, real experiments could be ethically delicate. For risk management, a model experiment can test the effects of different strategies on flood risk evolution. Different strategies can be virtually implemented and their effects on flood risk reduction can be analyzed. This would not be possible in the real worlds because a reset of a system to certain initial conditions is never possible. The virtual world can be exposed to a stress test. A virtual representation of a system can be forced into extreme conditions what is not possible with real situations. This reveals information about the sensitivity or vulnerability of the system. Furthermore, the effects of individual drivers of change on the overall system behavior can be studied in an isolated way. Thus, confounding variables can be virtually eliminated.

The model experiments shown in this book extended the methods for simulating the behavior of complex adaptive systems in flood risk research. The experiments allowed learning about the complexity inherent in the analysis of flood risk change. They showed methods for considering and modeling features of complexity in analyzing and modeling flood risk change. The modeling **framework of coupled component models** proved thereby to be a promising approach for conducting the model experiments. Coupling specialized disciplinary models together to represent a system is modular, and therefore it is flexible and robust. This modularity makes coupled component models more transparent and interpretable than lumped specific-purpose models. The model chain can be extended step by step with submodels that have already been validated in their specific domain of application. However, the coupling of already existing models remains a challenge, as they potentially address different scales in space and time. Coupled component models consider explicitly spatial phenomena, and thus they offer a prospective approach for coupling process models with agent-based models that simulate the interactions of individuals and institutions with the changing environment. This offers a huge potential for extending the capabilities for simulating complex adaptive systems such as floodplains. The combination of physics-based process models and agent-based models offers a thorough simulation of the spatiotemporal dynamics in floodplains, not only from the flood risk perspective but also from other disciplinary or interdisciplinary perspectives. The number of studies that couple agent-based models with process models is growing (Aerts, 2020; Aerts et al., 2018; Haer et al., 2019). To implement memory and societal learning in agent-based models or in multiagent models, further research is needed. New methods of machine learning can be implemented in model chains to represent societal learning mechanisms in these models (An et al., 2021). This would finally allow simulating adaptation processes in coupled human and natural systems or developing evolutionary models.

Another great challenge remains in coupling physics-based models with models and techniques for analyzing complex networks (Loreti et al., 2022; Papilloud et al., 2020). The analysis of spatial and temporal change of networks according to changes in their environment and their exposure to hazards is a remaining challenge. The case studies of this book focused mostly on flood risks for buildings and workplaces and only secondarily on networks or systemic risks. However, the presented methods or model experiment settings can also be adapted to study the impacts of floods on networks, both single-layer and multilayer networks. This might pave the way to develop experiments for analyzing cascading impacts of flood events to multiple sectors and human activities that result in systemic risks. These are characterized by high complexity, multiple uncertainties, major ambiguities, and influences of one system to other systems (Renn et al., 2020).

Coupled component models proved to be a viable approach for analyzing the state space of a floodplain system. The modeled systems can virtually be stressed by changing its boundary condition to absolute physical limits. This allows for searching the probable maximum flood impact event. This explorative approach to test the sensitivities of a system to external shocks can be adapted to discover future scenarios of extreme changes in flood risk (Moallemi et al., 2020). This allows defining and delimiting the future solution space, namely to explore in which range flood risk can potentially develop under the locally possible flood risk management strategies. This finally allows searching for unintended outcomes of selected flood risk management measures and simulating human adaptation. With this ability to explore scenarios of future system states and system development trajectories, coupled component models provide a meaningful way to generate and elaborate dynamic adaptation pathways.

A research gap remains in modeling structural change of systems. A system can change its behavior or by a structural adaptation after passing a threshold of a system state or a tipping point. If this system is represented by a model that is unable to simulate this structural change, we will never be able to explore the consequences of the structural adaptation of a system. The presented idea for incorporating multiple models to simulate the change from one system state to another might be promising to develop modeling frameworks that enable the simulation of structural change.

Flood risk management operates within a complex system. The coupled human and natural systems of floodplains show many features of complexity (e.g., as summarized by (Ladyman et al., 2013)), i.e., nonlinearity, feedback, spontaneous order, robustness and lack of control, emergence, hierarchical organization, and numerosity. The examples of this book showed how important it is to look at the interdependencies between the factors of risk when we aim at governing flood risk. Thus, a complexity perspective and a model coupling framework can aid flood risk management to look also at the interdependencies of flood risk components and not at the components only. The examples and methods described in this book furthermore help to extend flood risk management practice with the following:

- The presented model experiments focused on single factors of risk but also on interdependent factors and their connections. The coupling of specialized models and the development of a model coupler that simulates the connections and feedbacks between the model components allow to "connect the dots" (Zischg, 2018). Therefore, this modeling framework provides a toolbox for developing integrated assessment models and models for system-oriented analyses. It further enables the implementation of the conceptual ideas for modeling the resilience of coupled human and natural systems in mountain areas as suggested by (Hossain et al., 2020).
- The dynamic perspective on flood risks and the modularity of the proposed coupled component modeling framework creates the basis for developing information systems for monitoring flood risk change. Monitoring the change of risks as well as the changes of the drivers of risk is absolutely needed for the development of dynamic adaptation pathways and the operationalization of adaptive risk management strategies. Flood risk change monitoring systems are crucial to detect relevant change in time and to identify the emergence of new situations where flood risk management must intervene (Raso et al., 2019).
- The book aimed at contributing to exchange a static, mechanistic, and reductionist perspective with a dynamic perspective that considers the features of complexity in the topic of flood risk management. The presented network of interdependent components of risk is helping to navigate

through the complex journey of long-term flood risk monitoring and flood risk management. The network visualization allows identifying feedback and influences between the risk components. This systematically leads to find different pathways for influencing overall flood risk change.

- The dynamic perspective opens new ways to model transient scenarios of climate change impacts on flood risks. This allows to model adaptive flood risk management under climate change.
- A coupled component modeling framework paves the way forward to implement multimodel and multivulnerability frameworks and thus to analyze compound, cascading, interacting, interdependent, or accumulating risks or multirisks. This enables the development of causal maps and models for identifying and analyzing systemic risks (Ackermann et al., 2014; Renn, 2021; Schanze & Maleska, 2019; Schweizer, 2021). The deepening of knowledge on systemic risks in flood risk research enables governance of systemic risks (Schweizer & Renn, 2019).
- A modular framework enables the assessment of both direct and indirect impacts of flood events on society.
- The presented examples show first ideas of how to integrate feedback mechanisms between natural processes and human activities in spatially explicit models. These are needed to model coupled human and natural systems as well as socioecological systems and complex adaptive systems (Hull et al., 2015; Preiser et al., 2018).
- With such enhanced modeling frameworks, flood risk management practice is enabled to study the multifunctionality of infrastructure systems. For example, to analyze how road infrastructure can act as flood defense or flood retention systems or as a flood corridor and how this multifunctionality affects the systemic functioning of the infrastructure itself. Hence, coupled component modeling framework also allows studying the cobenefits of measures for a specific purpose to other topics, for example, the effect of green infrastructure for heat protection measures in cities on reducing surface water floods by increasing infiltration capacity.
- The proposed modeling framework is flexible; it can consider deterministic modeling for short-term flood risk change, probabilistic modeling for the simulation of medium-term flood risk change, and explorative modeling for confronting deep uncertainty issues. Moreover, we have seen that this framework is flexible to combine data-driven approaches with physics-based models. This leads to strengthen the ability to develop dynamical complex systems models of disaster resilience (Yabe et al., 2022).
- The complexity perspective provides a theoretical ground for participatory modeling processes. Model coupling eases the consideration of multiple perspectives of several stakeholders that are involved in a transdisciplinary analysis or in decision-making. Ideally, stakeholder can see their rules of thumb, heuristics, reasoning methods, or even models implemented in a coupled component modeling framework and experience or consider the

results of participatory modeling processes. Also stress tests can be complemented with a bottom-up procedure of vulnerability analysis by the stakeholder itself. The model coupling approach therefore allows implementing system-specific vulnerabilities into stress tests. This requires to develop plug-and-play components that can easily be coupled together to whole-system representations. Both, developers and users, would benefit from this programming technique (Verburg et al., 2016).

- Coupled modeling frameworks can implement policies of flood risk management and thus new policies can be tested for unintended consequences in the long term by considering feedback with other factors. This can help to generally better avoid unintended outcomes of a new policy.
- The examples of this book show how we can analyze flood risk with models that operate at different scales. Felder et al. (2018) demonstrated how models operating at the global scale can be coupled with models that operate at the local scale. This ability to bridge scales in modeling, together with the ability to link physical and socioeconomic drivers of risk, enables complex climate change risk assessments as proposed by Simpson et al. (2021).
- The integration of decision-making models with process models allows studying emergent patterns of environmental and social disparities. With integrated models we can analyze or identify possible unintended consequences of certain policies in flood risk management (Thaler et al., 2018).
- A multimodel environment moreover enables the simulation of cascading processes. In this case, one process model can feed another process model, for example, the hydrologic model drives a hydraulic and geomorphologic model and a subsequent large wood transport model. This can be implemented by elaborating pathway schemes as proposed by Schauwecker et al. (2019).
- The combination of physics-based models and network analysis methods leads to a more integrated assessment of the effects of local floods on large-scale networks (Rebally et al., 2021; Wang et al., 2020). It is essential for modeling the behavior of critical infrastructure during flood events (Fekete, 2019).
- Model chains from precipitation to flood impacts can also be applied for short-term forecasts of flood impacts. In this case, the meteorological input comes from a numerical weather forecast model that predicts the weather conditions for the next days. If computationally optimized, the model chains presented in this book can be fed by the weather forecasts. This leads to the prediction of the flooded areas and the flood impacts. Forecasting flood impacts is a key for developing impact-based flood early warning systems. These systems consider the system vulnerability (bottom-up sensitivity analysis) and the physical process (top-down model chain). If the predicted process magnitude exceeds the system vulnerability threshold, a user-targeted warning with behavioral recommendations can

be issued. This could principally drive location-based warning services. Therefore, this kind of early warning systems must base on spatially explicit simulation approaches.

Last but not least, the examples of interactive web tools and web-based applications showed methods to explain features of complexity and the effects of selected drivers of flood risk change to the overall flood risk evolution. Such interactive tools for learning and risk communication contribute to the understanding of a whole system's sensitivity and behavior. The interactivity of web-based applications allows developing model experiments that can be used by lay persons or students. With these tools, users can select the main levers with which decision-makers can influence flood risk change the most. According to Willamo et al. (2018), such approaches can foster the concept of differentiative thinking and thus to contribute to generalism, holism, and holarchism thinking in tertiary education. Such online tools can provide evidence or possibility for experiencing that the whole is greater than the sum of its part. Heinrich and Kupers (2019) furthermore stated that if the principles of complexity science can be made accessible for students, the awareness of students to engage in multidisciplinary learning is rising. This contributes to a general systems literacy and finally it increases the problem-solving competencies of future experts in an uncertain world full of wicked problems that need to be confronted. Therefore, a complexity perspective is useful to be implemented in problem-based learning (Lizier et al., 2018).

The presented model experiments and web-based tools show first attempts of learning tools for confronting complexity in flood risk research. Flood risk management can make use of coupled component models to overcome the reductionist and monodisciplinary perspective. Coupling existing models is similar to synthesizing knowledge. The synthesis of connected knowledge helps to work on the implementation of a holistic view on flood risk management. This comprehends the full set of risk factors and their change. In this sense, I hope that the book has aroused interest in the complexity perspective.

References

Ackermann, F., Howick, S., Quigley, J., Walls, L., & Houghton, T. (2014). Systemic risk elicitation: Using causal maps to engage stakeholders and build a comprehensive view of risks. *European Journal of Operational Research, 238*(1), 290–299. https://doi.org/10.1016/j.ejor.2014.03.035

Aerts, J. C. J. H. (2020). Integrating agent-based approaches with flood risk models: A review and perspective. *Water Security, 11*, 100076. https://doi.org/10.1016/j.wasec.2020.100076

Aerts, J. C. J. H., Botzen, W. J., Clarke, K. C., Cutter, S. L., Hall, J. W., Merz, B., Michel-Kerjan, E., Mysiak, J., Surminski, S., & Kunreuther, H. (2018). Integrating human behaviour dynamics into flood disaster risk assessment. *Nature Climate Change, 8*(3), 193–199. https://doi.org/10.1038/s41558-018-0085-1

An, L., Grimm, V., Sullivan, A., Turner, B. L., II, Malleson, N., Heppenstall, A., Vincenot, C., Robinson, D., Ye, X., Liu, J., Lindkvist, E., & Tang, W. (2021). Challenges, tasks, and opportunities in modeling agent-based complex systems. *Ecological Modelling, 457*, 109685. https://doi.org/10.1016/j.ecolmodel.2021.109685

Baldassarre, G. D., Kreibich, H., Vorogushyn, S., Aerts, J., Arnbjerg-Nielsen, K., Barendrecht, M., Bates, P., Borga, M., Botzen, W., Bubeck, P., De Marchi, B., Llasat, C., Mazzoleni, M., Molinari, D., Mondino, E., Märd, J., Petrucci, O., Scolobig, A., Viglione, A., & Ward, P. J. (2018). Hess opinions: An interdisciplinary research agenda to explore the unintended consequences of structural flood protection. *Hydrology and Earth System Sciences, 22*(11), 5629–5637. https://doi.org/10.5194/hess-22-5629-2018

Barth, N. C., & Döll, P. (2016). Assessing the ecosystem service flood protection of a riparian forest by applying a cascade approach. *Ecosystem Services, 21*, 39–52. https://doi.org/10.1016/j.ecoser.2016.07.012

Boulton, J. G., Allen, P. M., & Bowman, C. (2015). *Embracing complexity: Strategic perspectives for an age of turbulence*. Oxford University Press.

Delong, M. D., & Thoms, M. C. (2016). An ecosystem framework for river science and management. In *River science: Research and management for the 21st century* (pp. 12–36). Wiley Blackwell. https://doi.org/10.1002/9781118643525.ch2

Fekete, A. (2019). Critical infrastructure and flood resilience: Cascading effects beyond water. *WIREs Water, 6*(5). https://doi.org/10.1002/wat2.1370

Felder, G., Gómez-Navarro, J. J., Zischg, A. P., Raible, C. C., Röthlisberger, V., Bozhinova, D., Martius, O., & Weingartner, R. (2018). From global circulation to local flood loss: Coupling models across the scales. *Science of the Total Environment, 635*, 1225–1239. https://doi.org/10.1016/j.scitotenv.2018.04.170

Gilvear, D. J., Greenwood, M. T., Thoms, M. C., & Wood, P. J. (2016). River science: Research and management for the 21st century. In *River science: Research and management for the 21st century* (pp. 1–407). Wiley Blackwell. https://doi.org/10.1002/9781118643525

Haasnoot, M., Biesbroek, R., Lawrence, J., Muccione, V., Lempert, R., & Glavovic, B. (2020). Defining the solution space to accelerate climate change adaptation. *Regional Environmental Change, 20*(2). https://doi.org/10.1007/s10113-020-01623-8

Haasnoot, M., Kwakkel, J. H., Walker, W. E., & ter Maat, J. (2013). Dynamic adaptive policy pathways: A method for crafting robust decisions for a deeply uncertain world. *Global Environmental Change, 23*(2), 485–498. https://doi.org/10.1016/j.gloenvcha.2012.12.006

Haer, T., Botzen, W. J. W., & Aerts, J. C. J. H. (2019). Advancing disaster policies by integrating dynamic adaptive behaviour in risk assessments using an agent-based modelling approach. *Environmental Research Letters, 14*(4), 044022. https://doi.org/10.1088/1748-9326/ab0770

Heinrich, S., & Kupers, R. (2019). Complexity as a big idea for secondary education: Evaluating a complex systems curriculum. *Systems Research and Behavioral Science, 36*(1), 100–110. https://doi.org/10.1002/sres.2547

Hossain, M. S., Ramirez, J. A., Haisch, T., Speranza, C. I., Martius, O., Mayer, H., & Keiler, M. (2020). A coupled human and landscape conceptual model of risk and resilience in Swiss Alpine communities. *Science of the Total Environment, 730*. https://doi.org/10.1016/j.scitotenv.2020.138322

Hull, V., Tuanmu, M. N., & Liu, J. (2015). Synthesis of human-nature feedbacks. *Ecology and Society, 20*(3). https://doi.org/10.5751/ES-07404-200317

Ladyman, J., Lambert, J., & Wiesner, K. (2013). What is a complex system? *European Journal for Philosophy of Science, 3*(1), 33–67. https://doi.org/10.1007/s13194-012-0056-8

Lizier, J. T., Harré, M. S., Mitchell, M., DeDeo, S., Finn, C., Lindgren, K., Lizier, A. L., & Sayama, H. (2018). An interview-based study of pioneering experiences in teaching and learning complex systems in higher education. *Complexity, 2018*. https://doi.org/10.1155/2018/7306871

Loreti, S., Ser-Giacomi, E., Zischg, A., Keiler, M., & Barthelemy, M. (2022). Local impacts on road networks and access to critical locations during extreme floods. *Scientific Reports, 12*(1). https://doi.org/10.1038/s41598-022-04927-3

Mai, T., Mushtaq, S., Reardon-Smith, K., Webb, P., Stone, R., Kath, J., & An-Vo, D. A. (2020). Defining flood risk management strategies: A systems approach. *International Journal of Disaster Risk Reduction, 47*. https://doi.org/10.1016/j.ijdrr.2020.101550

Mitchell, M. (2009). *Complexity: A guided tour/melanie mitchell*. Oxford University Press.

Moallemi, E. A., Kwakkel, J., de Haan, F. J., & Bryan, B. A. (2020). Exploratory modeling for analyzing coupled human-natural systems under uncertainty. *Global Environmental Change, 65*. https://doi.org/10.1016/j.gloenvcha.2020.102186

Papilloud, T., Röthlisberger, V., Loreti, S., & Keiler, M. (2020). Flood exposure analysis of road infrastructure – comparison of different methods at national level. *International Journal of Disaster Risk Reduction, 47*, 101548. https://doi.org/10.1016/j.ijdrr.2020.101548

Preiser, R., Biggs, R., De Vos, A., & Folke, C. (2018). Social-ecological systems as complex adaptive systems: Organizing principles for advancing research methods and approaches. *Ecology and Society, 23*(4). https://doi.org/10.5751/ES-10558-230446

Raso, L., Kwakkel, J., Timmermans, J., & Panthou, G. (2019). How to evaluate a monitoring system for adaptive policies: Criteria for signposts selection and their model-based evaluation. *Climatic Change, 153*(1–2), 267–283. https://doi.org/10.1007/s10584-018-2355-3

Rebally, A., Valeo, C., He, J., & Saidi, S. (2021). Flood impact assessments on transportation networks: A review of methods and associated temporal and spatial scales. *Frontiers in Sustainable Cities, 3*. https://doi.org/10.3389/frsc.2021.732181

Renn, O. (2021). New challenges for risk analysis: Systemic risks. *Journal of Risk Research, 24*(1), 127–133. https://doi.org/10.1080/13669877.2020.1779787

Renn, O., Laubichler, M., Lucas, K., Kröger, W., Schanze, J., Scholz, R. W., & Schweizer, P. J. (2020). Systemic risks from different perspectives. *Risk Analysis*. https://doi.org/10.1111/risa.13657

Schanze, J., & Maleska, V. (2019). Towards a more systemic understanding of flood risks. *Journal of Flood Risk Management, 12*. https://doi.org/10.1111/jfr3.12569

Schauwecker, S., Gascón, E., Park, S., Ruiz-Villanueva, V., Schwarb, M., Sempere-Torres, D., Stoffel, M., Vitolo, C., & Rohrer, M. (2019). Anticipating cascading effects of extreme precipitation with pathway schemes - three case studies from Europe. *Environment International, 127*, 291–304. https://doi.org/10.1016/j.envint.2019.02.072

Schweizer, P. J. (2021). Systemic risks—concepts and challenges for risk governance. *Journal of Risk Research, 24*(1), 78–93. https://doi.org/10.1080/13669877.2019.1687574

Schweizer, P.-J., & Renn, O. (2019). Governance of systemic risks for disaster prevention and mitigation. *Disaster Prevention and Management: An International Journal, 28*(6), 862–874. https://doi.org/10.1108/dpm-09-2019-0282

Simpson, N. P., Mach, K. J., Constable, A., Hess, J., Hogarth, R., Howden, M., Lawrence, J., Lempert, R. J., Muccione, V., Mackey, B., New, M. G., O'Neill, B., Otto, F., Pörtner, H. O., Reisinger, A., Roberts, D., Schmidt, D. N., Seneviratne, S., Strongin, S., … Trisos, C. H. (2021). A framework for complex climate change risk assessment. *One Earth, 4*(4), 489–501. https://doi.org/10.1016/j.oneear.2021.03.005

Sterman, J. D. (2006). Learning from evidence in a complex world. *American Journal of Public Health, 96*(3), 505–514. https://doi.org/10.2105/AJPH.2005.066043

Thaler, T., Zischg, A., Keiler, M., & Fuchs, S. (2018). Allocation of risk and benefits—distributional justices in mountain hazard management. *Regional Environmental Change, 18*(2), 353–365. https://doi.org/10.1007/s10113-017-1229-y

Verburg, P. H., Dearing, J. A., Dyke, J. G., Leeuw, S.v. d., Seitzinger, S., Steffen, W., & Syvitski, J. (2016). Methods and approaches to modelling the Anthropocene. *Global Environmental Change, 39*, 328–340. https://doi.org/10.1016/j.gloenvcha.2015.08.007

Wang, W., Yang, S., Gao, J., Hu, F., Zhao, W., & Stanley, H. E. (2020). An integrated approach for assessing the impact of large-scale future floods on a highway transport system. *Risk Analysis, 40*(9), 1780–1794. https://doi.org/10.1111/risa.13507

Willamo, R., Helenius, L., Holmström, C., Haapanen, L., Sandström, V., Huotari, E., Kaarre, K., Värre, U., Nuotiomäki, A., Happonen, J., & Kolehmainen, L. (2018). Learning how to understand complexity and deal with sustainability challenges – a framework for a comprehensive approach and its application in university education. *Ecological Modelling, 370*, 1–13. https://doi.org/10.1016/j.ecolmodel.2017.12.011

Yabe, T., Rao, P. S. C., Ukkusuri, S. V., & Cutter, S. L. (2022). Toward data-driven, dynamical complex systems approaches to disaster resilience. *Proceedings of the National Academy of Sciences of the United States of America, 119*(8). https://doi.org/10.1073/pnas.2111997119

Zischg, A. (2012). *Participative planning processes in flood risk management and in integrated watershed management* (pp. 45–66). Nova Science Publishers.

Zischg, A. P. (2018). Floodplains and complex adaptive systems— perspectives on connecting the dots in flood risk assessment with coupled component models. *Systems, 6*(2). https://doi.org/10.3390/systems6020009

Zischg, A., Costa, R., Flury, C., & Schild, A. (2012). Einfluss der landwirtschaftlichen Bewirtschaftung auf alpine Naturgefahren - eine zusammenfassende Betrachtung. In *Conference proceedings INTERPRAEVENT 2012, Grenoble, France* (Vol. 2, pp. 833–844).

Index

Note: Page numbers followed by "f" indicate figures and "t" indicate tables.

E

F

G

H

FSC
www.fsc.org
MIX
Papier | Fördert gute Waldnutzung
FSC® C083411

Kolibri 360 GmbH
Ferdinand-Jühlke-Straße 7
99095 Erfurt, Deutschland
produktsicherheit@kolibri360.de